ADAS and Automa

A Practical Approach to Verification and Validation

ADAS and Automated Driving

A Practical Approach to Verification and Validation

BY

PLATO PATHROSE

Warrendale, Pennsylvania, USA

400 Commonwealth Drive
Warrendale, PA 15096-0001 USA
E-mail: CustomerService@sae.org
Phone: 877-606-7323 (inside USA and Canada)
724-776-4970 (outside USA)
Fax: 724-776-0790

Library of Congress Catalog Number 2022938723
http://dx.doi.org/10.4271/9781468604139

ISBN-Print 978-1-4686-0412-2
ISBN-PDF 978-1-4686-0413-9
ISBN-epub 978-1-4686-0414-6

To purchase bulk quantities, please contact: SAE Customer Service

E-mail: CustomerService@sae.org
Phone: 877-606-7323 (inside USA and Canada)
724-776-4970 (outside USA)
Fax: 724-776-0790

Visit the SAE International Bookstore at books.sae.org

Chief Growth Officer
Frank Menchaca

Publisher
Sherry Dickinson Nigam

Development Editor
Publishers Solutions, LCC
Albany, NY

Director of Content Management
Kelli Zilko

Production and Manufacturing Associate
Erin Mendicino

Dedication

This book is dedicated to

*My parents, **Pathrose Augustine Fernandez** and **Telma Fernandez**, for all the trouble they took for my education and for making me a better person.*

*My brothers **Plasbo** and **Pinto**, who were the guiding stars in my career,*

Sonia** and **Dew

Ava**, **Ethan**, and **Emma

My Friends, Teachers, Mentors, and Colleagues for their love, guidance, motivation, and support.

*My beloved wife **Teena** and son **Maximus** for their love, encouragement, and patience. It was greater than the number of words in this book.*

This would not have happened without your love and support, especially without that little helping hand that always switched ON the white lamp at my computer desk and kept me busy by placing all little toys on my desk.

Contents

Foreword

My first contact with Advanced Driver Assistance Systems (ADAS) and Autonomous Driving happened in 1991. As a young engineer, I became involved in the PROMETHEUS Program, a large European research initiative where OEMs, suppliers, and research institutes jointly worked on the future traffic. At that time, Daimler-Benz, my company, had built a large green van called VITA 1 that could drive autonomously on a well-marked route and was proudly presented to industry VIPs and worldwide press together with vehicles from other manufacturers in a huge event on a test track in Italy. There was a person behind the steering wheel who activated the system, and in the back of the van, in front of a huge computer rack, sat an operator who supervised the system. This van was soon replaced by a passenger car named VITA 2, which could be distinguished from an ordinary car by many cameras that stuck out of the bodywork in several places around the vehicle. We were incredibly proud when at the final event of PROMETHEUS in 1994, this vehicle drove autonomously on a motorway in the north of Paris.

In our naive thinking—we were nearly 30 years younger then—we believed that the problems of autonomous driving had been solved when VITA 2 drove in Paris, and only lack of computational power and the need for better sensors would hinder market introduction. In a brochure prepared on the occasion of the final event of PROMETHEUS, we even predicted that Autonomous Driving would be available on the market in the year 2000!

Functional safety was not an issue at that time. Our optimism did not even get damped with the introduction of CAN Bus technology and the dramatic increase in the number of ECUs inside the vehicle. The amount of software in vehicles became more and more unreliable toward the end of the last century and at the beginning of the new millennium. Fortunately, those problems have been overcome with the introduction of Automotive SPICE (ASPICE) in 2001, but the life of an automotive engineer became much more complicated through the requirements and processes that ASPICE brought with it.

In 2011, the functional safety standard ISO 26262 added further constraints to the development processes with its Automotive Safety Integrity Level (ASIL) recommendations for safety-relevant systems. Driver Assistance and Autonomous Driving fall in this category. ISO 26262 insisted on the safety of the people being more important than technology introduction and paved the way for today's ADAS systems. To release a new system, it was no longer sufficient that it had been tested for some 500,000 miles in test vehicles on public roads and test tracks in several places in the world. Instead, depending on the application, ISO 26262 recommends failure probabilities between one failure in one million hours for ASIL A and ten failures in one billion hours for ASIL D, where ASIL refers to the automotive safety integrity level classification by the ISO 26262 standard.

If you do your math, you will easily see that it is impossible to prove that those ASIL requirements are fulfilled with conventional vehicle testing. Even for ASIL A, the lowest ASIL level, more than 114 years of faultless driving would be needed. For ASIL D, this figure would increase to more than 11,400 years. Therefore, a new approach to system design, test, verification, and validation is needed, which is made even more complicated by the increasing use of artificial intelligence in ADAS and autonomous driving systems. Also, the upcoming ISO 21448 standard, which addresses the "Safety of the intended functionality" and is known shortly as SOTIF, needs to be considered. So far, only a few people know what this will mean. Still, it can be safely assumed that, like functional safety, SOTIF will heavily influence all aspects of system design, test, verification, and validation.

As already said above, safe functioning can no longer be proven only through vehicle testing for ADAS and Autonomous Driving systems. Even if they are still important, in-vehicle tests will play a minor role in the verification and validation of these kinds of systems. Most of the work will be done with simulations such as software-in-the-loop, hardware-in-the-loop, and driver-in-the-loop. The simulation requires data, and the data is used from the real world and the simulated data. The latter are particularly important for the test of Edge Cases, which rarely happen under usual driving conditions but are nevertheless crucial for safe functioning.

As you can see, the verification and validation of ADAS and Autonomous Driving systems is complex and time consuming, and even for somebody like me, who has spent more than 30 years in the ADAS and Autonomous Driving domain, it sometimes looks like "black magic." I am all the more happy with Plato Pathrose, one of the rare experts in this topic who lives and breathes testing of ADAS and Autonomous Driving and with whom I had the pleasure to work when I was the Head of ADAS Research and Development at Visteon, providing a comprehensive and comprehensible description of this challenging subject.

This book will be as useful for students who make their first steps in the fascinating world of ADAS and Autonomous Driving as it is for experts with a long working history in this field. It shows the way through the "methodological jungle" and creates the understanding of testing that developers of ADAS and Autonomous Driving systems need to have if they want their systems to be successful on the worldwide market.

Last but not least, I would like to thank Plato for the enormous amount of time and effort he invested into writing this important book, and I wish all readers that they not only find this book useful but enjoy reading as much as I did.

Matthias Schulze
Muehlacker, January 2022

Introduction

Tomorrow comes with a lot of surprises: one of those would be how mobility will change. There will be a day in the not too distant future that you can verbally communicate with the vehicle and instruct it to take you to a destination. The car will navigate through the street traffic by itself and successfully take you to your destination. This is the future, and the whole world is racing to have this kind of automated driving vehicle deployed on our roads.

Since we focus on future technology, no defined methods or practices act as a standard and can be used as a reference. The industry has a lot of nonstandard products developed and deployed in the field of automated driving. The research and development in the automobile industry are intensively and actively working on different autonomous driving projects across the world. However, these studies helped the industry build references for different methods, processes, and techniques in developing and deploying these complex products. The technology for fully autonomous vehicles for large-scale production and public usage is still under development, and it will take time to mature. Therefore, this is one of the highly innovative areas in the automotive industry. Today, different vehicle manufacturers consider their methods and approaches for bringing advanced technologies in Advanced Driver Assistance Systems (ADAS) and Automated Driving in their vehicles.

The methods and approaches described in this book were a few examples of how the outcome of research projects was adapted or tailored for the development and production of automated driving systems by different vehicle manufacturers and tier suppliers. Since the product development has a tight scope, time, and cost constraints, organizations have adapted the methods and approaches to fit the boundaries of these projects while ensuring a safe and quality product is developed. Mostly, it is unknown to the outside world what processes and methods were followed by various organizations while deploying these features in their vehicles or how they are evaluated for their quality and performance.

The examples covered in the book will provide a starting point for any novice to understand different concepts, methods, and use cases which would help to a quick start with the design and test of ADAS and Automated Driving systems with a system-level understanding.

This book brings a few details to the readers and some practical scenarios, examples, and methods used by Automotive OEMs and their suppliers in developing and deploying ADAS and Automated Driving systems in different vehicles. With this basic knowledge, you can kick-start your career in the field of ADAS and Automated Driving.

About This Book

ADAS and Automated Driving—A Practical Approach to Verification and Validation focuses on how an automated driving system can be developed from its concept to a product that can be deployed in the market. It covers some practically viable approaches, methods, and techniques with examples captured and collected from multiple production programs across different organizations. These concepts, methods, and techniques are used differently by various vehicle manufacturers and their suppliers as there is no uniform approach defined yet for deploying an autonomous driving system.

All information provided in this book is from industry experience and best practices in the automotive industry. This includes inputs and experiences from experts from vehicle manufacturers, their suppliers of the systems and tools, and engineering service organizations.

This book provides an overview of how different Advanced Driver Assistance Systems (ADAS) and Automated Driving systems are currently being developed and deployed in vehicles. Also, this will provide an understanding and act as a reference for any person interested in focusing their career on the verification and validation of ADAS and Automated Driving systems. Since no established standards and regulations for the deployment of autonomous driving road vehicles are available at the time of writing this book, every vehicle manufacturer has their approaches and methods tailored from research programs and studies. The examples here cover a few of those tailored methods, strategies, and techniques which will help the reader have foundational knowledge and follow best and proven practices from the industry.

Assumptions

I have only one assumption: you are interested to learn about ADAS and Automated Driving systems and their verification and validation methods and approaches that are currently being used in the industry. This book will serve as a reference for your product development to define the key performance indices and acceptance criteria compared to similar products in the market. You would consider the methods, techniques, and approaches here as references only, and necessary adaptations of these for your project will be considered as needed. You will follow all the required standards and applicable regulations based on different geographic locations for the deployment of your products based on its requirement and availability.

Availability of standards and regulatory guidelines will bring more light to the methods and approaches in deploying automated driving features in vehicles. You will adapt your methods and processes as required. The examples used in this book will help you have a good understanding of the methods used for product development across different organizations for specific cases. The contents of this book are not a complete set of methods and examples for full product development. Whenever you are working on a project, you should consider the state-of-the-art technology, processes, and methods as required based on the functions, features, and products you develop.

Acknowledgments

This book is the result of a discussion amid the global pandemic of 2020 as regards how my experience in ADAS and Automated Driving can be helpful for the younger generations. I want to thank my *Parents, Brothers, and Wife* for their support and motivation for the idea of writing this book. Thanks to my beloved wife *Teena* for inspiring me to write this book and for the patience, support, and help in finishing this book. You were always my first and the best critic who helped improve this book from a reader's perspective. I thank my son *Maximus* for his late-night companionship and for filling up my desk with toys while writing this book. It was the playtime with you I missed for many months.

I want to thank *Dhanya D. Rajeev* for her support, follow-ups, and reviews from the day I started writing this book. It would have been hard to achieve this without your support. It was a fantastic journey together over a long time, and we can be proud of it.

Many well-wishers supported me with great ideas, reviews, suggestions, and interviews. Thanks to *Dr. Tessy Theres Baby, Sinu Abraham, Reshmi Ann Varghese, Lavin Pottekkat, Asha Elizabeth* and *Sharika Kumar* for helping me with ideas, thoughts, and guidance. Many of my friends, acquaintances, and old colleagues contributed to this book with their knowledge and practical experience from various vehicle manufacturers and system suppliers. Thanks a lot to all those who contributed and supported me. It was hundreds of years of industry knowledge that enlightened me in writing this book. I tried my best to present it in a simpler way and appreciate all your support and contributions in realizing this book.

Special thanks to my friends and colleagues from *Tata Elxsi, JLR, McLaren, Magna, AVL, and Vinfast* and *my exceptional ADAS V&V team of Visteon*. The experiences with you helped me expand my knowledge and share it with the generations to come.

I want to thank my *Teachers, Mentors,* and *Friends* who supported and motivated me to writing this book.

Special thanks to *Matthias Schulze, my mentor and guide in the automotive field* and *Dr. Padmesh Parasuraman, my lifelong mentor and guide* for their support and motivation from the start of this project.

I want to thank *Sherry Nigam* and *Linda DeMasi* for their support in realizing this book. I thank the editors, the technical reviewers who spent time to review the manuscript and recommended updates, and the entire staff of SAE International for their support and guidance. It was a pleasant experience to work with you all, and the support I received was amazing.

Beyond all, I thank *God Almighty* for guiding me and helping me to do better in all difficult times.

/Autonomous
/Sensing
/Communication
/Battery
/Navigation
Autopilot mode

1

Introduction to Advanced Driver Assistance Systems and Automated Driving

Did you know that every year approximately 1.35 million people die in road accidents worldwide? As per the Association for Safe International Road Travel (ASIRT) report, almost 38,000 people die in the United States (U.S.) alone, and about 23,000 people die in Europe, with more than 4 million people getting injured in both regions [1.1, 1.2]. Globally this has grown to 1.35 million as per the World Health Organization (WHO) global status report on road safety [1.3]. What could be the way forward to avoid those? How could we help the people who are using vehicles and save them from being the victims of road accidents? There have been initiatives from both the European Commission (EC) and the U.S. government with a roadmap to reduce road accidents. The EC has set an ambitious plan to achieve zero deaths in road accidents by 2050 through the Vision Zero project [1.2].

The goal of Advanced Driver Assistance Systems (ADAS) and Automated Driving is to reduce road accidents, thereby reducing the associated fatalities across the world and providing more comfort and assistance for vehicle users. Nowadays, there is no vehicle available in the market that does not have basic safety systems. However, it is a pity that even the airbags in a vehicle are considered a luxury feature and not a mandatory life-saving component in some countries.

The main goal is to bring safe vehicles to market, which will be reactive along with additional comfort for its users and create a safe environment for the passengers of the vehicle and other road users. This is one of the ways of reducing accidents as we assume that a robotic system can be more robust and free from fatigue and can decide better than humans who are prone to emotions like bad mood, anger, and tiredness while driving a vehicle.

1.1. Sense Organs of a Vehicle

Like humans, vehicles also possess certain sense organs to understand the environment around them and to behave accordingly. These are the vehicle sensors, and we cannot imagine a vehicle these days without having multiple sensors in it. When we focus on ADAS and Automated Driving in vehicles, these sensors play vital roles from detecting the environment around the vehicle to executing certain actions and controlling the vehicle. This section will discuss various sensors in the vehicle, which help identify its surroundings and support in controlling the vehicle.

1.1.1. Camera

Cameras are one of the main sensors in a vehicle that has driver assistance or automated driving features. They can detect the surroundings based on the principle of light. A vehicle can have one or up to 13 cameras, sometimes even more, depending upon the different features in the vehicle. Cameras help in detecting various objects in the environment, which are classified using neural network-based algorithms. This helps the vehicle to take necessary actions and respond according to its environment. Depending on the position of the camera in the vehicle and its functionality, various names are used for the cameras, such as rearview cameras, front-facing cameras, surround-view cameras, pillar cameras, etc.

Each camera in the vehicle may differ in terms of its viewing angle, resolution, data format in which the image is generated, etc. One of the important camera assemblies in a vehicle is its front-facing camera assembly. This assembly usually consists of a single camera (mono camera), a dual-camera assembly (stereo camera), or a three-camera assembly (tri-camera system). Different vehicle manufacturers started using the front-facing camera with different resolutions from VGA (video graphics array) of 0.3 megapixels up to 12 megapixels. They also have a detection range from 60 to 150 m in distance. Usage of these cameras varies from simply detecting objects to even for distance calculation. Cameras perform well in almost all environmental conditions. As they rely purely on light, it will be challenging to depend only on cameras in poor lighting conditions.

1.1.2. Radar

Radar (Radio Detection and Ranging) works based on the principle of reflection of electromagnetic waves. It is mainly used to detect objects and calculate distance using the principle of the Doppler effect. A vehicle could have up to 8 radars or more depending on the different functions and features in the vehicle. Different radars in the vehicle are classified based on their operating frequency. Imaging radars are being used recently in vehicles, mainly to create a two-dimensional image of the environment like a camera. This will also act as a redundancy sensor for images. Radars can detect any object in any environmental conditions. For instance, in a night vision system, radars use infrared rays for detecting objects based on heat waves. In any environmental conditions, radars emit and capture the reflected electromagnetic waves and detect the presence of objects along with their azimuthal angle and relative velocity with the help of frequency shift. Radars in the vehicle are the main source of detection, positioning, and distance calculation for an object around the vehicle. They have a range of about 250 m of reliable detection. Radars are also given specific names depending on where they are placed in the vehicle and based on the frequency of the waves they emit, like short-range radar, front-facing radar, and long-range radar.

Radars are not influenced by light; hence, they are reliable sensors for object detection irrespective of lighting conditions. However, there are some disadvantages of radars as they are prone to noise. These noises are simple disturbances caused by unwanted reflections in the environment.

1.1.3. Lidar

Lidar (Light Detection and Ranging) works similarly to radar based on the principle of reflection of electromagnetic waves. A laser pulse is emitted instead of an electromagnetic wave of any other type. Prior to its entry into the automotive industry, lidars were used in other industrial areas like industrial automation and production lines. Lidar has an advantage due to its high precision and reliability, and it is becoming dominant in the automotive world. A vehicle with assisted or automated driving features may have no lidar used or can have up to six or more lidars depending on the features available in that vehicle. Lidars are available in the market based on different operating principles like mechanical rotating lidar or solid-state lidar. It has an operating distance of about 200 m, and it is one of the most reliable sensors in a vehicle. Like radar, lidar is also affected by noise; here, these noises are mostly due to the dispersion or scattering of laser pulses. Whenever there is moisture in the environment or during rain, the laser pulse gets scattered, which could be one noise source. Similarly, many reflecting objects present in the environment, for example, the presence of glasses or mirrors, can cause a lot of reflection, which could be another source of noise. We could experience these noises inside a city environment where there are numerous buildings with glass claddings.

1.1.4. Ultrasonic Sensors

Ultrasonic sensors have been used in vehicles for many years. They are one of the cheapest sensors for detection. They work based on the principle of reflection of electromagnetic waves. As the name suggests, it uses ultrasonic waves and is used to detect objects in the vicinity of the vehicle. They have an operating range of about 5 m and have been used mostly for parking assist features in the vehicle. A vehicle could have 4 to 12 or more ultrasonic sensors depending on the driver assistance features, which uses ultrasonic sensor inputs.

1.1.5. Inertial Measurement Unit Sensors

An inertial measurement unit, or IMU, is one of the important sensors in the vehicle that provides the state of motion of the vehicle in an environment. It mainly provides the orientation, motion, and angular position of the vehicle. In other words, the IMU will provide accurate vehicle data based on three axes (pitch, roll, yaw). These are pivotal information for the vehicle controls based on which the steering, braking, and acceleration can be decided. For example, the steering angle change will be different if the vehicle changes lanes on a straight and curved road. The IMU will provide the information and status of the vehicle with respect to the environment, and the vehicle controls can act accordingly. These sensors come with global positioning system (GPS) functionality, which will also provide the position of the vehicle in the geographical location.

1.1.6. High-Definition Maps

High-definition maps, or HD maps, are not sensors, but they are mentioned here as they play an important role in automated driving. Compared to a navigation map that we use in the vehicle, HD maps provide precision information about roads, markings, and structures in an environment with an accuracy of about 2 to 5 cm. Hence, this will provide additional and reliable position information as a redundant sensor to the vehicle about its surroundings and help the vehicle to navigate accurately. The drawback, however, with maps is that they require regular updates and may cost the user.

1.2. ADAS and Automated Driving

SAE International (formerly named the Society of Automotive Engineers) has defined the levels of automation in vehicles in its publication "SAE J3016™: Taxonomy and Definitions for Terms Related to Driving Automation systems for On-Road Motor Vehicles" [1.4]. The classification of automation levels is based on the automation capabilities of the vehicle and the level of involvement required by a human for control and management of those functionalities. Based on these, the features are classified into two groups: driver support features and automated driving features [1.4] (Figure 1.1).

The systems that provide driver support features in vehicles which are classified up to SAE Level 2 are collectively called Advanced Driver Assistance Systems, or simply ADAS. This includes support features for safety like lane keep assist or comfort features like adaptive cruise control. During driving, these features help impart additional safety in the form of information or response and comfort to the driver. The driver is in full control of the vehicle

FIGURE 1.1 Levels of driving automation as defined by SAE J3016™.

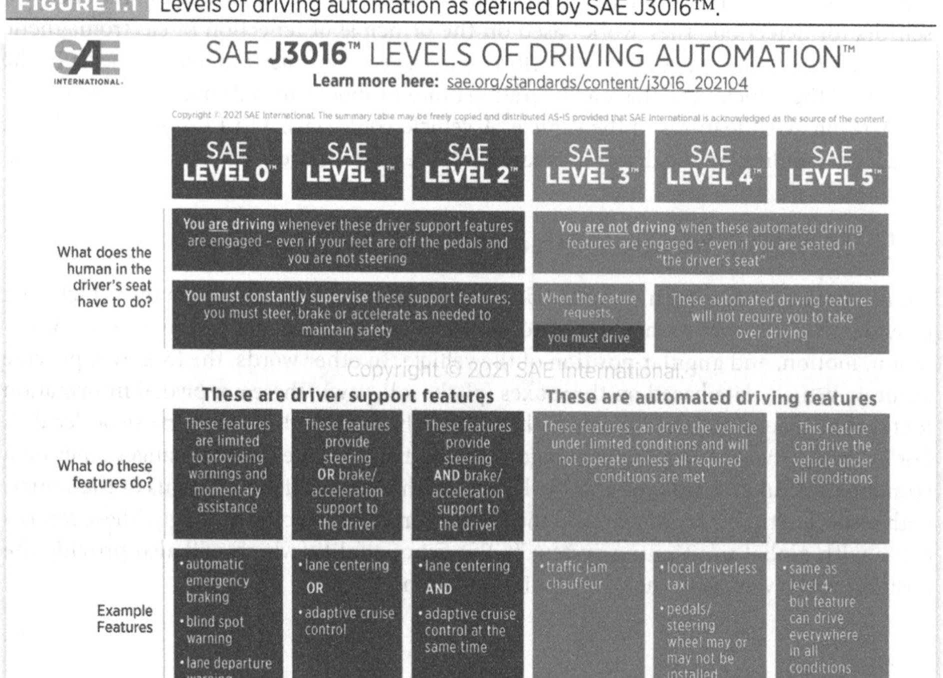

during the entire course of operation. Whenever there is a malfunction or a condition in which the support feature is not behaving as intended, the driver is still in control and can override the support functionality or disable it [1.4].

Automated driving features are those features in the vehicle which can take full control of the vehicle when certain conditions are met for which it is designed. During these times, the vehicle can move without any supervision of the driver. These features can control the vehicle and navigate through traffic while the driver can focus on other activities. Whenever there is a state in which the conditions change or the vehicle cannot navigate due to error or any confused state, the system can request the driver to take back control without any delay.

Certain vehicle manufacturers implement these automated driving features with a predefined threshold time for the driver to take over the control of the vehicle. Multiple systems are available in the market which have been designed with 10 sec, 8 sec, or 12 sec takeover time before the safe maneuver is performed by the vehicle itself as the driver cannot take control of the vehicle. This approach is slightly different from the definition of SAE standard classification of automated driving features; however, different vehicle manufacturers adopt this approach for product deployment. Some examples of these automated driving features include traffic jam chauffeur, highway chauffeur, urban pilot, highway pilot, valet parking, etc.

Description of certain automated driving features is given below with its SAE International automation level classifications. In product development, automotive OEMs adapt these features in terms of their operation boundaries like speed ranges and behavior to a practically feasible and safe level based on vehicles. These descriptions are only for reference and to understand different features. Some of the functional descriptions below were reused with permission from the European Road Transport Research Advisory Council (ERTRAC) Roadmap for Connected Automated Driving 2019 [1.5].

1.2.1. Highway Assist and Traffic Jam Assist (Level 2)

Assisted Driving up to 150 km/h (Highway Assist) on highways or similar roads; from the entrance to exit, on all lanes, including overtaking. The driver must deliberately activate the system and must monitor the system continuously. The driver can, at any time, override or switch off the system as he wishes [1.5]. During traffic conditions on highways, Traffic Jam Assist functionality assists the driver in driving at speeds up to 60 km/h utilizing a combination of other assistance functions and helps the vehicle navigate in the traffic situation on the highway. The driving task is completely executed here by the driver, and these assist features only assume to assist the driver for smooth navigation in these conditions when enabled.

1.2.2. Remote Parking (Level 2)

This is a feature that is operational with a speed limit of approximately 5-10 km/h. The vehicle can be parked remotely through a mobile device to a designated parking area. The driver must monitor and will be in control remotely at all times and can override or switch off the system remotely. This is mostly designed to control and perform short-distance parking maneuvers for the vehicle, mostly in a controlled environment.

1.2.3. Traffic Jam Chauffeur (Level 3)

This feature allows for conditional automated driving in traffic jams with an operational speed of up to 60 km/h on motorways or similar roads. The system can be activated in case of a traffic jam scenario. It detects a slow driving vehicle in front and then automatically controls both the longitudinal and lateral motions of the vehicle [1.5]. The later version of this functionality might include automatic lane change functionality. The driver must deliberately activate the system, but he/she does not have to monitor the system constantly. The driver can override or switch off the system at any time. In case of a takeover request to the driver from the system, the driver has sufficient time reserved to orient himself and take over the driving task. If in case the driver fails to perform the takeover, the system will move to a reduced risk condition. For example, bringing the vehicle to a safe stop.

1.2.4. Highway Chauffeur (Level 3)

This is a conditional automated driving feature with an operational speed of up to 130 km/h on motorways or similar roads from entrance to exit, on all lanes, including overtaking. The driver must deliberately activate the system but does not have to monitor the system at all times [1.5]. The driver can at any time override or switch off the system if he wishes. In case of a takeover request to the driver from the system, the driver has sufficient time reserved to orient himself and take over the driving task. In case the driver does not take over, the system will go to a reduced risk condition, i.e., bring the vehicle to a safe stop. If possible, depending on the traffic situation and system capabilities, the reduced risk conditions will include necessary lane changes to stop at the hard shoulder of the highway. For example, stopping at the emergency lane or side of the road.

1.2.5. Urban and Suburban Pilot (Level 4)

This is a highly automated driving feature with an operational speed of up to 50 km/h or as defined in urban and suburban areas. The feature can be activated by the driver in all traffic conditions [1.5]. The driver can override or switch off the system anytime during its operation.

1.2.6. Highway Autopilot (Level 4)

This is a highly automated driving feature with an operational speed of up to 130 km/h on motorways or motorway similar roads from entrance to exit, on all lanes, including overtaking and lane change capabilities as required. The driver must deliberately activate the system but does not have to monitor the system constantly. The driver can at any time override or switch off the system [1.5]. There will not be any request from the system to the driver to take over when the system is in a normal operation area (i.e., on the motorway). If a situation occurs that requires the driver to take control, and if the driver does not take over the control, the system has the capability to leave the highway and park the vehicle safely as a risk mitigation response.

FIGURE 1.2 ERTRAC Roadmap 2019 for features in passenger vehicles.

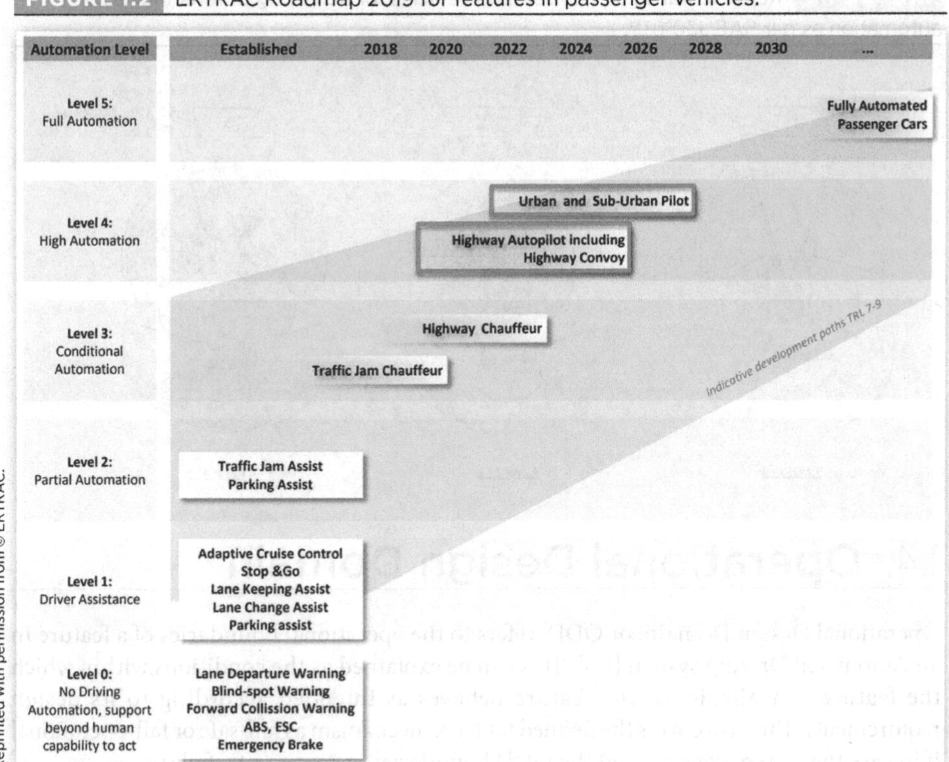

1.2.7. Valet Parking (Level 4)

This is a parking feature with an operational speed of approximately 5-10 km/h. The driver can move out of the vehicle and engage the valet parking function. The vehicle will navigate itself to the destined parking location and take necessary maneuvers to reach the parking destination. The vehicle will also handle by itself most of the failure conditions. A driver can override or switch off the system remotely, or the vehicle can inform the driver if it cannot complete its parking maneuver due to some specific reason (Figure 1.2).

Many vehicle manufacturers have published Level 2.5, Level 2.9, or Level 2+ automated driving. This is still assisted driving and classified as SAE Level 2. It is done for marketing and advertising purposes and can be seen mostly as an effort to create a perception that there are certain add-ons beyond a normal SAE Level 2 feature.

1.3. Level 5: Full Automation

SAE classification of Level 5 fully automated driving refers to vehicles that are capable of handling all required tasks of a driver by itself without any supervision. These vehicles do not require a driver or control interfaces like steering wheels or brake pedals available for any human interaction. They are fully autonomous and capable of handling any situation and failures with a fail-operational response by itself. These are normally pods or smart vehicles deployed in a controlled environment and with predefined paths. Until now, they are not planned for public road usage or as mass production vehicles. In the future, we could expect to see these kinds of vehicles on our roads too (Figure 1.3).

FIGURE 1.3 Representative sensor setup in vehicles for achieving different levels of automation as per SAE J3016™.

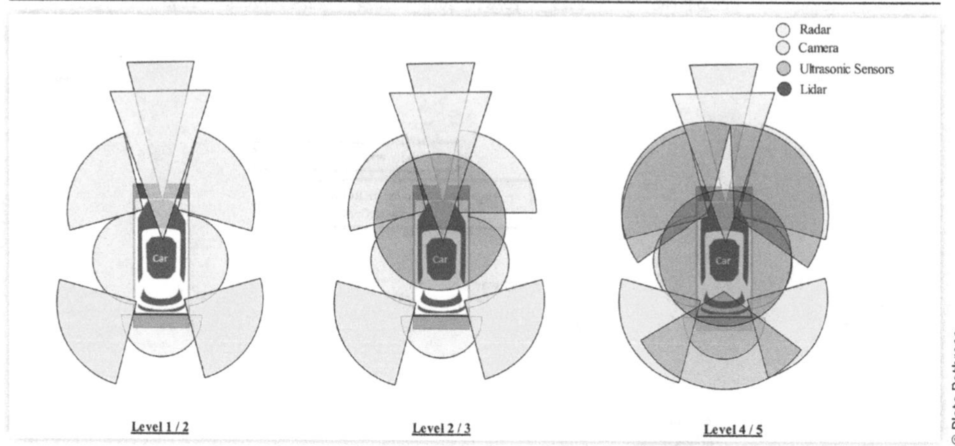

1.4. Operational Design Domain

Operational Design Domain, or ODD, refers to the operational boundaries of a feature in an Automated Driving system [1.6]. This can be explained as the conditions within which the feature is available, or the feature behaves as intended according to its design requirements. This also covers the defined fallback mechanism as fail-safe or fail-operational if in case the system goes beyond the ODD boundaries or in case of a failure.

While designing any system, the operational requirements of a system are defined for its successful operation and optimal performance. When the system is exposed to conditions beyond these defined operational boundaries for which it was designed, there could be system malfunctions, errors, or failures. Similarly, the ODD for an automated driving feature must be defined to provide its operational boundaries for optimal performance and keep the system safe, thereby protecting the users.

The following are some of the benefits of defining the ODD for a feature in its definition and concept phase:

- The ODD describes the operating conditions under which an ADAS or Automated Driving feature is specifically designed to function. This is to be decided by the vehicle manufacturer. Thus, all stakeholders involved in the concept, definition, design, and development phases will have a clear understanding of it.
- Verification and validation of the feature should be performed with the proper understanding of the ODD. This will clarify whether the feature is behaving as intended within its boundaries and the possibility of testing the fallback mechanism if the system moves beyond those boundary conditions.
- The defined ODD for a feature can be utilized by the authorities while performing safety checks, assessments, and audits. Ample proof needs to be provided that the feature behaves as expected within and the fallback mechanism comes into effect beyond the defined ODD.
- The ODD is also used for sharing information about the features of the vehicle to the end user, although extracts from the ODD of a feature from a vehicle manufacturer go into the vehicle user guide to give basic feature details and its operating limits to the end user.

Any tool can be used to capture the ODD requirements and can also use diagrammatic representations to have a better understanding for all stakeholders. The main goal for defining an ODD is that anyone reading it should have a clear understanding of the feature or the system and its boundaries. This will prevent wrong interpretations during different phases of product development and testing.

Note: The vehicle manufacturer is responsible for defining the ODD of a planned feature for a vehicle. This should act as input and reference to any tier suppliers.

1.5. Dynamic Driving Task

Dynamic Driving Task (DDT) refers to the task of driving and controlling the vehicle in real time. This can be done by a human driver or by an Automated Driving system while the vehicle is in motion [1.6, 1.7]. For features up to SAE Level 2, the vehicle is controlled by the driver, and the human driver performs the DDT during its operation. During an event of failure of the system, the driver will perform the fail-safe maneuver. During all operational times when the feature is engaged, the driver will decide and control the vehicle.

In an Automated Driving system or a vehicle with conditional or high automation features, the DDT is performed by the vehicle itself. Even in the event of failure, the vehicle will decide on what should be the response and execute it. During certain situations, switching the DDT back to the human driver is considered in SAE Level 3 and Level 4 features. In fully automated vehicles (SAE Level 5), the vehicle will handle the driving task even if there is a failure without switching it back to a human. Depending on who performs the DDT, features are classified to various automation levels defined by the SAE standard.

1.6. Object and Event Detection and Response

Object and Event Detection and Response (OEDR) refers to the detection and response during a driving task or during the operational condition of the vehicle. This is like a driver detecting the objects and events on the road while driving and responding to those to avoid accidents or to handle specific situations to reach his/her destination without any trouble [1.5, 1.6, 1.7].

In automated driving features, this refers to the detection of objects using sensors, identifying, and responding to them. For example, a camera detects the lanes on a multilane road and helps the vehicle keep in its driving lane without barging into other lanes by controlling the steering wheel of the vehicle. This is the lane-keeping functionality which is a driver assistance feature. Similarly, in adaptive cruise control, a vehicle uses radar or camera or both to detect the vehicle in front of it to follow or to maintain a definite speed. In case of an event like the vehicle in the front reduces the speed or breaks abruptly, the sensors in the vehicle will detect that event, which will be processed by the system and provide a response through the vehicle actuators by applying brakes and slowing down. The driver can control the vehicle by applying brakes or moving to a different lane to prevent a hazard, which is a response toward that event. In both cases, the driver is completing the OEDR subtask while being supported by other automated functions.

For conditional and highly automated driving features, which are from SAE Level 3 and above, specific ODDs are defined for the feature for its operation and handling of the OEDR. During its operation, the feature is expected to detect the objects and events and adapt itself by responding to those to avoid any hazards [1.6, 1.7].

For example, in the traffic jam chauffeur feature (SAE Level 3), the vehicle will navigate by itself when the feature is in operation on highway traffic and execute certain maneuvers required. This is performed by detecting the objects around the vehicle and adapting its movements accordingly in case of events that are affecting the normal course of the vehicle. In this case, there may be multiple sensors in operation, including cameras, radars, and lidars. The system has been designed to detect those events and control the longitudinal and lateral movements of the vehicle by itself without any support from the driver. Even during a failure, the vehicle will respond to those with a fallback mechanism, either performing a safe maneuver or bringing the driver back to take control of the vehicle, thereby preventing any hazard.

1.7. Summary

In this chapter, an overview of ADAS and Automated Driving systems and different levels of automation have been covered as classified by the SAE standard J3016. Some examples of various features that are classified to higher levels of automation are also familiarized. Certain common terms used for defining features in an automated driving vehicle are explained. These include Operational Design Domain (ODD), Dynamic Driving Task (DDT), and Object and Event Detection and Response (OEDR), which act as the basic terms for defining a feature. The next chapter will cover various methods and approaches to consider in designing and developing ADAS and Automated Driving systems. It will also take you through certain methodologies used for the development of robust and safe systems.

References

1.1. Association for Safe International Road Travel (ASIRT), "Annual Global Road Crash Statistics 2020," accessed February 7, 2021, https://www.asirt.org/safe-travel/road-safety-facts.

1.2. European Commission, "EU Road Safety Policy Framework 2021-2030—Next Steps towards Vision Zero," accessed February 7, 2021, https://ec.europa.eu/transport/themes/strategies/news/2019-06-19-vision-zero_en.

1.3. World Health Organization, "Global Status Report on Road Safety 2018," July 17, 2018, accessed February 7, 2021, https://www.who.int/publications/i/item/9789241565684.

1.4. SAE International, "Taxonomy and Definitions for Terms Related to Driving Automation Systems for On-Road Motor Vehicles," SAE Standard J3016_202104, April 30, 2021.

1.5. ERTRAC Working Group, "'Connectivity and Automated Driving,' 'Connected Automated Driving Roadmap'," Version 8, March 8, 2019, https://www.ertrac.org/index.php?page=ertrac-publications.

1.6. PAS 1883:2020, "Operational Design Domain (ODD) Taxonomy for an Automated Driving System (ADS)—Specification," 2020.

1.7. PAS 1880:2020, "Guidelines for Developing and Assessing Control Systems for Automated Vehicles," April 30, 2020.

/Autonomous
/Sensing
/Communication
/Battery
/Navigation
Autopilot mode

2

Design Approaches for Automated Driving Systems

"Good product requires good design." This is the framework of any product available in the market. When we speak about having automated driving features in the vehicles, we also expect those systems to be reliable, robust, and safe. This chapter will cover certain methods and approaches followed in the industry for the design and development of ADAS and Automated Driving systems so that they are robust, safe, and secure. Some of these methods will help in designing validation frameworks and test cases for qualifying these systems. Readers will get basic knowledge on how to start with designing a product from scratch if there are not enough references.

2.1. Product Development

Every product development starts with a vision. A product manager has a vision of what the product should be and how he could sell it. This is based on his analysis of the market and analysis about the acceptance of that product from customers. In the ADAS and Automated Driving market, these visions act as the foundation for a new product or a feature inside the vehicle that would finally focus on how these products and features will benefit the customer in reducing accidents. Along with any driver assistance features, every manufacturer would like to project themself as superior to their competitors by bringing in some special features on their products.

FIGURE 2.1 A Kano model.

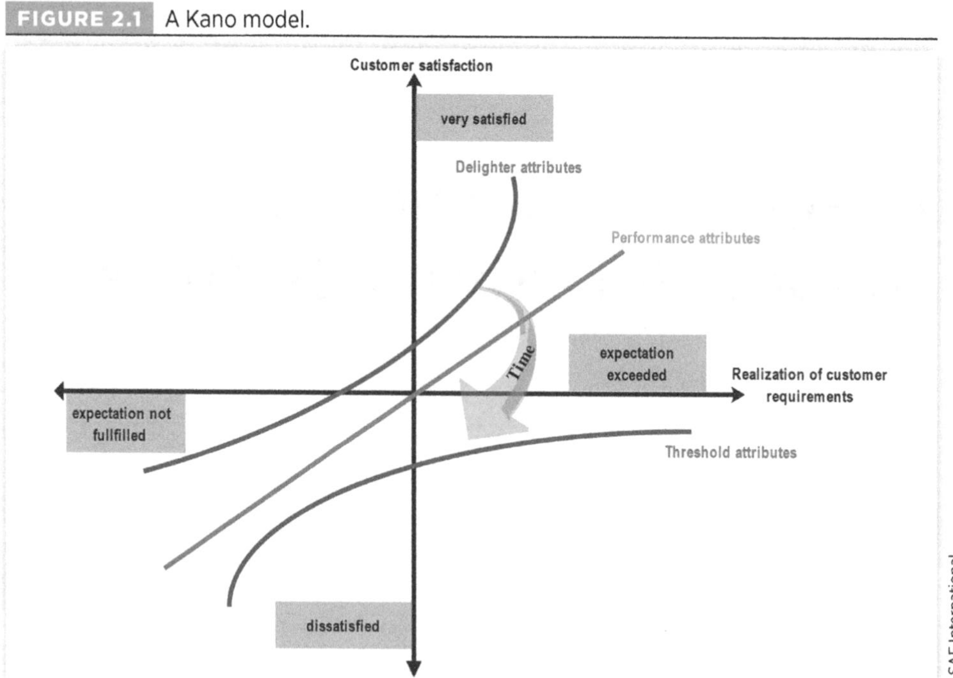

Any new product feature can be explained with the help of the "Kano model." The Kano Model of product development and customer satisfaction was published in 1984 by Dr. Noriaki Kano, professor of quality management at the Tokyo University of Science [2.1]. This model will help a product manager to define features, which will add value to the product and help in maintaining competition in the market. It also helps him to bring in delighters in the product. Delighters are WOW factors or features which give unexpected benefits that exceed customer expectations, thereby resulting in higher customer satisfaction. Delighters play an important role for any new market entrants to penetrate the market (Figure 2.1).

A good example of a "delighter" in a product, when compared with the Kano model, is the first touch screen device from Apple, the iPod Touch, which was launched in 2007. IBM Simon was the first communicator with a touch screen launched in 1992 [2.2], but the technology was so early that the market was not ready to accept that [2.1]. The commercialization of touch screen technology was successful with the iPod Touch from Apple and was further extended to iPhones. Apple launched the touch screen as a delighter in their 2007 product, which was widely accepted in the market, and now those delighters have become an inevitable requirement in any mobile device. We cannot imagine a mobile phone without a touch screen these days. This is an example of how a delighter gained market acceptance when launched and later became an integral part of the product requirements in the future. Similarly, ADAS and Automated Driving features are highlighted by different vehicle manufacturers and suppliers to have a competitive advantage as well as market acceptance. When a new product is planned in the area of automated driving, the features should be planned to have commercial benefits. Moreover, they should be safe for their users.

2.2. Distributed Architecture versus Centralized Architecture

These days, an average vehicle in the market has a good amount of electronics inside them. It is tough to think about a vehicle without having any electronics in it. The building blocks of these electronic features are called Electronic Control Units (ECUs), which are interconnected with harnesses based on specific communication protocols. We can correlate these with a network of computers inside the vehicle. As the number of ECUs inside the vehicle keeps on increasing and getting more complex, there is a need for advanced harnesses and protocols for their communication.

Early on, vehicle electrical and electronic architecture (E/E architecture) was designed in a distributed manner, where the overall functions of the vehicle were distributed among different ECUs where each ECU is dedicated to specific functionalities. Technology has changed in recent years, and this evolution of technology made the things inside the vehicle complex, especially with the introduction of ADAS and Automated Driving features. Vehicle manufacturers started looking for a solution to reduce the overall weight of the vehicle and to make the architecture inside the vehicle simpler. Thanks to the electronic revolution, today there are processors with multiple teraflops of processing capacity compared to what we had ten years before. Similarly, there are many communication protocols that found their way inside the vehicle with less complexity and more bandwidth. Hence, to reduce the harness inside the vehicle and also to reduce power dissipation across all different components, a much simpler architecture with advanced networking called Domain-Centralized, or simply Centralized, E/E architecture was introduced (Figure 2.2).

FIGURE 2.2 Distributed versus domain-centralized E/E architecture.

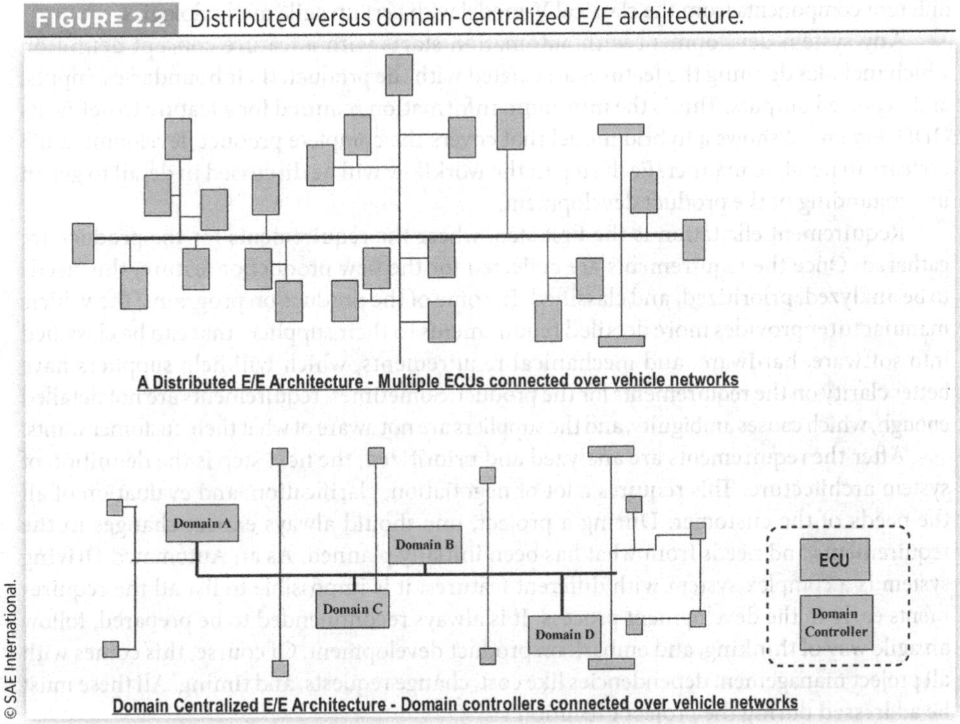

In a domain-centralized architecture, domain controllers are used instead of individual ECUs. Domain controllers are electronic systems with multiple processors in them, each running with the same operating system or different operating systems. This is configurable and can work together among themselves with high-speed interfaces and with safe and secure software. As the name indicates, the domain controllers are called so because they take over the major functionalities associated with a specific domain in the vehicle. For example, a domain controller can take over the whole cockpit functionality (instrument cluster, head-up display system, infotainment), while another domain controller is responsible for all the ADAS and Automated Driving features in the vehicle. Future vehicles will have only a few of those domain controllers and software nodes that communicate through high-speed interfaces. This will reduce the harness inside the vehicle and thereby reduce the overall weight of the vehicle. An average luxury vehicle with distributed E/E architecture can have a harnesses weight reaching up to 100 kg. The reduction in weight and the number of ECUs will keep the vehicle networks simple as well as help with the vehicle mileage. Especially moving toward electrification, mileage requirements are important for any vehicle manufacturer.

2.3. Developing an Automated Driving System

Any product development focused on large-scale production should follow certain processes. There are specific processes to reduce wastage and to improve the quality of the product [2.3, 2.4]. This section will cover various system engineering processes that are followed for the development of Automated Driving systems. This can be correlated with the classical V-model with certain agile methodologies in development. Nowhere in the industry is following a classical V-model of product development life cycle these days. It is always a hybrid model with different components from the classical V-model with certain agile methodologies.

Any system development with automation starts with a feature concept definition which includes defining the features associated with the product, their boundaries, inputs, and expected outputs. This is the minimum information required for a feature to define its ODD. Figure 2.3 shows a hybrid model that covers the complete product development life cycle in an iterative manner. Each step in the workflow will be discussed in detail to get an understanding of the product development.

Requirement elicitation is the first step, where the requirements for the product are gathered. Once the requirements are collected for the new product or feature, this needs to be analyzed, prioritized, and classified. In some of the production programs, the vehicle manufacturer provides more detailed requirements to their suppliers that can be classified into software, hardware, and mechanical requirements, which will help suppliers have better clarity on the requirements for the product. Sometimes, requirements are not detailed enough, which causes ambiguity, and the suppliers are not aware of what their customer wants.

After the requirements are analyzed and prioritized, the next step is the definition of system architecture. This requires a lot of negotiation, clarification, and evaluation of all the needs of the customer. During a project, one should always expect changes in the requirements and needs from what has been initially planned. As an Automated Driving system is a complex system with different features, it is impossible to list all the requirements early in the development process. It is always recommended to be prepared, follow an agile way of thinking, and embark on product development. Of course, this comes with all project management dependencies like cost, change requests, and timing. All these must be addressed during the project execution.

FIGURE 2.3 A product development workflow.

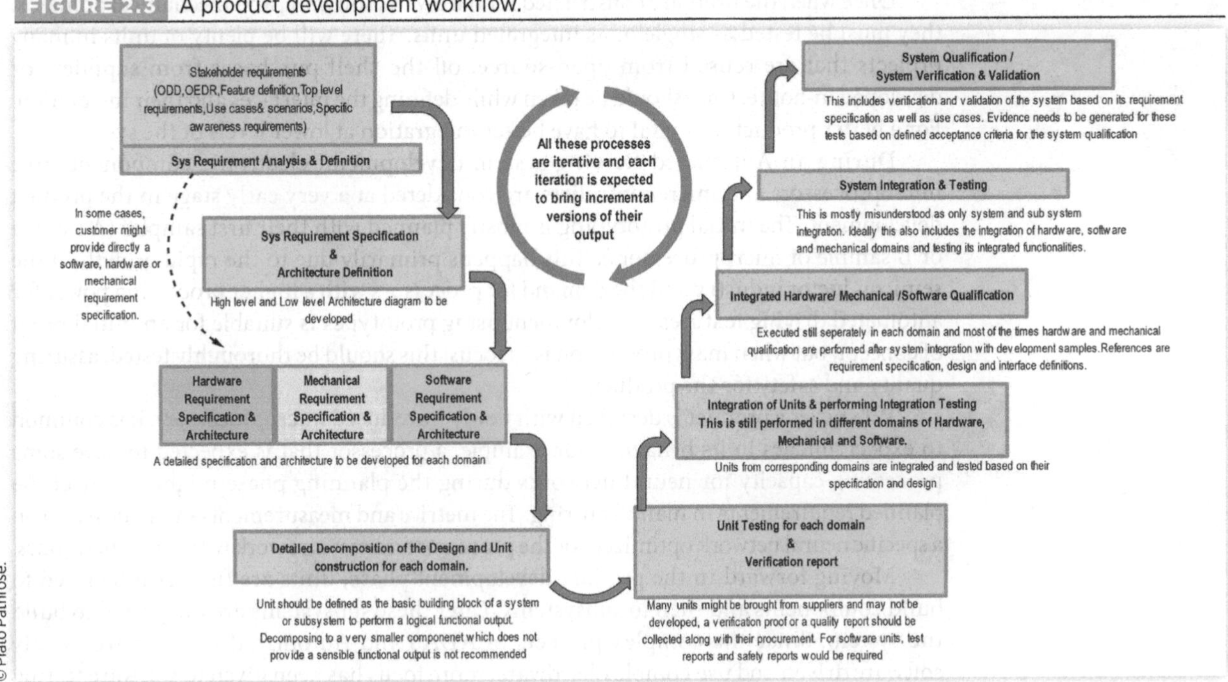

Defining a system architecture for the product starts by creating a high-level block diagram and its boundary diagram. These will usually be the logical and functional blocks of the system. These blocks will be further decomposed to define a more detailed low-level architecture, including smaller components of the system with its inter-relationships and interfaces. Having a detailed architecture of the system is very important to understand the system better to be developed.

The next step is to decompose the architecture and system requirements into hardware, software, and mechanical domains. Having separate requirements for hardware, software, and mechanical areas is an ideal approach. This will help focus on each domain and help not only in the development but also in the verification of those requirements. As there is no ideal approach, one could expect in the industry, in most projects, a master requirement specification is created without any bifurcation on domains and is used for the development and testing. The challenge of maintaining separate requirement specifications for each domain is usually attributed to time constraints or organizational reasons. If a good requirement management tool or a life-cycle management tool is used, it will be easier for requirement management and traceability until validation. This will also help in managing the complexity of the product as well as throw light on the number of requirements that have to be considered for the entire duration of the project. This is usually one of the metrics used in some organizations for measuring product complexity.

The requirements identified for hardware, software, and mechanical domains would act as the foundation for developing their corresponding architecture. Here, the hardware, software, and mechanical requirements will be listed out to the smallest possible level, called a unit, for enhanced clarity. Having clarity about each unit of the product and its functionality is recommended for a good product design. There may be doubts about how to define a unit in the software, or how small or big it should be. A unit is defined as the smallest possible piece of software or hardware that would act as an independent entity with a defined set of functions.

Once when the units are constructed for hardware, software, and mechanical domains, they must be tested as single or as integrated units. There will be plenty of units in many projects that are reused from open-source, off-the-shelf purchases from suppliers or developed in-house. Care should be taken while defining the interfaces and their integration. For a better product, it is vital to have better integration at lower levels of the system.

During an Automated Driving system development, electronic components like microprocessors and microcontrollers are considered at a very early stage in the product development. The initial prototyping is mostly planned with their first samples (A sample or B sample of microprocessors). This happens primarily due to the rapid growth in the semiconductor industry and the demand for processors with a higher processing power for automated driving features. Development using prototypes is suitable for an initial proof of concept, but when mass production is in focus, this should be thoroughly tested, assuring quality and safety for the product.

Whenever a product is designed with newly introduced microprocessors, it is common to expect glitches in its behavior. For example, a processor that is expected to have some processing capacity for neural networks during the planning phase might not meet the planned requirements in manufacturing. The metrics and measurements might be based on a specific neural network optimized for the processor and may not work with any generic ones.

Moving forward in the product development phase, units are further integrated to build components and then to subsystems. Later these subsystems are integrated to build the system. Since the complex projects of ADAS and Automated Driving are mostly software driven and use complex hardware, more focus has been given to software testing both in the unit and the integrated software in the system. Software qualification is a standard term used for integrated software testing, which comes from the International Organization for Standardization/International Electrotechnical Commission (ISO/IEC) 15504 standard and revised standard ISO/IEC TS 33061 [2.5]. Once when the software components are integrated, it undergoes a verification phase. Environments used for these kinds of integrated testing are either reference hardware or a simulated environment. Similarly, hardware and mechanical parts will undergo verification and qualification once they are integrated.

In many production projects, cross-domain integration and testing are not clearly defined or executed. For example, how do the hardware and its components behave during a software flashing? What would be the voltage change in a pull-up resistor during a flag being set or during the software flashing process, read/write cycles to memory timings, etc.? Many of these would get tested during hardware tests, and in some projects, these are executed during the board bring-up activity. But in many projects, no evidence clearly shows a structured testing approach for cross-domain integration testing.

Once the subsystems are integrated, a subsystem-level integration testing is performed, including its interfaces, action-response between subsystems, error conditions, fail-safe or fail-operational behaviors, etc. This can be executed as part of functional testing or verification and covers the functional architecture compatibility of the implemented functions. This is part of verification, and various subsystems are checked for their successful integration at this phase.

System qualification is the next step where the system verification and validation are performed. Verification of the system refers to the compatibility check of the system with the defined system requirements, which could be from the system requirement specification, including applicable standards and regulations. System validation refers to the compatibility check of the system and its functionalities for specific use cases for which it was designed. This is the acceptance criteria for vehicle manufacturers to check if the system provided to them from a supplier with a particular feature is performing as expected and serving all the use cases for which it is designed. System verification and validation are collectively called system qualification, where both requirements-based tests and various use cases are evaluated.

In many organizations, there was a tremendous push to move to agile development and testing in the last few years. Surprisingly the product development processes were always iterative from before itself. Unlike present days, agile principles or values were not followed or facilitated with the help of any tools. Ten to fifteen years back, the development was also followed in an iterative fashion, including iterative development of hardware and software. Those were also part of continuous improvement. These days, following agile principles in product development adds a lot of advantages, especially in complex projects handling ambiguity and changing requirements while providing incremental improvements in a regular fashion.

2.4. Requirement Elicitation

Requirement elicitation is the process of gathering requirements. Once there is a product vision established and a concept is developed with different features for the product and its boundaries, the next step is to collect the requirements for the product. Many assume that requirements for a new product or a feature come only from customers. Requirement elicitation is the process of collecting requirements. This should not be from the customer or a vehicle manufacturer alone but also from various other stakeholders. Requirements received from the customer or a vehicle manufacturer are just one of those sources.

In many texts, the term stakeholder requirement is used as part of requirement elicitation. The term stakeholder requirement does not need to mean that the customer is the only stakeholder, and the source of requirement is only from a customer for whom the product is being developed. When a new product is planned and developed, different stakeholders have to be considered, out of which customer is one of them and the most important because he/she is the end user who is paying for the product. However, other stakeholders are from different areas which provide input for the product development and launch (Figure 2.4).

Requirements for a product can come from its customers, which is the major source of requirements in terms of use cases from the vehicle perspective. For new product development, the product management team and the research team would have some requirement inputs, which can be the result of a market study or from the analysis of competitor products existing in the market.

As per the Kano model [2.1], any new product development might consider having certain delighters as part of the product. This will be in addition to the mandatory requirements for that product. There will be requirements that need to be considered from standards and regulations, depending on the location of the product to be launched and sold. In addition to those, there will be requirements coming from the platform, which are generally known as platform requirements. Platform requirements can be provided by a vehicle manufacturer or even from a supplier. If a vehicle manufacturer wanted to share the platform details, they would provide details of the vehicle as well as the interacting systems to the newly developing system. This will help the supplier to understand how the new product needs to be developed as well as its operational ecosystem. Some of the platform requirements from the vehicle manufacturers would be like, whether the vehicle is an electric vehicle—Availability and compatibility of different vehicle networks to the newly planned system, vehicle details like the gear change managed through switches or levers, etc. Similar to these platform requirements

FIGURE 2.4 Requirement elicitation.

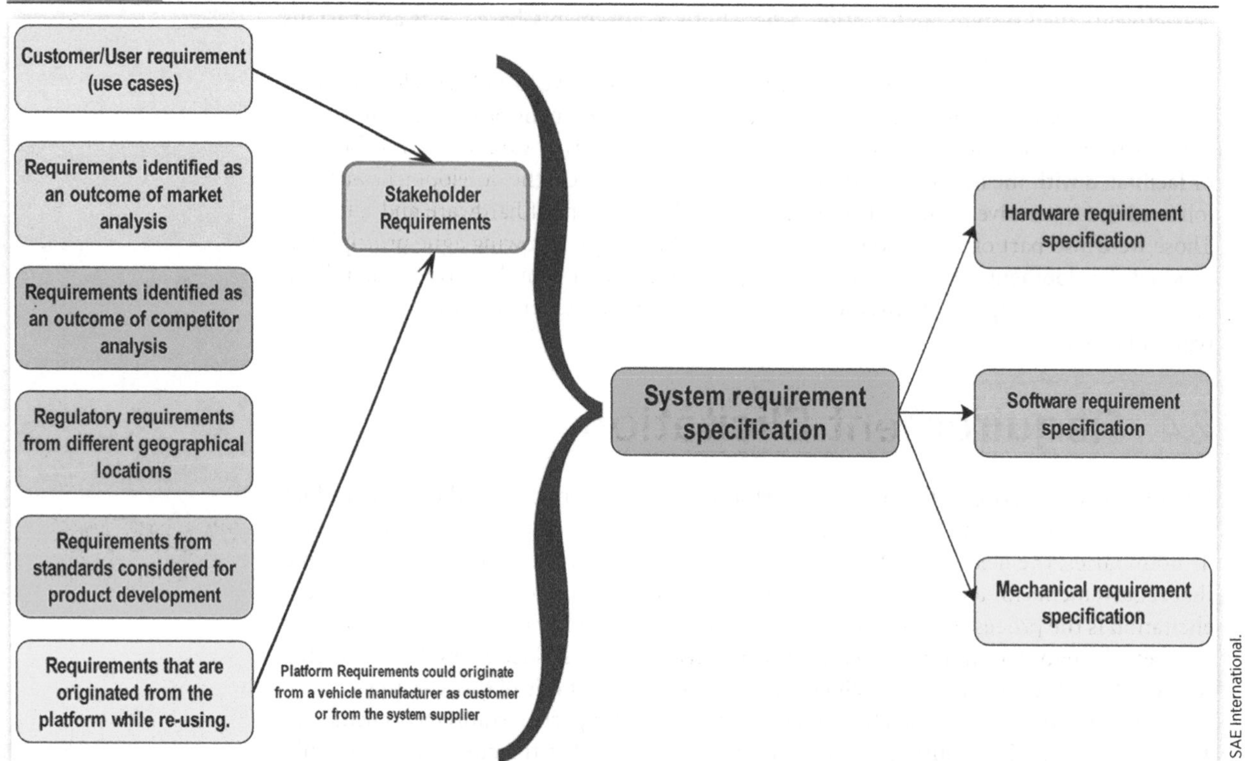

from a vehicle manufacturer, there could be platform requirements from the supplier of the system as well. This is mainly seen when a system supplier has an existing system or a prototype and wants to reuse most of it for product development. Both these are platform requirements, and they must be evaluated during the product to establish the system requirements.

All these requirements are consolidated to generate system requirement specifications. There are many other tools and methods used in the industry to develop further and derive requirements for a system. Some of those include Design Failure Mode and Effect Analysis (DFMEA), Parameter Diagram (P-Diagram), Quality Function Deployment (QFD), etc. Applying these tools and methods makes it possible to generate further technical requirements and detailed requirements for the system during its lifetime. When a new system is developed, requirement elicitation is one of the critical phases. Although it is time consuming, there must be enough effort put into establishing requirements for the product early in the development phase.

In product development, it is impossible to implement all the requirements which come from different stakeholders. How do we manage those? A simple and easiest approach that is followed in the industry is to classify all those requirements into three different groups such as:

1. Must have requirements
2. Good to have requirements
3. Nice to have requirements

After classifying the requirements to these three groups, the focus should be given in covering the first two groups of requirements in the product. If the time and budget permit, the requirements from the last group should be considered depending upon the technical feasibility and the added value it would bring to the customer.

How to classify requirements to these three groups? That should be the call of the requirements manager along with the rest of the project management team. This also involves seeking agreement from various stakeholders of the project.

2.5. Quality Function Deployment

Quality Function Deployment, or QFD, is a Japanese technique for capturing customer or stakeholders' requirements and converting them into measurable technical requirements with detailed analysis and evaluations [2.5, 2.6]. QFD is the preferred method for creating and capturing the majority of requirements before or at the start of product development. Using QFD will help capture the Voice of the Customer (VOC) and convert them to technical specifications for the product.

The QFD method is a quality management technique developed in Japan. QFD is considered here because for new product development, knowing the wants and needs of the customer is an important step. A product developed using the QFD method will result in increased customer benefits that are economical and suitable for production and can be developed within the stipulated time and cost plan of the project.

The QFD method has four different phases. Phase one is executed utilizing the "House of Quality matrix." This is a tool that captures the customer needs into product design characteristics or attributes with the help of a relationship matrix, as shown in Figure 2.5. The matrix gives the relationship between the customer's needs or wants to the product design parameters, which is to be developed. This is a data-intensive matrix developed with expert knowledge of the system and the technology and its relationship with customer needs. This is not a single-person activity; instead, it is performed by an expert group in various phases. The matrix got its name as "House of Quality" because of its shape, where the top portion of the matrix looks like the roof of a house. Once the QFD Phase one is completed from the system level, it can be handed over to different departments for further analysis specific to those domains. The outcome of the relationship matrix would be a high-level technical requirement based on customer needs from the system and its subdomains.

Phase two QFD is applied during the design phase of the product. It will help in identifying the systems, subsystems, and components of the product under development. The data available in the House of Quality matrix will be classified based on its dependency and interactions from the system, subsystems, and component levels. This information will act as the input for performing DFMEA. It will also help with the Production Part Approval Process (PPAP) of the components while working together with various suppliers. Several subdomain-level experts should be involved in this phase for analysis and establishing results.

QFD Phase three is about process development. As mentioned earlier in this chapter, the need to have a process in product development is to reduce wastage. Process analysis and classification can identify the characteristics expected in a sub-system or at the component level. The outcome would indicate which processes will get impacted by having a specific characteristic for sub-systems or the components. This information will help in fine-tuning the processes involved and will be able to identify the critical processes in the development workflow and to define the acceptance criteria for those processes. Even though

FIGURE 2.5 An indicative House of Quality matrix for requirement analysis.

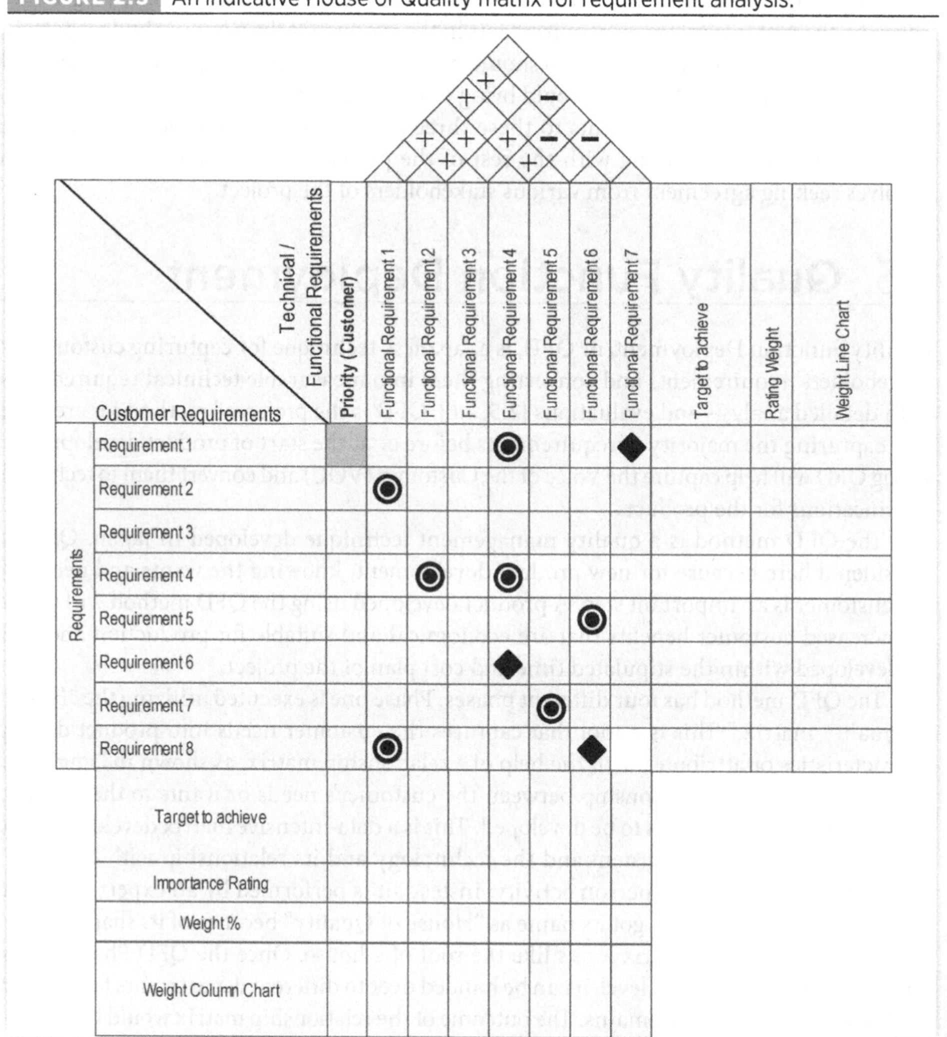

this phase looks like a quality assurance activity, bringing it into the development phase will help implement a structured way of product development.

The last and final fourth phase of QFD is the process improvement area, where the analysis of existing design and development activities are evaluated for their drawbacks. Based on the analysis results, they are improved to achieve the required quality product to match customer requirements. This is similar to the continuous improvement approach in process developments. The scope is to develop a better-quality product by improving the processes followed during the development. This phase usually is not performed widely for any new product development. It is not practical to completely change the existing organizational processes to design and develop a product for a particular customer project. The organizational process improvement teams usually take care of process improvements with long-term plans that extend beyond a single project. The fourth phase is included here as a piece of information and is not generally executed in projects because of the time and effort required.

2.6. Designing a Robust Product

Once the requirements for a product are finalized or have achieved a certain maturity, the next step is to start with the design and development of the product. How can one design a robust and quality product? One of the well-known methods for designing robust and quality products was introduced by Dr. Genichi Taguchi, and those were certain engineering methods for quality improvement. They are collectively called Taguchi methods for robust product design.

Robust product design is a disciplined engineering process that defines the best way to design a product. The term "best" refers to the lowest cost solution for a product design. Cost is not considered only from a manufacturing perspective but also the expenses involved in the whole life cycle of product development [2.8]. ADAS and Automated Driving products, with these complex features which are cost-effective and of better quality, are vital in today's commercial markets. Any organization that considers these topics as separate aspects and not integrated into their product development risks losing business to their competitors who have better delivery time, low cost of operation, and minimal startup problems. This plays a significant role in the current startup ecosystem, where one could see many young organizations coming up with solutions and products for ADAS and Automated Driving.

Designing a robust product focuses mainly on the concept selection (for a particular feature implementation) as the first step and then the parameter optimization as the next one. Both of these are achieved by reducing the variation of the quality characteristics and taking a control and management approach to keep those characteristics within the required range. Here the quality characteristics are those associated with the system which is getting designed and its features. Identifying quality characteristics is a way to measure the overall system and the features for its quality and robustness [2.7, 2.8]. This product design approach will help so that it will be less sensitive to variations and thereby provide better performance and quality to its customers. Additionally, this will also remove the need for stringent quality control activities during the deployment and operation of the product.

Dr. Taguchi proposed "Signal to noise ratios," which acts as the central metric for a robust product design [2.8]. When a product is in operation, due to many reasons, the quality of the product or the functionality may get degraded. We can call these factors that affect the quality or the functionality as Noise factors. There are mainly three types of noise factors which are:

1. External noise factor or noise in the environment
2. Unit-to-unit noise factors or system to system/subsystem level noise
3. Deterioration noise factors or noise caused due to aging or deterioration.

There are two ways in which we can make our product robust. This can be done either by eliminating the source of the noise factors or by designing our product to be insensitive or less sensitive to these noise factors. For a product design like an Automated Driving system in the vehicle, the first option is to reduce the noise by controlling the nature or the environment from where the inputs are received through sensors and networks. There are a lot of limitations here, and the cost associated is very high if we try to control the environment where the product is operated. The second option is to design the product to be less sensitive or insensitive to those noise factors the system is exposed over its lifetime. Since we expect that the customer will operate the system in an environment where the noise factors are present and if the system is designed accordingly to mitigate the influence of these noise factors, there will be better acceptance for the products by following the second option.

The application of the metric "Signal to Noise Ratios" (SNR) method can be executed in three steps, the first being the selection of a concept. For instance, while designing an Automated Driving system and its features, there could be multiple concepts one could think of to achieve a particular solution. How do we select one? The best option would be to select the concept already proven in the market and is superior and optimum in terms of its functionality and cost. Here the decision is purely based on historical data or the market data available for that particular concept. Since automated driving is a new area, it is impossible to predetermine a superior concept already proven in the market. Thus, the only option left is to select a technology that has its legacy from existing products and can be easily adapted for further improvements. There are many tools that can be utilized for the concept selection based on the needs and the application; QFD, Design of Experiments (DoE), Theory of Inventive problem solving (TRIZ), etc. are widely used for these purposes.

The second step is the design of parameters. In this phase, the product development team identifies the associated input parameters based on the selected concept with inputs from the user requirements as well as the control parameters and their setpoints for the selected concept. A setpoint is a control system term that means the desired value or target value for an input parameter or an input variable while the system is in operation. This will help keep the selected concept less sensitive to the noise parameters, and it will be robust. This helps identify and create a stable set of control parameters and will later aid in the calibration or tuning of the product and its features.

One of the most important tools used here is the P-Diagram or the Parameter Diagram [2.7]. This is a powerful tool that helps in representing and characterizing the various parameters that affect the system's output and performance. In this phase, one should identify the parameters which influence the system both in its static and dynamic conditions. Once those parameters are identified, Signal to noise ratios are evaluated for these different parameters at the output of the system, which will provide the overall functional quality of the system for particular parameter input. Selection of control parameters and possible variation ranges are performed based on their effect on controlling the system from the impact of different noise parameters. Optimization of the system is required to make it either insensitive or less sensitive to different noise parameters, which is achieved by using the control parameters of the system. Control parameters can be considered as the calibration agents of the system.

The world has changed, and so does technology. As per the original Taguchi methods, a Parameter diagram is used for parameter identification and its characterization considering the external noises a system could experience during its operation over time [2.7]. Since the technology has evolved, most of the systems are software and network-dependent. A vehicle with ADAS or Automated Driving features has complex cyber-physical systems that are always connected to the external environment by different means. These include connectivity to sensors within and outside the vehicle, connection to the infrastructure elements which are part of the environment, or connection to an external server for data transfer. As the connectivity and networks inside the vehicle increased, cybersecurity threats have also increased. Cybersecurity threats have been more common in the computer world before. Nowadays, vehicle systems have become high-power computers with multiple software's inside them, making them more prone to cyber threats.

This calls for adaptation and improvisation of the parameter diagram used for parameter identification, like in Figure 2.6. The product design must not only consider the noise parameters alone but should also consider the cyber security attacks. Any robust system design should consider possible cyber security threats and design the system to be insensitive or less sensitive to those threats. This could also help in considering a risk mitigation

FIGURE 2.6 Parameter diagram and its improved version with cybersecurity attacks.

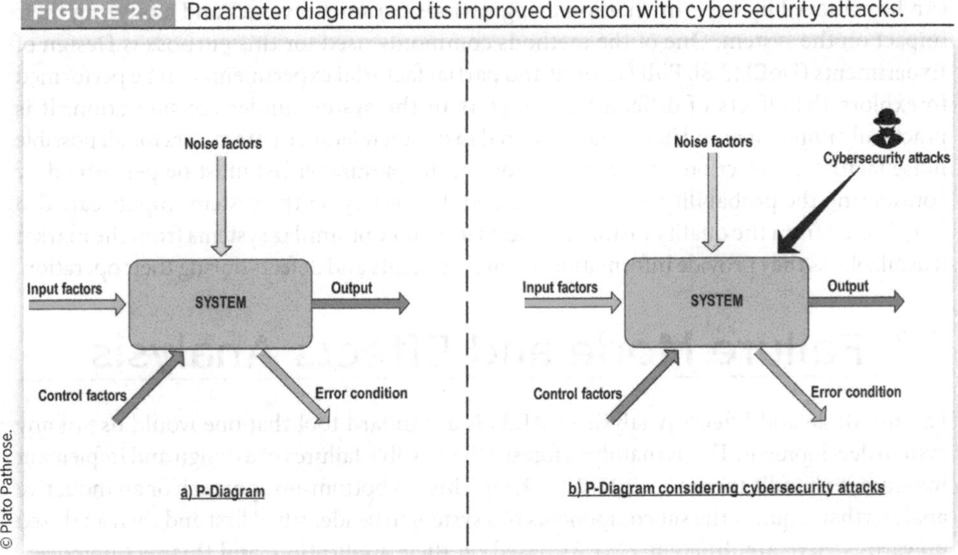

mechanism in the system like encryption or with security keys to handle attacks. If one could foresee the threat by means of cybersecurity attacks and plan mitigation actions during the system development, it would help the system to continue its operation without failure and with no loss of quality during its operation, even with these attacks. These security attacks can occur in any form [2.9]. Hence all possible cybersecurity attacks and threats should be analyzed, and the system should be designed accordingly. Common attacks include spoofing, tampering, Denial of service (DoS), Repudiation attacks or denial of true information, etc. There are certain standards that propose cybersecurity requirements and the processes to be followed for the system development and deployment in the automotive industry [2.9, 2.10]. SAE J3061, ISO/SAE FDIS 21434 are a few to mention.

The third step in Taguchi's method is the tolerance design phase. During this phase, the tolerance levels and range are calculated based on their impact on the system for each control parameter. This is the phase in which the acceptable threshold or limits of the noise parameters in the system are defined based on quality requirements. For example, one can decide in this phase for the acceptable quality of sensor data input to the system in terms of data fidelity to an Automated Driving system to operate. Or it can be a value of the highest input voltage an ECU should consider to remain in its normal operation mode. Identification of tolerance limits are important, as this will also decide and influence the ODD of an automated driving feature in the vehicle. Since it is not practical to expect a system to be operational always in an ideal environment, one way to provide better quality products and features is by defining the limits of its operating conditions where the system provides optimal performance.

Considering the complexity of systems inside the vehicle these days with multiple features like an Automated Driving system, defining tolerance limits for all identified parameters will be challenging. A practical approach would be to consider the tolerance designed for the control parameters, which would affect the basic functions and their features. As the basic ADAS functions act as the building blocks for higher levels of automation, this is a practically feasible approach for better control and management.

How do we define the tolerance for each parameter? This is not an easy question to answer and requires certain tools to determine that. The tolerance design of the parameters can be achieved by experimentation using various parameter values and analyzing their impact on the system. One of the methods commonly used for this purpose is Design of Experiments (DoE) [2.8]. Full factorial and partial factorial experiments can be performed to explore the effects of different parameters in the system under consideration. It is practically impossible and hence, not required to design tolerance parameters for all possible noise factors. A selection and prioritization on the parameter list must be performed by considering the probability of occurrence and its severity on the system. Inputs can also be collected from the quality history and service history of similar systems from the market if available, as they provide information about their faults and defects during their operation.

2.7. Failure Mode and Effects Analysis

Failure Mode and Effects Analysis (FMEA) is a standard tool that one would use in any system development. This is mainly to foresee the possible failures of a design and implement measures to handle them [2.3, 2.11, 2.12, 2.13]. This is a bottom-up approach or an inductive analysis that requires the subcomponents of a system to be identified first and then analyzed upwards. There are different FMEAs based on their application, and this section covers only the Design FMEA. FMEA can be started early as soon as requirements are identified and classified to various technical requirements of the system. This can be started directly after performing the QFD Phase one, explained earlier in this chapter.

The starting point for FMEA can be toward the end of the concept phase in product development and can extend up to its production. If there are multiple iterations for the system development as A sample, B sample, or C sample, FMEA should cover those as well. Each version of the system is improved from its previous version, and this needs to be checked and adapted based on the failures identified and measures implemented. For an Automated Driving system, the system concept and the high-level architecture developed after the concept phase can be the starting point for performing FMEA. Before starting with FMEA, one has to remember that it is a team activity that requires the participation of experts from different subdomains. In many projects, it is a common mistake that the responsibility of performing FMEA is assigned to one or two people without the actual participation of experts. This is not an effective way for using this tool as it works based on knowledge and brainstorming from different areas of product design.

FMEA can be executed with a minimum five-step process [2.12], or with seven steps. Here the five-step process is explained and that which is widely followed in the industry. The seven-step process was introduced by AIAG (Automotive Industry Action group) and VDA (Verband der Automobilindustrie e.V.) in their recent publication [2.11] (Figure 2.7):

1. Structural Analysis
2. Functional Analysis
3. Failure Analysis
4. Countermeasure Analysis
5. Optimization of the System

System structural analysis starts with analyzing the system for failures, considering the structural decomposition of the system, and identifying possible failures associated. A minimum of three levels of decomposition are required to perform FMEA, which are

Failure effect → Failure mode → Failure cause

FIGURE 2.7 Five-step process of performing FMEA.

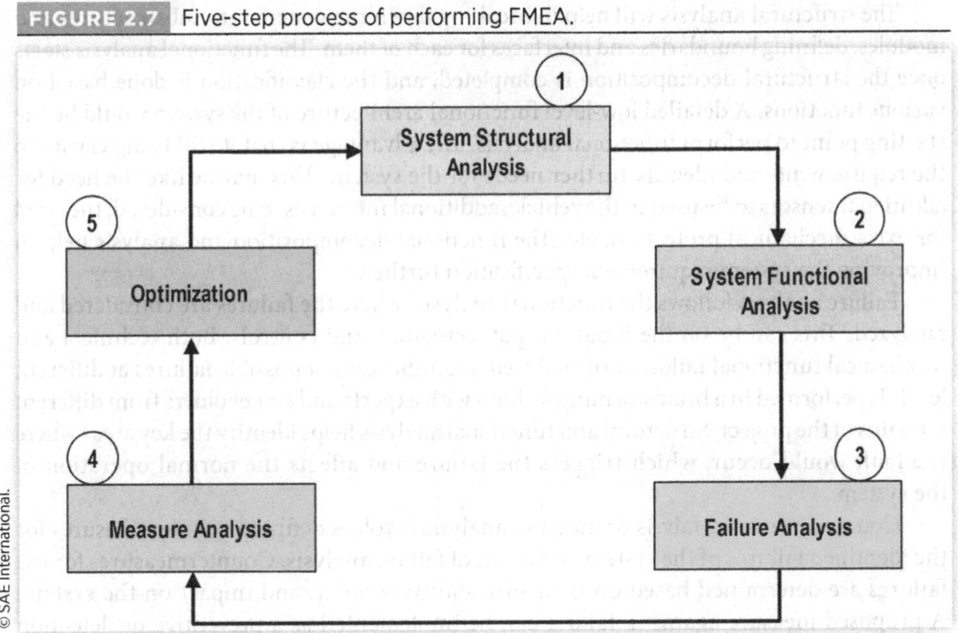

As an Automated Driving system is a safety-critical product, a different approach of classification needs to be considered and followed as per the functional safety standard: Level 4 (Hardware/Software function failure cause) → Level 3 (Failure mode, this is comparable with technical safety concept in functional safety) → Level 2/Level 1 (Failure effect, this is comparable to functional safety concept or safety goal) (Figure 2.8).

FIGURE 2.8 Structural analysis in FMEA.

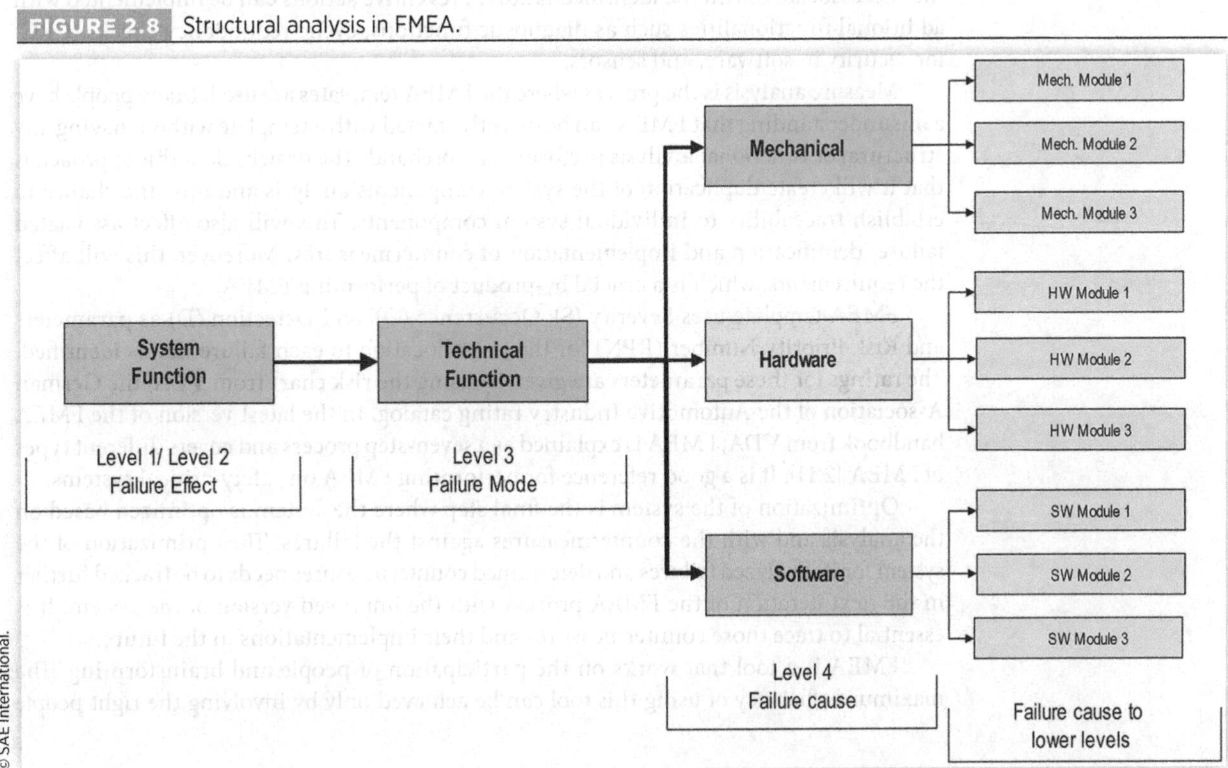

The structural analysis will help in breaking down the system into substructures and modules, defining boundaries and interfaces for each of them. The functional analysis starts once the structural decomposition is completed, and the classification is done based on various functions. A detailed low-level functional architecture of the system would be the starting point to perform functional analysis. The advantage is that it will bring clarity to the requirements and identify further needs for the system. This may be like the need for additional sensors to be used in the vehicle, additional interfaces to be considered, the need for extra mechanical protection, etc. The functional decomposition and analysis help in improving the system requirement specification further.

Failure analysis follows the functional analysis, where the failures are considered and analyzed. This can be on the input, output, actuators, and controls. Both technical and mechanical functional failures are analyzed. Identification of possible failures at different levels is performed in a brainstorming fashion with experts and stakeholders from different domains of the project. Structural and functional analysis helps identify the key areas where the fault would occur, which triggers the failure and affects the normal operation of the system.

Countermeasure analysis or measure analysis involves defining countermeasures for the identified failures of the system as a result of failure analysis. Countermeasures for the failures are determined based on their probability, severity, and impact on the system. A proposed measure against a failure can be implemented as a preventive or detection measure. That solely depends on the phase of the project in which it should be considered. Preventive measures are considered when in the system development phase, for example, reviews and worst scenario calculation. The detection measures are considered while the product is in the testing phase or on a finished product, for example, hardware tests, integration tests, and image quality tests of the camera. The actions for the abovementioned measures should add additional requirements and must be implemented depending upon the risk associated with the identified failure. Preventive actions can be implemented with additional functionalities such as diagnostic functions, cyclic redundancy checks (CRC) for security in software, and sensors.

Measure analysis is the process where the FMEA templates are used. Many people have a misunderstanding that FMEA can be directly started with a template without having any structural or functional analysis performed beforehand. The drawback of this approach is that it will create duplication of the system components analysis and miss the chance to establish traceability to individual system components. This will also affect associated failure identification and implementation of countermeasures. Moreover, this will affect the requirements, which is a crucial by-product of performing FMEA.

FMEA template uses Severity (S), Occurrence (O), and Detection (D) as parameters and Risk Priority Number (RPN) for the risk allocation to each failure that is identified. The ratings for these parameters are given by using the risk chart from VDA, the German Association of the Automotive Industry rating catalog. In the latest version of the FMEA handbook from VDA, FMEA is explained as a seven-step process and covers different types of FMEA [2.11]. It is a good reference for performing FMEA on safety-critical systems.

Optimization of the system is the final step where the system is optimized based on the analysis and with the countermeasures against the failures. The optimization of the system for its analyzed failures and determined countermeasures needs to be tracked further in the next iteration of the FMEA process with the improved version of the system. It is essential to trace those countermeasures and their implementations in the future.

FMEA is a tool that works on the participation of people and brainstorming. The maximum efficiency of using this tool can be achieved only by involving the right people from different domains during product development. In any product development, one can have an overview of the product maturity by referring to the FMEA performed on that product. In most of the projects in the industry, FMEA is not performed in its right sense but only for achieving process compatibility rather than improving the overall quality of the product.

2.8. Summary

Designing a product is an important stage that determines product quality. In this chapter, we have covered some of the different methods and tools used for designing a robust product. When developing a complex system like ADAS or Automated Driving systems, it is important to follow structured methods for design and development. There are hidden advantages in that which will improve the overall quality of the product and the customer satisfaction. In most projects, either due to time constraints or other reasons, standard methods and tools are not used. This will affect the quality of the final product and its features and even dissatisfaction at the customer end. Moreover, this will add up additional expenses in the long term for production, service, and maintenance in the aftermarket scenario. This chapter aims to familiarize the readers with certain best practices and methods to design and develop an economical product with good quality.

As there is a lot of startup ecosystem and agility around ADAS and Automated Driving systems in the market, there is a misunderstanding in the industry that once a prototype is developed, it can be easily put into production for a vehicle program. Getting a system production-ready is a complex process, and one should expect a lot more challenges while doing so to deploy a quality and safe product. Some may even fail because of not following structured methods and processes and trying to take shortcuts in making the system production-ready. A piece of advice to the reader is that if you are developing a complex or even safety-critical product, you must follow those standard methods and processes using the right tools. This would be complex, but it would be advantageous for building a quality and safe product for the users in the long run. Sacrificing on the product safety due to the project time constraint is not an excuse.

Developing a complex safety-critical system like an Automated Driving system requires robust verification and validation methods. Testing is a way to measure the quality of your product. If you cannot measure, you cannot improve. The next chapter will cover different verification and validation methods used and the types of test environments commonly used in the industry.

References

2.1. Kano, N., Seraku, N., Takahashi, F., and Tsuji, S., "Attractive Quality and Must-Be Quality," *Journal of the Japanese Society for Quality Control* 14, no. 2 (1984): 39-48 (in Japanese).

2.2. Sager, I., "Before IPhone and Android Came Simon, the First Smartphone," *Bloomberg Businessweek*, June 29, 2012.

2.3. Hanington, B. and Martin, B., *Universal Methods of Design* (Beverly, MA: Rockport Publishers, 2012), ISBN:9781610581998.

2.4. ISO/IEC TS 33061:2021, "Information Technology—Process Assessment—Process Assessment Model for Software Life Cycle Processes," 2021.

2.5. Ficalora, J.P. and Cohen, L., *Quality Function Deployment and Six Sigma: A QFD Handbook*, 2nd ed. (New Jersey: Addison Wesley Engineering, 2009), ISBN:978-0135138359.

2.6. QFD, "Introduction to Quality Function Deployment (QFD)," accessed March 3, 2021, https://quality-one.com/qfd/.

2.7. Juran, J.M. and Gryna, F.M., *Quality Planning and Analysis from Product Development through Use*, 3rd ed. (New York: McGraw-Hill, 1993), 256.

2.8. Folkes, W.Y. and Creveling, C.M., *Engineering Methods for Robust Product Design Using Taguchi Methods in Technology and Product Development* (Reading, MA: Addison-Wesley Publishing Company, 1995), ISBN:0201633671.

2.9. Sommer, F., Dürrwang, J., and Kriesten, R., "Survey and Classification of Automotive Security Attacks," MDPI Article, April 19, 2019.

2.10. SAE International, "Cybersecurity Guidebook for Cyber-Physical Vehicle Systems," SAE Standard J3061, 2016.

2.11. Automotive Industry Action Group, *AIAG & VDA FMEA Handbook* (Automotive Industry Action Group, 2019), ISBN:978-1605343679.

2.12. Carlson, C., *Effective FMEAs: Achieving Safe, Reliable, and Economical Products and Processes Using Failure Mode and Effects Analysis*, 1st ed. (Hoboken, NJ: Wiley, 2012), ISBN:978-1118007433.

2.13. Chrysler LLC, Ford Motor Co, General Motors Corp, *Potential Failure Mode and Effects Analysis (FMEA)* (Southfield, MI: Chrysler LLC, Ford Motor Co., General Motors Corp., 2008), 21.

/Autonomous
/Sensing
/Communication
/Battery
/Navigation
Autopilot mode

3

Different Test Approaches

Testing is required of any product to evaluate and measure the anticipated quality and functionality of the designed product. Generally, in the industry, testing is not considered an attractive job, unlike development. It is either because of misunderstanding or because of lack of knowledge. Appropriate testing needs to be performed from time to time, especially for complex systems like ADAS and Automated Driving systems. This is to prove that the system is functional and is performing as desired, and the users are safe to use the system without any faults that could cause hazards.

This chapter will discuss different test approaches used to develop and deploy ADAS and Automated Driving systems and features in the vehicle. Most of the testing approaches are the same as those used for other automotive systems. The main difference here is the complexity of many subsystems involved and the processing of huge amounts of data. Unlike other classical automotive products, it becomes more complex and challenging because of the amount of software and neural networks used for realizing functions. The test methodologies and approaches need to be adapted and improved to test and evaluate those subsystems, making the testing complex and challenging compared to classical automotive systems.

3.1. Verification and Validation

Verification and Validation are two standard terms used by a test engineer quite often. How would you differentiate them and plan to execute them for a project?

Verification refers to the evaluation of a system based on a defined set of requirements. Those requirements are defined for a specific product during the requirement elicitation phase and were used as the foundation to generate subrequirements for development. In other words, one could describe verification as the compatibility check of a system with its specifications to see if it meets the requirements defined for its development.

Validation refers to the evaluation of a system based on its use cases. It checks whether the system which was developed addresses the use cases for its customers. The evaluation here focuses on the end user and the expectations of the end user. Specific use cases are captured during the requirement elicitation phase for ADAS and Automated Driving systems before converting them into technical requirements. These use cases are further decomposed to functional and nonfunctional requirements with defined acceptance criteria. It is sometimes very difficult to define acceptance criteria during the validation phase. For example, the use cases of adaptive cruise control, one of the comfort features of ADAS, have been defined so that the behavior of those features should have cruise mode, the vehicle follow mode, etc. Those use cases can be evaluated with defined test cases, but a human factor is associated with "comfort," which is subjective to the end user. How can that be measured? Because it is subjective to the end user, it cannot be measured with a single test case and a single tester but with multiple test cases and different testers. It is also possible to have a data-driven approach for measuring the comfort of a particular feature using attribute agreement analysis. Usually, verification and validation will not go to that depth of evaluating the comfort of the feature by many vehicle manufacturers or their suppliers in the ADAS and Automated Driving.

Quality assurance is another term that is used frequently in the industry with different meanings. Quality assurance refers to the processes followed to have a quality product as output. This is part of quality management rather than verification and validation. As explained in the previous chapter, the primary purpose of having processes is to reduce wastage. Having a good process to develop and evaluate the system will help reduce wastage in terms of rework, effort, etc., thereby saving cost for the product deployment and the project. This means quality assurance comes into play during the design and development phase and the verification and validation phase. This is a proactive approach to reducing defects. Quality assurance, along with verification and validation, helps boost the confidence in the developed system for its quality and performance and makes it ready for market deployment.

3.2. Agility in Verification and Validation

Agile or being agile is a common term used these days in the automotive industry. What is its actual relevance in today's world? Being agile means moving forward even in the presence of ambiguities and bringing improvements continuously, which is acceptable and satisfies the end customer. Taking an agile approach will help move forward even when one has to deal with complex projects like the ADAS and Automated Driving systems. How would one develop and test a product based on ambiguous inputs and requirements? The answer is to be agile in verification and validation following the processes and methods. Bringing agility to verification and validation will help inspect and adapt the system development, performance, and quality with continuous outputs even in the middle of ambiguities.

In the automotive industry, the commonly followed approach for system development is V-model. V-model refers to the graphical representation of the system development life cycle in the shape of the English alphabet "V." The requirement elicitation starts from the top left-hand side followed by different development steps toward the bottom corner. The right-hand side consists of processes and steps related to testing from the smallest component to the system level while traversing toward the top. The need for adapting to agile verification and validation processes over the classical V-model approach has many reasons, especially when dealing with complex features in a complex system like an Automated Driving system. It is impossible to list out all the requirements for the system for its development early in the development phase. Instead, these requirements can be defined iteratively. The system development can incorporate these requirements as its features in a continuously improving fashion and fix the gaps in the system behavior with precise requirements. Previously, a vehicle development program from the concept phase to its production took around five to six years, but today, this time got reduced to two or three years, and many platform-based approaches are used to bring the products quickly to the market. Because development happens in an iterative manner starting from the requirements, the following testing, also done in an iterative manner, would help bring value in the product development and quality of the product.

3.3. Different Levels of Testing—A Reference from V-Model

This section discusses different levels of testing that have to be taken care of in the development and deployment of ADAS and Automated Driving systems. It also explains different types of testing that should be considered at each level and their benefits. When comparing various test types at different levels and relating them with the process assessment model recommended by "ISO/IEC 15504 Information technology—Process assessment" in the form of Automotive SPICE (Software Process Improvement and Capability Determination) [3.1, 3.2], a life-cycle model specific for testing can be defined as shown in Figure 3.1. It is adapted from the classical product life-cycle V-model considering various test types and associated confidence levels in verification and validation.

For the verification and validation part, which lies on the right-hand side of the V-model, software unit testing is the building block where one would start with the test activities. This refers to verifying the software units against their requirements derived from the system and software requirements. These software units will also be checked against specific guidelines and standards that are to be followed for the software development, such as the Motor Industry Software Reliability Association (MISRA) [3.3, 3.4]. Also this can be based on specific standards or coding guidelines defined internally by an organization.

Almost all organizations have certain guidelines defined for software programming. These include specific rules that need to be followed to guarantee readability and to have quality software being developed. As an example, some of those guidelines include the number of maximum characters in a line of code, number of lines of code in a file, maximum number of functions which can be defined in a single file, types of variables defined, function calls, compatibility with MISRA standard, etc.

Once all the developed software units are tested, they need to be integrated. For the integration of these software units, the interfaces need to be tested along with the evaluation of cross-unit function calls and data transfers among these software units. Usually, requirements need to be defined for the integrations of software units, their functionality, and performance. Software units are integrated to build software components. These components are tested and integrated to build the complete software package. Tests need to be performed at all these different levels to evaluate the quality of the software, its functionality, and performance based on the requirements and the use cases.

The process and methods used for integration testing at the system level are some of the most misunderstood areas in product development. In many projects, system integration

FIGURE 3.1 An iterative V-model of development and test processes.

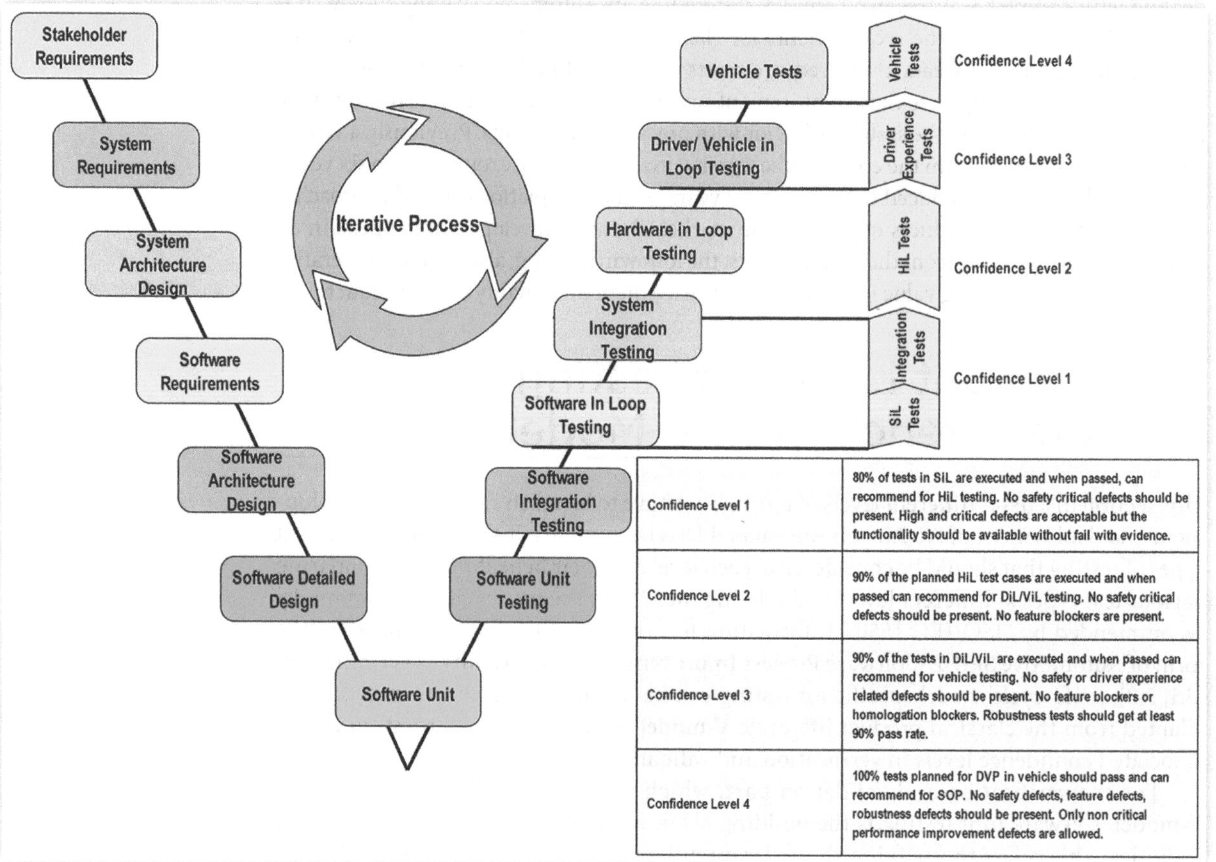

refers to the integration of different system components or subsystems to build a system. However, in certain other projects, the integration tests cover hardware and software integration and associated tests. Both types of tests are required for a complex and safety-critical system like Automated Driving systems, and there is no choice to select one over the other. The decision on what needs to be tested should be defined based on the system boundary and what lies within the organization's control.

Once the integration tests are completed, the next step is to perform system-level verification and validation. Different types of tests are executed during these phases to ensure that the system meets the requirements and the use cases.

3.4. Defects at Different Levels of Testing

Testing is usually performed at different stages of product development. This can be considered as different levels when comparing it with the right-hand side of the V-model. It would help to identify the defects as early as possible, thereby reducing the cost of fixing those defects if they were moved to the final product. A generic representation of the defect

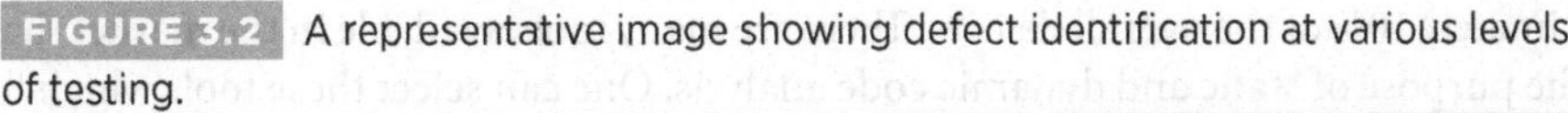

FIGURE 3.2 A representative image showing defect identification at various levels of testing.

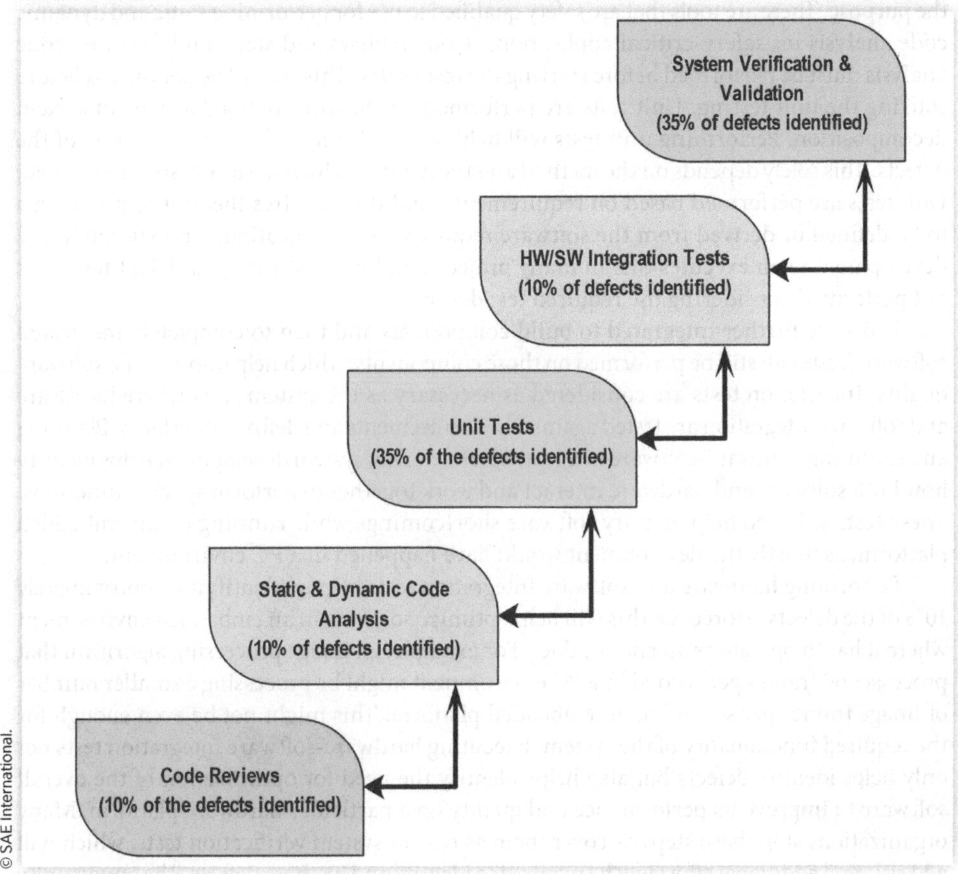

distribution over a complete project when mapped to different test levels can be represented as in Figure 3.2. This is a generalized view of defect distribution collected from multiple production programs of ADAS products. These can vary based on the projects and the processes followed for development and testing. The number of defects identified at each test level can vary depending on the team's skillset, test depth considered in each level, methods used, and the organizational process maturity. In many organizations, all these test levels are not executed for the test life cycle for product development.

Walking through different steps, as shown in Figure 3.2, it can be found that adequate reviews performed during the software development help in identifying at least 10% of the overall defects in a product during its development phase. Reviews can be of any type, such as self-reviews, peer reviews, or even group reviews. Reviews should be performed on the technical content of the code by technical experts. Otherwise, there will always be a drawback of having a beautifully written code following all guidelines and standards but not functioning as intended. This is one of the most common challenges various project teams face during product development. The reviews are usually performed for compatibility with coding guidelines and not on its technical content.

Using tools to perform static and dynamic code analysis can help in identifying the behavior of the code during its operation. This analysis helps in identifying and reducing approximately 10% of the overall defects. There are many tools available in the market that serve the purpose of static and dynamic code analysis. One can select these tools which fit the purpose. There are tools that are safety qualified to use for performing static and dynamic code analysis for safety-critical applications. Code reviews and static and dynamic code analysis must be performed before starting the test cycles. This should be completed before starting the unit testing. Unit tests are performed on the units defined as part of system decomposition. Performing unit tests will help in identifying at least another 35% of the defects. This solely depends on the method and the depth at which the unit tests are executed. Unit tests are performed based on requirements, and this requires the unit requirements to be defined or derived from the software requirement specification. Unfortunately, the development team executes this in many projects, and in most cases, sufficient testing is not performed considering the required test depth.

Units are further integrated to build components and then to completely integrated software. Tests can still be performed on those components, which help improve the software quality. Integration tests are considered as necessary as the system tests where hardware and software integration are tested against the requirements and defined interfaces. Planning and executing hardware-software integration tests during system development helps identify how both software and hardware interact and work together to perform specific functions. These tests will also help identify software shortcomings while running on an embedded platform, as mostly the development would have happened in a PC environment.

Performing hardware and software integration tests help in identifying approximately 10% of the defects. Moreover, this will help optimize software in an embedded environment where it has to operate as an end product. For example, an image processing algorithm that processes 60 frames per second in a PC environment might be processing a smaller number of image frames per second in an embedded platform. This might not be even enough for the required functionality of the system. Executing hardware-software integration tests not only helps identify defects but also helps identify the need for optimization of the overall software to improve its performance and quality on a particular hardware platform. Many organizations skip these steps or cover them as part of system verification tests, which will add risk, and additional effort will be required to identify and fix those defects. The opportunity to optimize and calibrate is a by-product of performing hardware-software integration tests.

System verification and validation come as the last steps of product testing. A system can be defined here as a complete integration of all subsystems that must be verified according to the defined set of requirements. These requirements can be detailed technical specifications of the system or use cases. Approximately 35% of the defects are expected to be identified at this phase of testing. This is usually performed independently from the product development team and will influence a lot in the product development. This is the phase where the evaluation of the product is made to check if it fits the end user.

3.5. Simulation and Testing

Simulation refers to imitating the real-world operation of a system in a virtual environment. Simulations can be performed early in the development phase of a system to understand its behavior when it is exposed to the real world. There are many tools available in the market that could replicate a real-world environment and operational conditions of systems. One cannot ignore the usage of simulation tools these days when developing and testing complex systems like ADAS and Automated Driving systems. While selecting simulation tools for different purposes, one should be cautious in identifying and selecting qualified tools for those purposes. Qualification of the simulation tools measures its correlation with the real world and its performance and known limitations, including the probability of output error values. There is no guarantee that a simulation tool matches exactly the same as the real world. The major part of simulation tools is software and is used for both development and testing phases in the product development. For the development of ADAS and Automated Driving systems, there are many types of simulation tools used. This starts from the simulation of sensors in the vehicle to the whole environment around the vehicle and extends to vehicle characteristics and attributes.

Simulation is the most common and widely used method for testing ADAS and Automated Driving systems [3.5]. They provide flexibility for both the development and the testing teams to simulate the real-world behavior in a laboratory or an artificial environment rather than building a system and evaluating it in the real world. Such an environment will help run early testing with models and software components even before the actual system is built. This helps to evaluate the design and the implementation before even producing any of those parts. There are a lot of tools and integrated toolchains used for different types of simulation done in the industry [3.6]. Some of the commonly used simulation methods in ADAS and Automated Driving system development are explained in this section.

3.5.1. Model-in-the-Loop Simulation

As the name suggests, model-in-the-loop simulation is the simulation of models of a system as in the real world by replicating the external environment and associated input vectors. The concept comes from model-based development of a system or software where it is easy to identify and fix the defects or bugs even before developing and writing codes. A model-based development approach is used mainly to prove a concept and design by creating models and evaluating its operation in the real world by replicating the real-world environment [3.7]. Simulation is not only used for testing but also for prototyping and evaluating concepts and designs. Models can be developed based on a concept or design for a complete system or the smallest system unit. There is no need to have a defined set of requirements if the model is developed to prove certain concepts or designs which have not been evaluated before. The advantage of using model-based development is that the development team is aware of all the interfaces and the parameters associated with a system or software even before the actual development of the software and writing lines of codes. At the same time, the testing team can use the outcome of the model simulation to design their test cases, considering all possible inputs and test vectors.

There are many model-based development and testing tools available in the market, which help model the system and replicate the real-world environment [3.7, 3.8]. These simulation tools can also be integrated as part of toolchains used to automate the input vectors to these models and convert those models to actual lines of code. Model-based development and testing are not a standard process followed in all organizations, even though few organizations insist on using it. This might be because of the time for the project execution or the nonavailability of the models for the subsystems and components that match the product. It is recommended to consider model-in-the-loop simulation for developing complex systems like ADAS and Automated Driving systems where all the requirements cannot be listed early in the development phase, and the cost associated with testing the actual product is very high.

3.5.2. Software-in-the-Loop Simulation

Software-in-the-loop simulation refers to the simulation of the integrated software or its components and its operation in the real world utilizing simulation tools [3.7]. Even though the software needs to run in an embedded environment in the real world, the simulation can be performed in a PC environment in a laboratory. The simulation environment can replicate the operating system, middleware, and the application layers like the embedded environment for the software to operate. A set of simulation tools and frameworks are used as part of the toolchain for the software-in-loop simulation. These tools can also be established in a scalable manner for an individual software unit or completely integrated software.

Environment simulation and vehicle simulation tools are pervasive for the software-in-the-loop simulation for ADAS and Automated Driving. With the help of these tools, it is possible to perform software simulations replicating the real-world environment where it would operate and evaluate different ADAS and Automated Driving features. This will be beneficial, especially for the algorithms utilizing input from sensors exposed to the physical environment. There are many tools available in the market capable of providing environment simulations, sensor simulations, vehicle simulations, etc. A software-in-the-loop simulation environment would include one or more of these simulation tools integrated as part of the simulation toolchain.

For the simulation environment to achieve better correlation with the real world, an approach of co-simulation is used in the industry [3.7, 3.9]. This is achieved by using multiple simulation tools together over a middleware, thereby utilizing the benefits from multiple tools that are part of the simulation environment. Using co-simulation for establishing the simulation environment is based on the approach that each simulation tool has its strengths and weaknesses and using multiple tools to utilize their strengths together in replicating the real-world operating conditions. Some simulation tools have their strength in providing better environment simulation but at the same time not good in the area of vehicle simulation. Some others provide better vehicle dynamics simulation but average or below average environment simulation or sensor simulation. In a co-simulation environment, one could use multiple simulation tools operating together at the same time and exchanging information with the system software under test. Usually, a co-simulation environment consists of two or more different simulation tools to provide input vectors to a system model or the system software. The response of the system software is visualized and analyzed for its compatibility, performance, and correctness for the requirements and use cases (Figure 3.3).

3.5.3. Hardware-in-the-Loop Simulation

A Hardware-in-the-Loop (HiL) simulation simulates the complete system, including hardware and software, in a laboratory environment for its real-world operations. This includes a set of hardware and software tools and infrastructure to replicate the real-world operation of the system in a laboratory environment. This is usually performed as part of the development and system-level verification and validation [3.10]. The test infrastructure is complex, which includes real-time hardware and software components to generate inputs to replicate the real-world operations of the system. HiL simulation is one of the most critical tests considered for ADAS and Automated Driving systems. These tests help to evaluate the system-level functionalities and their robustness in a laboratory environment. For the ADAS and Automated Driving applications, there are commonly three types of configurations used.

In the first configuration, as shown in Figure 3.4, the HiL environment consists of an ECU with all the sensors attached to it as in the actual vehicle. This can be set up in a laboratory

FIGURE 3.3 A software-in-the-loop co-simulation environment for scenario-based testing.

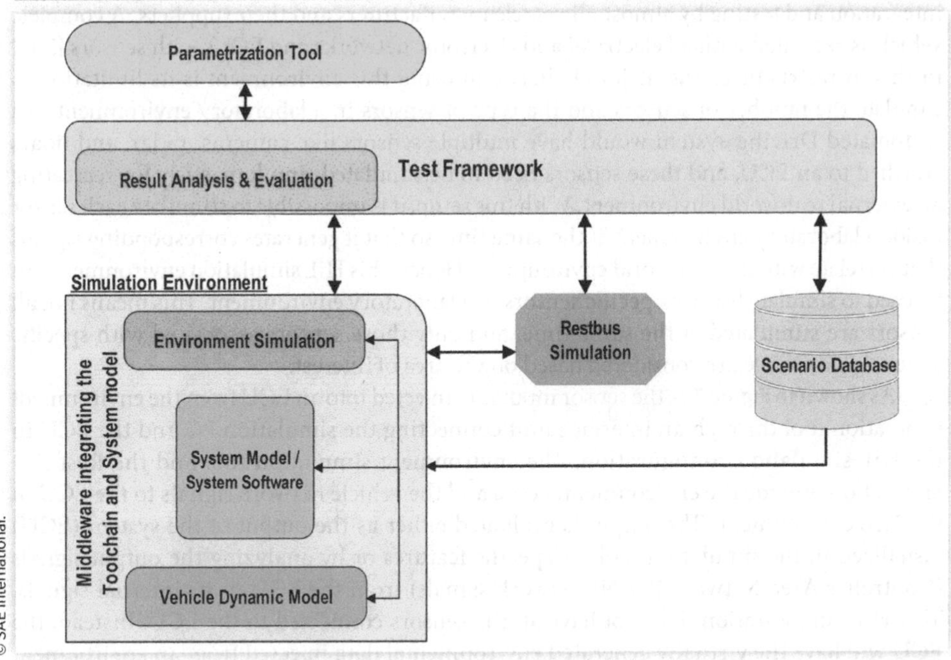

FIGURE 3.4 A HiL simulation environment using original sensors.

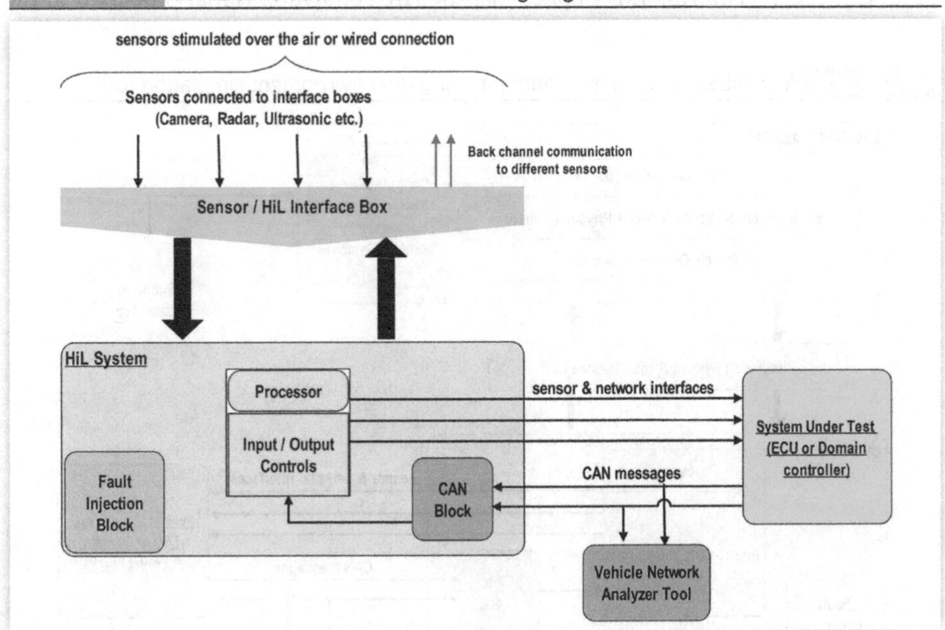

environment as a portable test environment or even scaled to the full vehicle as a vehicle integration platform commonly called Labcar. Labcar is a common platform used for vehicle integration and testing by almost all vehicle manufacturers and their suppliers. A complete vehicle is recreated with all electrical and electronic networks and ECUs with sensors [3.11] in this infrastructure. The major challenge in using this environment is its limitation to simulate the number of sensors and the type of sensors in a laboratory environment. An Automated Driving system would have multiple sensors like cameras, radar, and lidars attached to an ECU, and these sensors need to be simulated simultaneously for recreating an external real-world environment. With this setup, it is impossible to stimulate each sensor inside a laboratory environment at the same time so that it generates corresponding signals that correlate with the real-world environment. Hence, this HiL simulation environment can be used to simulate feature-specific sensors in a laboratory environment. This means not all sensors are stimulated at the same time, and only those sensors associated with specific features in a vehicle are considered based on the area of interest.

As shown in Figure 3.5, the sensor inputs are injected into an ECU from the environment simulation tool through an interface unit connecting the simulation PC and the ECU in the HiL simulation configuration. The environment simulation tool and the Rest Bus Simulation provide the environmental data and the vehicle network signals to the ECU in the HiL environment. The output is evaluated either as the output of the system (ECU) visualized in the simulation tool for specific features or by analyzing the output signals (Controller Area Network [CAN] network signals) from the ECU or its internal signals. This HiL configuration does not have all the sensors connected to the ECU. Instead, the ECU will have the virtually generated environmental data injected from an environment simulation tool. This configuration is usually used for software and system verification and validation. In software testing, this configuration is mostly used to test and optimize the software components early in the development phase and evaluate the software performance in an embedded platform. This implies that the HiL simulation can also be used in the

FIGURE 3.5 A HiL simulation environment using the environment simulation tool.

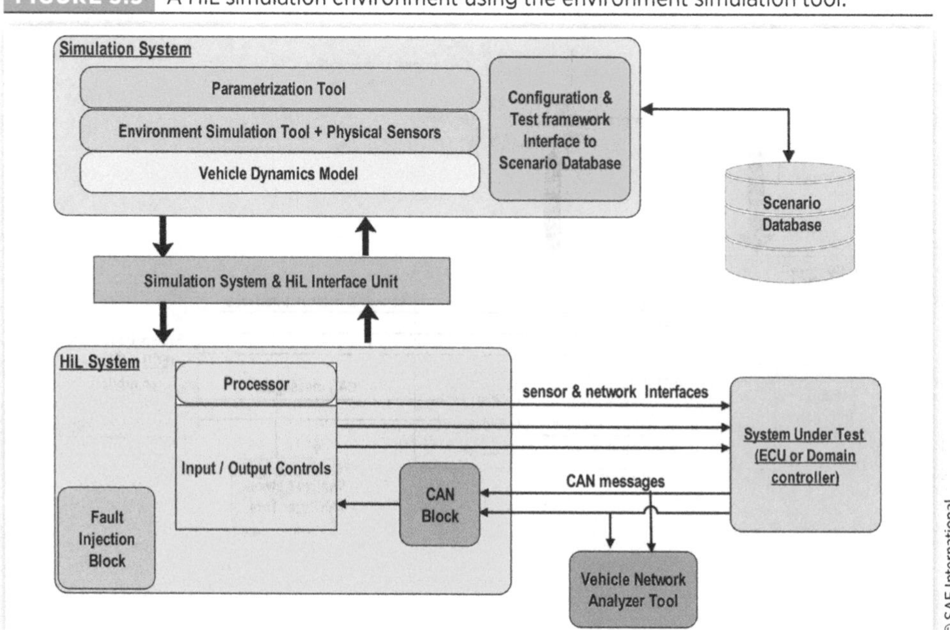

development phase as a software testing platform using prototype hardware or even with development boards where the operational software can be flashed and evaluated.

Meanwhile, system verification is performed with this configuration to evaluate the functionalities and features in a vehicle. This uses either a prototype or production-intended ECU and software for simulation. In this configuration, usually, the sensor data that needs to be injected into an ECU is planned based on the features to be tested as it is challenging to virtually generate and inject sensor data from more than ten sensors (camera, radar, lidars, etc.) to an ECU in a time-synchronized manner. There are many solutions available in the market that claim to support the injection of a higher number of sensor channels, but they need to be evaluated. The major challenge is the need for high-end infrastructures like simulation PCs and interface cards and the limitations caused by the simulation tools when multiple systems are used in parallel.

There are multiple sensors connected over different types of harnesses for an Automated Driving system to provide environmental data. The HiL simulation should have the whole system integrated with its subcomponents over its harnesses. It is impossible to generate realistic sensor data as required for any robust test approach like a real-world environment in a laboratory environment. This can be addressed either by using a simulation tool to simulate the environmental data or by using precaptured environmental data injected into the system or the ECU.

Scenario-based testing is one of the critical testing types for an Automated Driving system. One of the approaches used for scenario-based testing will be discussed in the next chapter. Here scenarios act as inputs for a vehicle from the environment. The sensors in the vehicle should capture those from the environment and transfer them to the Automated Driving system. The evaluation is made based on the system behavior for those scenarios, checking if it matches with what has been expected based on the system design. The most common approach used in the HiL simulation uses an environment simulation tool to generate scenarios and feed the corresponding environmental data to the ECU under test.

Another configuration used in the HiL simulation is the sensor data re-simulation, as shown in Figure 3.6. Here the real-world environmental data from the vehicle is precaptured using data loggers, and later these data are injected into the ECU in the HiL test environment. One of the significant challenges in following this approach is capturing environmental data for many scenarios. This is an expensive approach considering the expenses associated with data capturing and re-simulation.

However, having a HiL environment with re-simulation will provide the real-world data from the vehicle as input to the ECU than any virtual environment data. This will help the system experience the real-world environment and corresponding sensor data in the simulation environment, which would correlate to reality and is much better than the simulated environmental data provided by any of those environment simulation tools. In most of the production projects, multiple HiL configurations are used to achieve the required coverage in testing using scenarios and evaluate the robustness of the software and the system when exposed to the real world and synthetic data in a laboratory environment.

A HiL simulation configuration using the re-simulation approach requires data loggers to be integrated into the vehicles, capable of recording environmental data from all the sensors in a time-synchronized manner [3.10]. The HiL environment should be capable of replaying these data to the ECU in a simulation environment also in a time-synchronized fashion like in the actual vehicle. This is one of the major challenges faced by almost all organizations using the re-simulation approach in the HiL environment. The re-simulation in HiL will get complex with the type of data loggers used and the data required to be captured and replayed. The simulation system should handle huge bandwidth of data

FIGURE 3.6 A HiL simulation environment for re-simulation of sensor data.

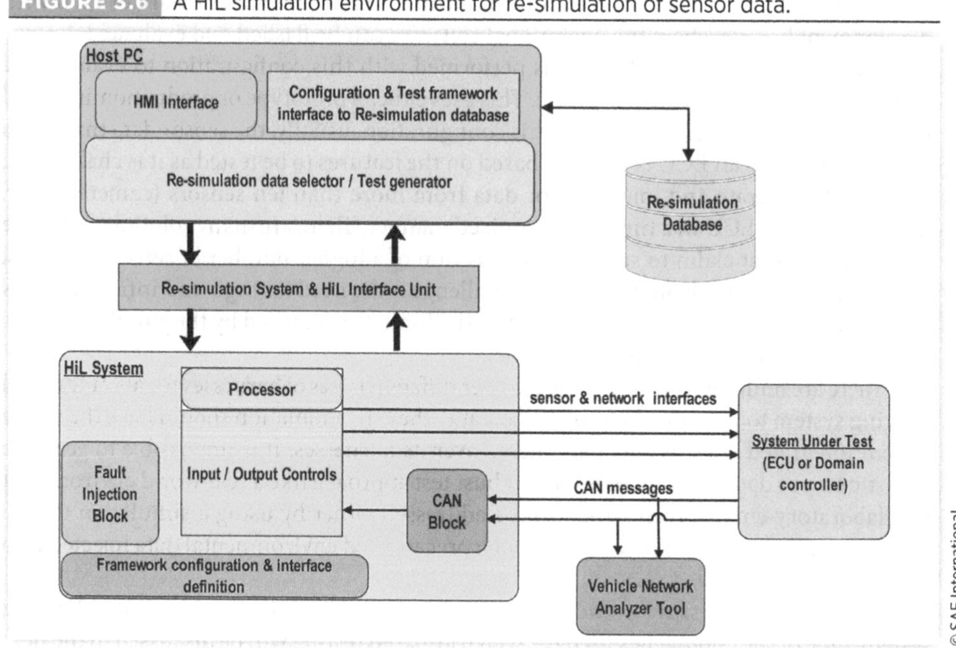

from multiple sensors and from different scenarios, which will be in the size of Terabytes or Petabytes of data. Data logging and its usage will be explained in detail in further chapters.

3.5.4. Driver-in-the-Loop Simulation

The Driver-in-the-Loop (DiL) simulation brings the human into the loop for simulation and testing purposes in a laboratory environment. In all the ADAS and Automated Driving features in the vehicle, the driver is involved in a monitoring role or in controlling mode when these features are engaged and are in operation [3.12]. The DiL simulation helps in simulating these conditions with a real driver inside a laboratory environment so that the influence of human factors during the operation of these features can be evaluated, and the system can be designed accordingly. DiL simulation is primarily used to test driver experience in different ADAS and Automated Driving features in a vehicle environment. The same test environment can evaluate the effect of human factors in different assisted and automated driving features and test the mode-switching functionality with the vehicle and the driver in assisted and automated driving. Mode switching is switching the control of the vehicle from the Automated Driving system and the human driver when the vehicle goes beyond any predefined ODD or if the system encounters any failures [3.13].

DiL simulation includes a simulation toolchain, a movable platform that replicates the vehicle dynamics, Human-Machine Interface (HMI), steering wheel, and other actuators used to control the vehicle, as in Figure 3.7. A DiL simulation is also used to calibrate ADAS or Automated Driving features by fine-tuning the control signals fed into the vehicle network from the ECU [3.12]. DiL simulation is more widely used by the ADAS and Automated Driving system suppliers than the vehicle manufacturers as these system suppliers have to work with different vehicle manufacturers to deliver the system with various features. This test environment is a handy simulation infrastructure to evaluate and calibrate the features, including vehicle dynamics inside the laboratory.

FIGURE 3.7 A DiL simulation environment.

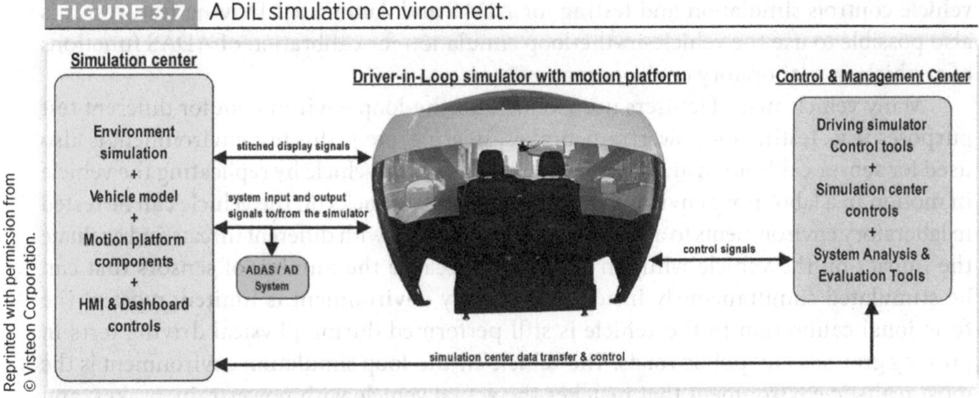

3.5.5. Vehicle-in-the-Loop Simulation

Simulation of the whole vehicle operation as in the real world with virtually generated or prerecorded environmental data for different scenarios in a laboratory environment is executed with the help of a vehicle-in-the-loop simulation environment. The primary purpose of vehicle-in-the-loop simulation is to evaluate the behavior of the complete vehicle in different scenarios inside a laboratory environment [3.14]. The input for the sensors in a vehicle would be generated virtually using an environment simulation tool along with different road models, which provide the input for vehicle dynamics in the simulation environment. The behavior of the vehicle in different scenarios, in different road conditions for the ADAS and Automated Driving features, can be evaluated with this simulation environment. This is a test environment mainly used at vehicle manufacturers' sites. The vehicle-in-the-loop simulation environment is widely used as an integration and test platform for the entire vehicle by many vehicle manufacturers. The sensor inputs are either simulated, stimulated, or directly injected into the ECU of a vehicle. The response is evaluated from the vehicle level. There are different configurations possible for the vehicle-in-the-loop simulation. One of them is shown in Figure 3.8 using the test environment for

FIGURE 3.8 A vehicle-in-the-loop simulation environment.

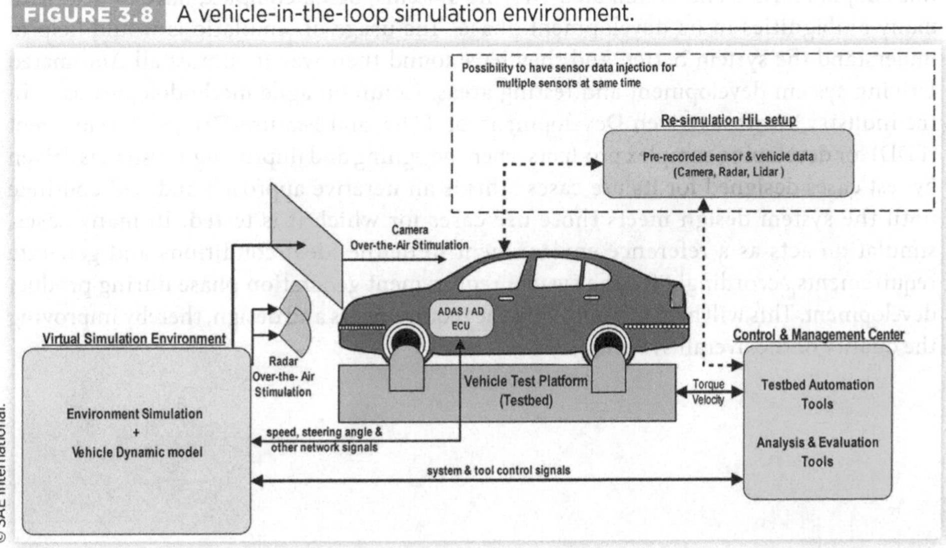

vehicle controls simulation and testing for ADAS and Automated Driving features. It is also possible to use the vehicle-in-the-loop simulation for calibration of ADAS functions of a vehicle in a laboratory environment.

Many vehicle manufacturers use a vehicle-in-the-loop environment for different test purposes (e.g., testing for powertrain, brakes, steering, etc.). This test environment is also used for sensor calibration and fault injection tests in the vehicle by replicating the vehicle in motion in a laboratory environment. Cybersecurity attacks on the vehicle can be tested in laboratory environments to analyze the risks associated with different threats and evaluate the impact on the vehicle while in operation. Because the number of sensors that can be stimulated simultaneously inside a laboratory environment is limited, most of the functional calibration in the vehicle is still performed during physical driving tests in proving grounds and public roads. The vehicle-in-the-loop simulation environment is the most realistic environment that matches the actual vehicle with powertrain, brakes, and steering wheel components in it [3.15]. The whole data pipeline for an automated driving feature from the sensors to actuator components can be tested in a laboratory environment with this.

3.6. Summary

Different types of testing are considered at different phases of the product test life cycle to evaluate the quality of the software and the system. A few of these test types and the different levels of testing considered for the ADAS and Automated Driving system are discussed in this chapter. In addition, different types of tests, use cases, and configurations used in the industry for performing different tests were covered with a few examples. How the agile methodologies and their influence can help product development and testing was explained in detail. Later, the chapter describes the importance of simulation in developing and testing ADAS and Automated Driving systems. Various examples regarding the usage of simulation environments and their configurations for testing were discussed. For new product development, how the simulation helps generate requirements and identify the possible reuse of simulation in testing was one of the key takeaways from this chapter. ADAS and Automated Driving systems, being complex, have to deal with many ambiguities in its development phase. The usage of simulations would help to understand the system better, and they have found their way in almost all Automated Driving system development and testing areas. Common agile methodologies used in the industry are Test-Driven Development, or TDD, and Feature-Driven Development (FDD) for developing complex products where designing and improving a system is driven by test cases designed for its use cases. This is an iterative approach and will continue until the system design meets those use cases for which it is tested. In many cases, simulation acts as a reference environment to define ideal conditions and generate requirements accordingly for the system requirement generation phase during product development. This will help in improving the requirements and design, thereby improving the quality of the overall system.

References

3.1. ISO/IEC 15504-5:2012, "Information Technology—Process Assessment—Part 5: An Exemplar Software Life Cycle Process Assessment Model," 2012.

3.2. ISO/IEC TS 33061:2021, "Information Technology—Process Assessment—Process Assessment Model for Software Life Cycle Processes," 2021.

3.3. MISRA C, "Guidelines for the Use of the C Language in Critical Systems," HORIBA MIRA Limited, 2008.

3.4. MISRA C++, "Guidelines for the Use of the C++ Language in Critical Systems," HORIBA MIRA Limited, 2008.

3.5. Winner, H., Hakuli, S., Lotz, F., and Singer, C., *Handbook of Driver Assistance Systems* (Amsterdam, the Netherlands: Springer International Publishing, 2016), doi:https://doi.org/10.1007/978-3-319-12352-3.

3.6. Yiming, X., Zou, Y., and Sun, J., "Accelerated Testing for Automated Vehicles Safety Evaluation in Cut-In Scenarios Based on Importance Sampling, Genetic Algorithm and Simulation Applications," *Journal of Intelligent and Connected Vehicles* 1, no. 2 (2018): 28, doi:https://doi.org/10.1108/JICV-01-2018-0002.

3.7. Meng, X., Gan, H., and Yan, Y., "Simulation Test of Driving Assistance System Based on Virtual Scene," *Journal of Physics: Conference Series.* 1419 (2019): 012047, doi:https://doi.org/10.1088/1742-6596/1419/1/012047.

3.8. Feilhauer, M. and Häring, J., "A Real-Time Capable Multi-Sensor Model to Validate ADAS in a Virtual Environment," in Springer Fachmedien Wiesbaden GmbH (Eds), *Fahrerassistenzsysteme 2017* (Wiesbaden, Germany: Springer Vieweg, 2017), 227-256, doi:https://doi.org/10.1007/978-3-658-19059-0_14.

3.9. Benedikt, M., Watzenig, D., and Zehetner, J., "Functional Development of Modern Control Units through Co-Simulation and Model Libraries," *ATZ Electronic Worldwide* 10 (2015): 30-33.

3.10. Feilhauer, M., Haering, J., and Wyatt, S., "Current Approaches in HiL-Based ADAS Testing," *SAE Int. J. Commer. Veh.* 9, no. 2 (2016): 63-69, doi:https://doi.org/10.4271/2016-01-8013.

3.11. Reway, F., Huber, W., and Ribeiro, E., "Test Methodology for Vision-Based ADAS Algorithms with an Automotive Camera-in-the-Loop," in *2018 IEEE International Conference on Vehicular Electronics and Safety (ICVES)*, Madrid, Spain, 2018, 1-7, https://doi.org/10.1109/ICVES.2018.8519598.

3.12. Plato, P., Kavita, K., Jithin, S., and Rohit, B., "HAD Validation Using Driving Simulator," in *DSC 2019 Conference*, Strasbourg, France, September 2019.

3.13. Morse, P., "Driving Simulator Hardware and Software Compatibility," Ansible Motion Blog, January 20, 2017.

3.14. Bock, T., "Bewertung von Fahrerassistenzsystemen Mittels der Vehicle in the Loop-Simulation," in Winner, H., Hakuli, S., and Wolf, G. (Eds), *Handbuch Fahrerassistenzsysteme* (Wiesbaden, Germany: Vieweg + Teubner Verlag, 2012), 76-83.

3.15. Albers, A. and Düser, T., "Implementation of a Vehicle-in-the-Loop Development and Validation Platform," in *FISITA World Automotive Congress*, Budapest, Hungary, 2010.

/Autonomous
/Sensing
/Communication
/Battery
/Navigation
Autopilot mode

4

Scenario-Based Testing

Scenario-based testing is a test type that is commonly used for testing ADAS and Automated Driving features. Here, scenarios are used as the reference to generate test vectors for evaluating the software and the behavior of the feature in the vehicle [4.1]. Since the requirements keep on incrementing during the development of an automated driving feature, the development and testing are usually performed based on scenarios as inputs. It is impossible to have a complete set of requirements or scenarios listed out before developing any feature. As scenarios keep on incrementing, so does grows the database for scenarios to consider for the development and testing. This chapter will cover the details and complete workflow of scenario definition, generation, implementation, and how scenarios can be used for testing. Although there are many different methods followed in the industry, this chapter discusses one of those methods that have been used for production programs for the deployment of automated driving functions.

4.1. Scenario Elicitation, Description, and Structuring

There are many references available in the literature regarding scenario-based testing for automotive applications. One of those interesting papers will be discussed in detail here as it was considered for developing the concept and executing the scenario-based testing for the deployment of automated driving features in a few vehicle programs [4.2]. The scenario elicitation process is the process of collecting scenarios that are applicable for a particular feature under consideration. As discussed in the second chapter, scenarios have to be collected from various sources similar to requirement elicitation. In general, one could classify those scenario elicitations as an outcome of at least four different analyses, such as:

1. Accident data analysis: Here the scenario source is the accident database of various countries or from different geographical locations. One could access it from the regulatory organizations, private watchdog organizations, or information available with the vehicle manufacturers.
2. Technology analysis: Scenarios extracted from scenes that are based on the feature and the technology used for features that are under development. Scenarios specifically focused on the technology which is getting implemented.
3. Scenarios from use cases: The scenarios within and beyond the operational design domain of a particular feature and its use case under consideration. Here various use cases of the feature from the end user are considered as the source for deriving scenarios.
4. Scenarios from regulations: The scenarios that are defined as part of specific regulations that must satisfy the requirements of vehicle type approvals. For example, the scenarios defined in the United Nations Economic Commission for Europe (UNECE) Regulation R157e for the Automated Lane-Keeping System.

Scenarios collected from these sources can be used to generate a scenario database for further implementation or replication and utilize it to develop and test the Automated Driving system and its features. It is essential to follow a structured approach during the elicitation of scenarios as this will act as the input for identifying those corner cases where the system might misbehave or fail during its operation. In general, no scenario should be left out, assuming it to be not so significant or difficult to implement. These are the common errors with various teams during the scenario elicitation process. All these scenarios contribute to the scenario database. One may think that it is possible to cover all possible scenarios associated with a particular feature during its development and testing. Even though it is a wish for many, the reality is different. Especially for a vehicle program that has constraints with time, scope, and cost. It is impossible to consider every scenario for which the feature can be made compatible. Hence, it is essential to follow suitable methods, processes, and tools in situations where constraints need to be considered.

The scenarios from the technology area can be captured using brainstorming sessions and analysis methods similar to that of performing FMEA. These can be further prioritized based on risk weightage. Accident databases can be reused from different geographic locations around the globe and the vehicle manufacturer's internal database if it is available. Use case-based scenarios are generated considering the feature concept and ODD defined during the concept phase. A detailed explanation of the process will be covered further by taking the Highway Chauffeur feature as an example. Different infrastructure elements that should be considered are classified as scenes that a feature would experience during its operation. These are taken as the basis for deriving the use case-based scenarios and their relationships in the scenario implementation. For the Highway Chauffeur feature, the following scenes can be considered based on the feature concept and operation as represented in Figure 4.1:

- Entry and exit to highways
- Congestion (traffic on the highway)
- Driving through the tunnel
- Lane keeping and lane change

FIGURE 4.1 Scene classification for Highway Chauffeur feature.

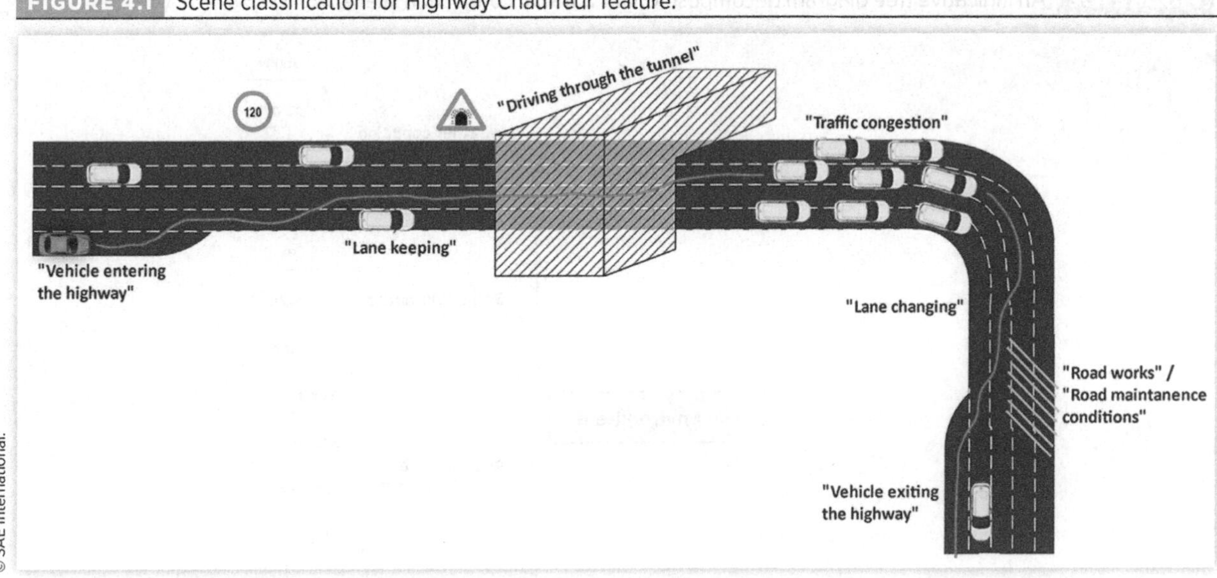

The applicable scenes for a particular feature can be described and defined together with all the stakeholders involved in developing the system and the feature. This helps both the vehicle manufacturer and its suppliers to have a clear understanding of the feature and its expected operational domain.

Scenarios can be generated by further decomposition of the scenes which were identified for each feature. Different tools can be used to decompose the scenes. One would be a simple approach of using a tree diagram to decompose every scene, like in Figure 4.2. A tree diagram is a diagram that is typically used to represent the hierarchy in a tree-like structure. Here the tree diagram is used to decompose the scene to its subcomponents further, keeping the hierarchy.

A scene that was decomposed, as shown in Figure 4.2, includes the static components or the Scenery, the Dynamic components, and the Attributes for each, along with the various possible state of the user. In this case, for the decomposition, it is the whole vehicle one should consider. This is referred to as the "Subject vehicle." A scenario can be derived by considering each path defined in the tree diagram while considering different attributes for the vehicle.

Scenario description starts with defining pseudo scenarios for each branch of the tree diagram starting from its root. Here the tree diagram acts as a reference to generate pseudo scenarios by identifying all possible components applicable in a particular scene. Those components that act as the branches and leaves in the tree diagram are identified by the method of brainstorming with experts and different stakeholders. This activity is similar to identifying possible failures in an FMEA.

The drawback with this method is the time consumed to derive all the possible combinations from the tree diagram and to measure the completeness of the defined pseudo scenarios for a particular feature. This keeps the scenario generation process iterative, which has to be revisited and improved for further scenarios. Then the next question is "How many scenarios are expected to be generated from a tree diagram?" One should open up ones' mind to generate as many pseudo scenarios based on the tree diagram.

FIGURE 4.2 An indicative tree diagram decomposition for a traffic congestion scene.

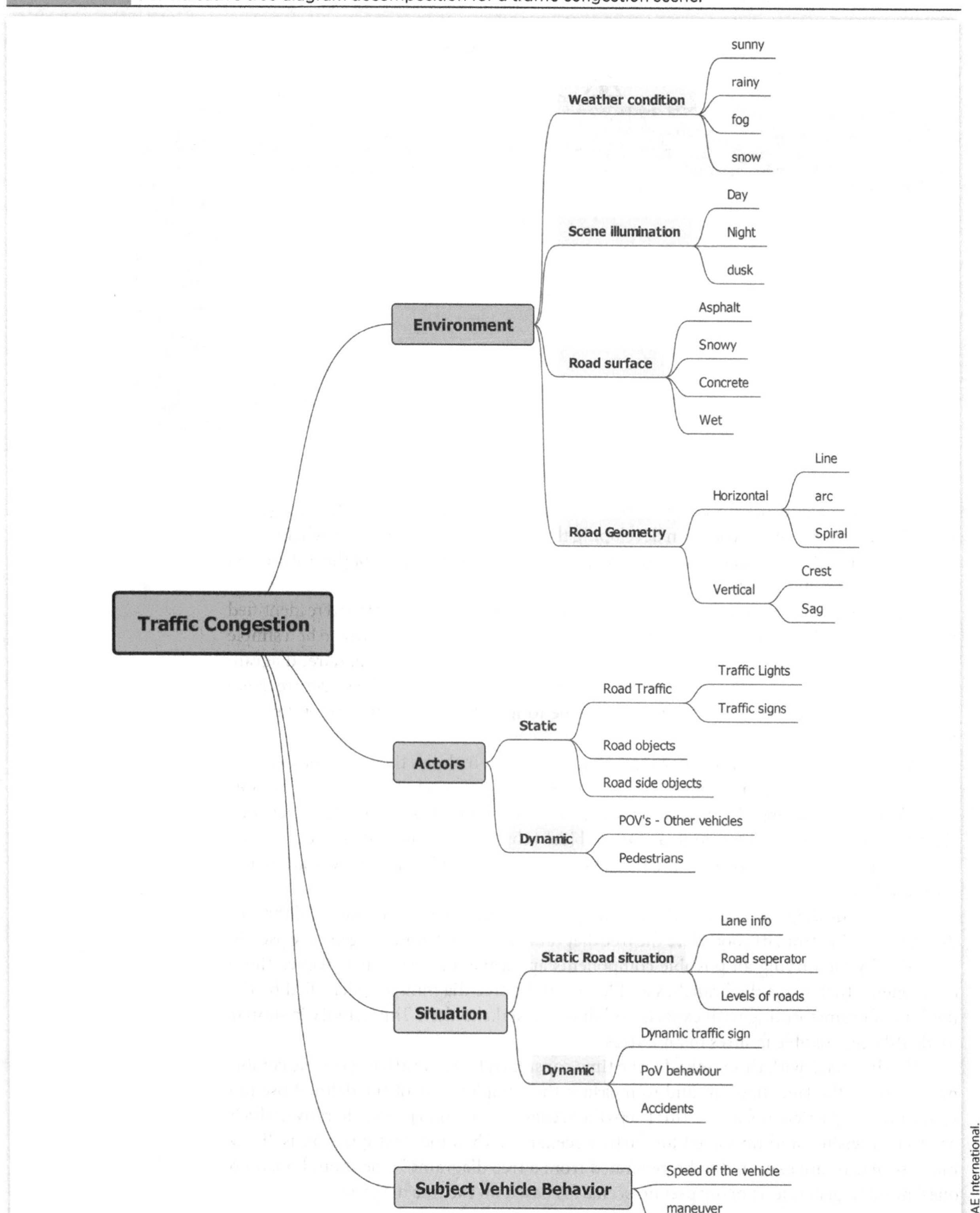

Pseudo scenarios from a tree diagram can be captured using various tools. This can be done with the help of a spreadsheet or can even use custom tools for defining pseudo scenarios. Once the team is confident enough that all possible pseudo scenarios are captured from the defined tree diagrams for different scenes, the next process is to take it through a selection process. From a tree diagram, one could generate hundreds of scenario combinations considering different parameters associated with each branch and leaves and the status of the subject vehicle. For product development and testing, especially for a vehicle program, all the pseudo scenarios cannot be considered as valid scenarios for which the system and its features should be compatible. There might be many scenarios that are beyond the defined ODD of the feature, and there might be even scenarios with a very low probability such that they might not even occur. The selection of pseudo scenarios from the list must be made carefully. Otherwise, there is a chance that valid scenarios might be removed, affecting the overall quality and coverage of scenario-based testing. Until the occurrence of a scenario, one might not recognize the importance of that scenario for development and testing. The approach that was followed here to avoid and to minimize the loss of identified scenarios was to make sure that the selected pseudo scenarios should cover the whole tree diagram, including all the branches and leaves of the tree structure, with a minimum of three overlapping pseudo scenarios in the same branch to its leaves. This approach was only considered to make sure to have scenarios selected so that the probability with which one covers the scenarios associated with each branch of the tree with more than a 0.5 probability value as each path will be covered at least three times in the tree. The filtering of pseudo scenarios should be done as a team activity with technical experts and should never be done by a single person.

Another challenge one would face while generating pseudo scenarios using the tree diagram approach would be maintaining traceability of scenarios with the tree structure. This can be addressed by classifying the tree diagram from their roots to branches, sub-branches, and leaves as different levels and numbering them to retain the hierarchy. It is required to use some method to establish traceability. Else it would become complex to associate the pseudo scenarios to each branch and path later. One could consider using a numbering system or an automatic code generation with alphanumeric values for each level of the tree diagram and its path for traceability purposes.

Once when the pseudo scenarios are screened and selected, they are further described as scenarios with details. Natural or Programming languages can be used for the description of these scenarios. A Scenario Description Language has been utilized for scenario description by many organizations these days. The description of the scenarios can be further elaborated with specific scenario diagrams and charts. Different organizations use different methods, and the ultimate goal of using any method for scenario description is to have a clear understanding of the scenario and various events and actors involved in it. The process of generating a scenario from the tree diagram is explained in Figure 4.3.

Description of all the scenarios from different sources is accumulated together to create the scenario database. This acts as the baseline for any further development and testing, which makes use of scenarios. This scenario database should not be a closed box. Instead, this should be a continuously growing database. The filtered-out scenarios from the pseudo scenario database should be gradually taken into the scenario database in phases, and any new scenario coming from any of the abovementioned sources should be included in this growing database.

To avoid difficulties in selecting and using scenarios at a later period, a classification of the scenarios can also be included in the scenario description file. This is simply a grouping of scenarios that may or may not be applicable or used for a particular feature but with

FIGURE 4.3 Scenario development process from a tree diagram.

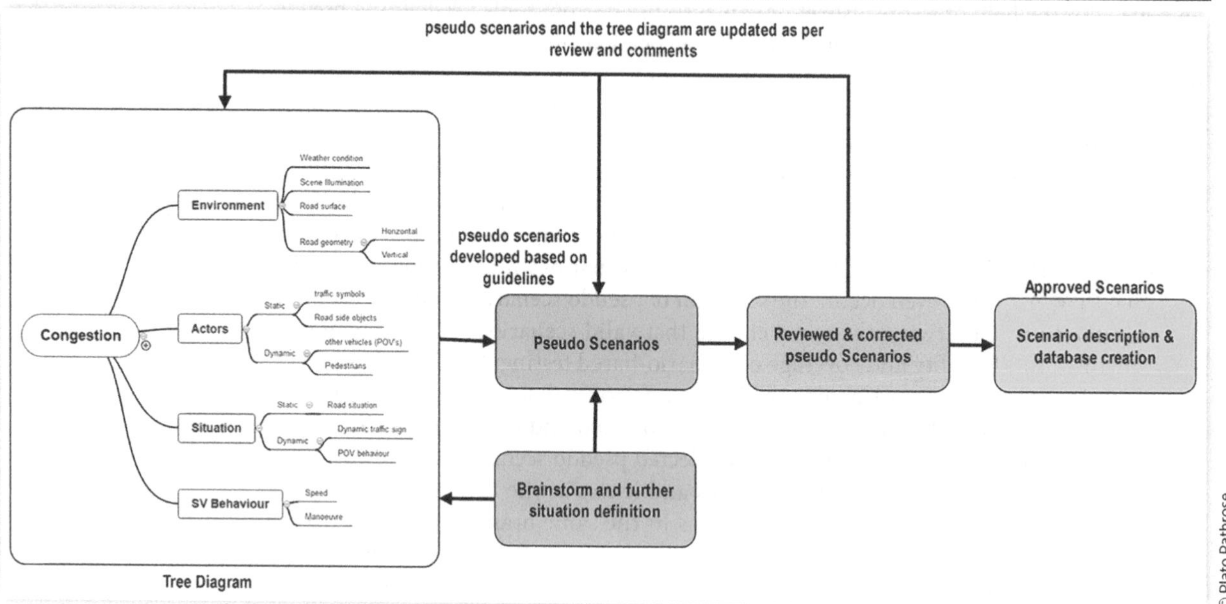

traceability. Input for classification of scenarios and allocation of those to different ADAS or Automated Driving features can be taken from the feature definition and ODD defined for each of the features. There are different ways in which scenario classification can be done. This could be a one-to-many classification where one scenario can be used to test multiple automated driving features, or this could also be that one scenario is applicable for only one feature. Figure 4.4 shows the classification of scenarios in three different levels that were used for a project. The scenarios are grouped based on features, an alternate grouping based on the motion of the subject vehicle, and the grouping based on scenes. One can select the applicable scenarios based on any of these classifications for test executions. It is also possible

FIGURE 4.4 Different classifications in a scenario database.

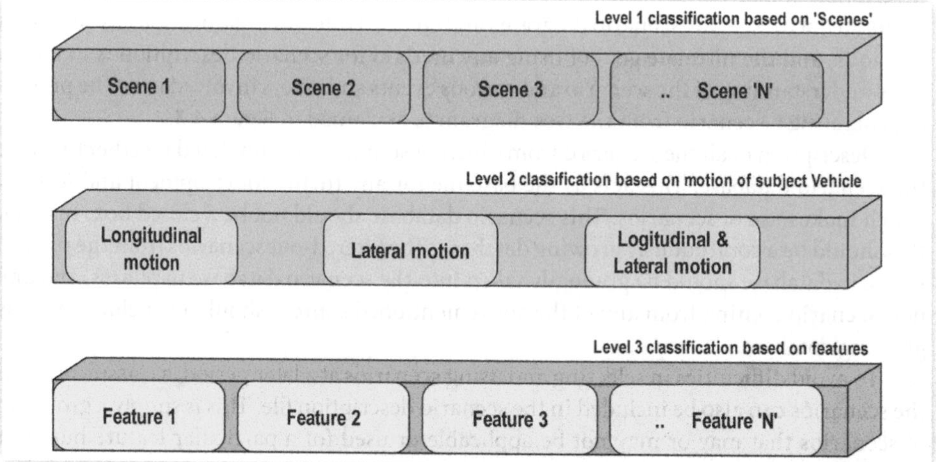

to have multiple selections at different levels. Scenario classification should be performed whenever a new scenario is defined based on a tree diagram or other input sources.

It is impossible to consider and plan a response for each and every available scenario for a feature in the database in a production program. The next steps will explain how to move forward from the scenario database and how these scenarios are further used for testing.

4.2. Scenario Implementation and Parameterization

Scenario implementation is the next step where the defined scenarios are being implemented using certain simulation tools. Depending on the need, one could select from the several environment simulations tools available in the market as the purpose of these tools is to provide a virtual environment where the behavior of an ADAS or an Automated Driving system can be simulated in a virtual environment. Scenario implementation involves translating the defined scenarios (which are either in natural language or scenario description language) to virtual and synthetic scenarios in an environment simulation tool.

In the current industry, there were also different methods followed where the scenario description process only covers up to the generation of pseudo scenarios using any method. The implementation of scenarios follows a different process like pseudo scenarios are defined using scenario description language and then converted to scenarios using an environment simulation tool that understands the scenario description language or by means of any other tools. One could adapt any of these methods as long as the scenarios are traceable and are replicated as defined. There are also standardization efforts in the industry regarding how scenarios need to be defined and how they could be generated [4.4, 4.5].

Implementing scenarios using a simulation tool is not easy in the case of poor planning and lack of strict guidelines. Each of the scenarios is to be implemented using a simulation tool considering different environmental layers [4.4, 4.6, 4.7]. For implementing a scenario, a minimum of at least four layers is to be considered, as described in Figure 4.5. These four layers are as follows:

1. Road: This layer covers the road geometry, elevation, structure, and other characteristics of the roads. Here the parameters are controllable later after the implementation.
2. Traffic infrastructure: This layer covers the infrastructure associated with the traffic flow. This could be traffic signs, bridges, barriers during road works, lamp posts, etc.
3. Static and dynamic objects: This layer considers all the static and dynamic objects interacting with the other two layers. This could be from other vehicles, persons, animals, etc.
4. Environment: This layer is about the environment, which needs to be set up including below three layers. The environment layer covers all the characteristics of the environment like lighting conditions, weather, wind, etc.

Out of these four layers, the first two layers can be grouped as layers that are part of the infrastructure and, therefore, can be considered as permanent layers in a scenario implementation. The second and third layers are variable and are frequently changing ones. All these four layers come with characteristics and associated parameters that vary and are critical for a simulation environment.

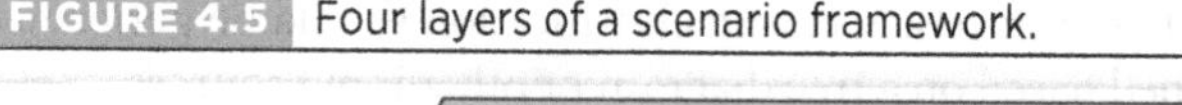

FIGURE 4.5 Four layers of a scenario framework.

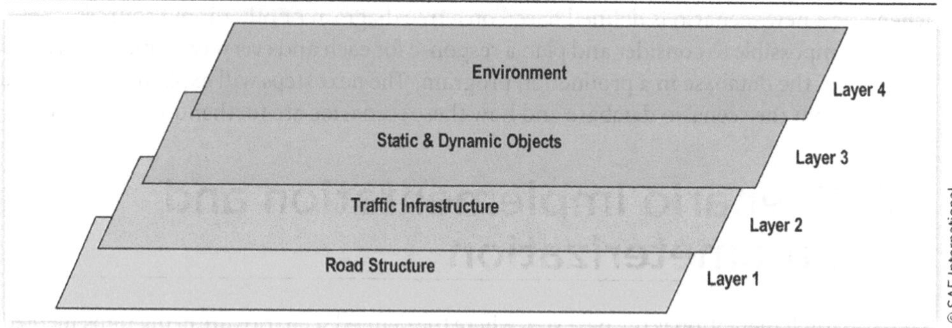

The research project PEGASUS (Project for the Establishment of Generally Accepted quality criteria, tools and methods as well as Scenarios and Situations) considers scenario implementation in six different layers. It is an excellent reference to scenario development and implementation [4.3].

As mentioned before, scenario implementation involves a lot of manual and automatic tasks, and the ease of implementing also depends on the tool selected. Beyond all these, the major challenge will be regarding the guidelines required for scenario implementation. Any guidelines should be defined before any implementation activity to keep it traceable and measurable. Mainly, these guidelines will cover unique naming conventions of the scenario files such as how the naming should be for each variant of it, defined length for the road layer used in each scenario implementation, the minimum number of Principle Other Vehicles (POVs) required at any point of time in a scenario, etc. This is crucial for the traceability of the implemented scenario files and their different variants while establishing a scenario database. This will also help measure the simulation mileage when speaking about the maturity of an automated driving feature in terms of kilometers driven in a virtual simulation. After implementing scenarios, a detailed analysis needs to be done again compared to the description database which was created. The analysis identifies the scenarios applicable for each feature that intend to simulate and the parameters of these four layers considered for implementation and their relationships. Along with identifying applicable parameters for a particular feature, a prioritization of parameters should also be done.

The reason for analyzing the parameters based on different features for simulation is because every parameter in a scenario might not have an equally important effect on different features. For example, an ADAS feature like lane keep assist has a higher dependency on the markings and lanes in the road and, therefore, needs to classify the road markings to have a higher effect. At the same time, when considering a cruise control or an emergency braking feature, road markings or lanes do not have much importance. Instead, the object in front of the vehicle and speed has a greater influence on the feature than lanes. When speaking about higher levels of automation where the basic ADAS features are acting as the building blocks, one can find that there are dependencies on almost every parameter identified. Having the ranking of parameters done on the ADAS features will help prioritize the parameters in automated driving features. This will also help perform automated tests, which is part of complete test execution, or a regression test as part of the software and system testing.

A unique approach was taken in this project for parametrization and prioritization of parameters associated with each feature and scenario. The approach from Six Sigma was used for identifying the parameters as well as to rank them based on their impact on

FIGURE 4.6 A simple mathematical representation of a system.

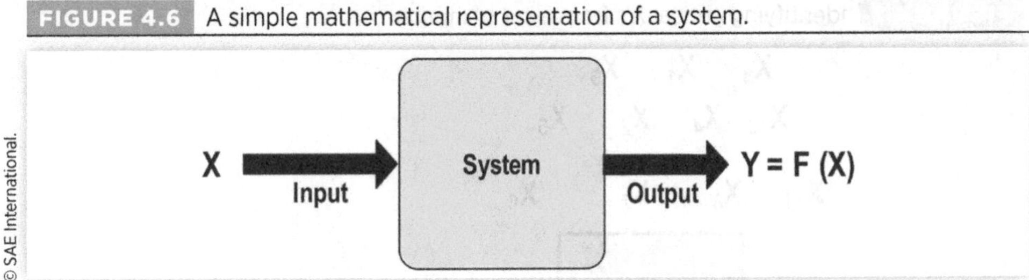

each feature. In the Six Sigma approach, any process or a system can be considered as a simple function "F." All the inputs to that process can be considered as "X," and the output can be considered as "Y." The output can be mathematically defined as a function of input such as "Y = F(X)" [4.8]. This can be graphically represented as in Figure 4.6.

Here, each ADAS feature and the parameters on which they depend are considered as inputs to the function. To identify those input parameters, one should have a good understanding of the functional architecture and may also need support from the development team to gain insights into the requirements and functional descriptions. This means the parameter identification for a feature and ranking of those parameters are part of teamwork rather than an individual activity. This is an activity that can be considered similar to performing an FMEA or risk analysis with the scenario description database [4.9, 4.10].

The approach taken here is Six Sigma methods and tools for the classification and prioritization of parameters. At the same time, assuming that one feature is considered and all the parameters are identified for that feature. Suppose the number of parameters is considerable, for example, above ten. It would become difficult to manage and prioritize its impact and influence on the feature. The approach taken here is the application of the Six Sigma methodology of identifying all applicable X's, which are the inputs to the function is as represented in Figure 4.6. As all the input parameters do not have the same influence or impact and are critical for the feature, the critical parameters should be identified [4.8].

The critical parameters to the function or the critical X's can be identified using different tools and approaches like as FMEA or Pareto principle. Here the focus is not only to identify the failures associated but also the parameters and their severity. A statistical tool can be used to identify critical parameters associated with a feature if the data for all parameters are collected for certain iterations. Identifying critical parameters of a feature will help focus the simulation and parametrization on those parameters specifically rather than the whole set of parameters for the feature. This will help a lot in the development as well as for testing those features (Figure 4.7).

Analyzing the parameters to identify which one has a more significant influence than others requires sample data generated by multiple trial runs of the feature in a simulation environment. The classification and ranking of these parameters were done with the help of a statistical tool Minitab. One would think that all these analyses could have been completed when describing the scenario itself. With this approach described here, it is impossible to identify all critical parameters earlier. This method requires sample data and parameter rankings as input to identify the critical parameters. The sample data for the parameters were collected by simulation using a golden sample of the feature. This golden sample of software model was already a proven feature in the market as an ADAS feature with years of history. That was used to generate sample data for parameters of the ADAS features. Since the automated driving features did not have any golden samples, the feature dependency diagram analysis and the fault tree analysis and FMEA will help identify critical

FIGURE 4.7 Identifying critical X's from a set of inputs.

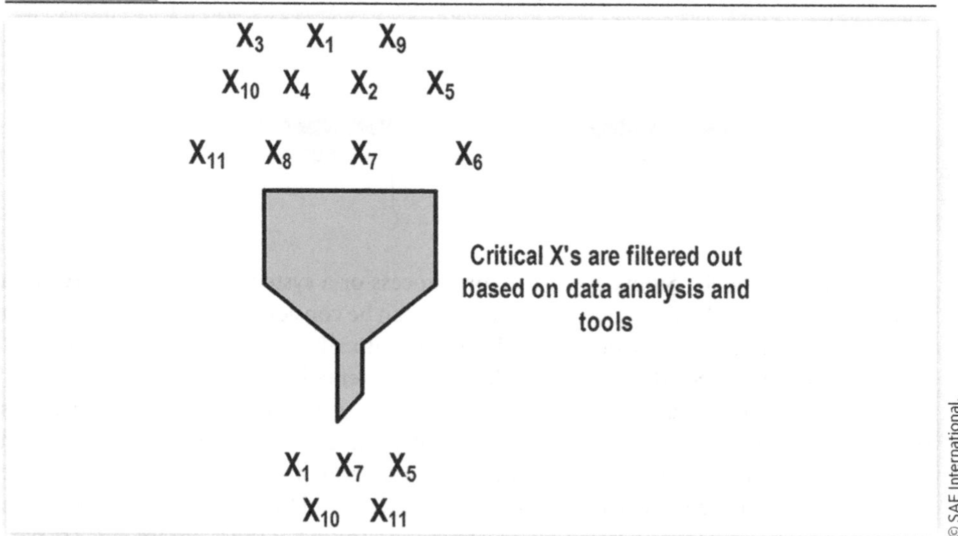

parameters of those. This needs to be performed for all the features which are planned for scenario-based simulation and testing. The process of scenario description and parameter identification and prioritization with sample data generation, like the process defined in Figure 4.8, took about one year when this was started from scratch.

After all the parameters are identified and classified for each feature, the next step is to define the data input range for each of those parameters. The input range is about identifying the values which can be passed through each of those parameters during the operation of the system and its features. These values need to be identified and listed to apply different test techniques while performing scenario-based testing. For example, while using an environment simulation tool, one could pass the value for vehicle speed from 0 to 250 km/h, or the lighting conditions can be varied as the morning, noon, evening dusk, etc., which varies based on luminous intensity. These possible parameter inputs need to

FIGURE 4.8 Parametrization and testing workflow for scenario-based testing.

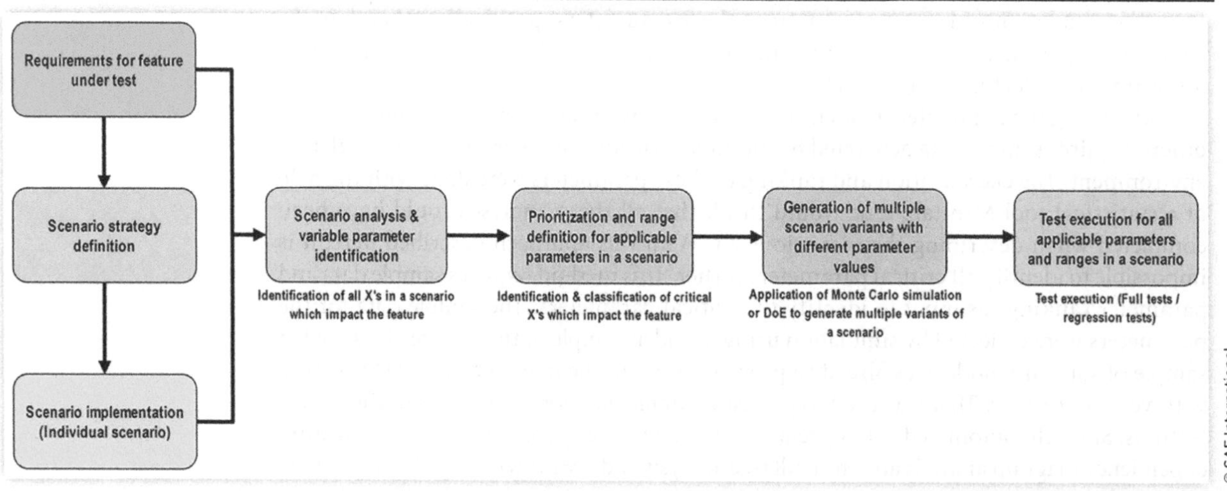

be listed along with each parameter in the scenario database. Depending upon the feature under test, the parameter values vary and have some possible minimum and maximum values that would consider the ODD of the features. For example, while defining the parameters for the feature Traffic Jam Pilot, which is designed to be functional up to a maximum speed of 60 km/h, the vehicle speed parameter should be covering up to the value of 60 km/h and a few more as part of boundary condition tests [4.11, 4.12, 4.13, 4.14, 4.15]. Exceptions exist if someone wants to apply different test techniques while testing the feature like negative testing, fault injection, etc.

When all the required parameters and their input value ranges are identified for each feature, the next step is to apply these in simulation. Depending upon the selected tool, some can generate multiple scenarios based on a combination of these identified parameters and their ranges. This is one way of executing multiple scenarios from an already defined scenario with different parameter values applied for each scenario run. One could use different simulation methods to pass these parameter values to generate scenarios. Suppose the selected environment simulation tool has a built-in scenario generator. In that case, it might not always be powerful and capable enough in providing flexibility to apply your method of scenario-based simulation with parametrization. An alternate way to handle this challenge is by integrating a scenario generator tool available in the market to your existing validation framework or to develop a scenario generator fitting with your validation framework on your own. The scenario generator should satisfy the needs of the scenario generation for the usage of both development and testing teams. If a scenario generator is planned to be developed, it is possible to utilize different simulation methods as you wish to pass the parameter values for each iteration while executing a scenario in a simulation tool. Markov chain Monte Carlo simulation, Monte Carlo simulation, DoE, Active DoE with symbolic regression, multifactorial combinatorial method, etc., are a few methods used in the industry for parameterization [4.12, 4.13]. The advantage of using these methods is that a set of reliable input values are generated for each parameter and can be used to generate different scenarios. The advantage of using the DoE and Active DoE method is that it helps generate a reduced set of parameter combinations if the same scenario database is used for the calibration and tuning of the software for a feature [4.8, 4.16]. This will also help in the regression tests when the scenarios are used for testing only a particular feature or a part of it after a bug fix.

4.3. Scenario-Based Simulation and Testing

Once when the parameters are identified after scenario analysis, the input value range for each parameter is defined. The next step is to execute these scenarios with those identified parameter values as inputs. Scenarios are used for simulation purposes during the development and testing phases to evaluate the behavior and performance of a software model or the software itself. During the development phase, scenarios are designed and used for identifying the optimal threshold values for many features. These can be calculated based on mathematical calculations as well as simulation using an environmental simulation tool. For example, a scenario applicable for the feature Automatic Emergency Braking (AEB) can be used to simulate and calculate the braking distance at different speeds of the subject vehicle. These measurements can be used for adapting the requirements of the parameters associated with the feature and the software. Simulation is also a method to generate requirements and the associated values on physics-based calculations.

When a scenario is defined and implemented and is used for testing, there are certain things one should keep in mind:

- Scope of testing
- Control and management of scenario parameters
- An approach to measure and evaluate results of scenario-based testing

Using scenarios for testing a particular function model or software is not an easy task. A good test environment with the aforementioned conditions is essential in performing scenario-based testing for any features. Depending on the need for scenario-based simulation, an interface is required to pass through specific values for the parameters associated with a scenario. Most of the environment simulation tools provide these interfaces for passing specific values for parameters for a particular scenario. This is helpful during development testing and for analyzing the corner cases associated with a feature and checking how the software behaves in those cases.

Executing scenarios for large-scale testing purposes requires a flexible and user-friendly interface to the environment simulation tool. With this, one could select all the applicable parameters for scenarios depending upon the feature under test. This will also provide flexibility to the user to define and pass the range of values for each applicable parameter in a scenario while it is being executed in multiple iterations. This can be implemented as part of the validation framework with which the scenario-based testing can be controlled. The parameter values can be passed as input for each scenario and can be structured in steps for increment or decrement based on the user. The step splitting or step-size definition for the values associated with a parameter varies upon each parameter in a scenario. For example, consider the parameter "vehicle speed" to test a feature with its ODD defined from 0 to 150 km/h. A step size of 1 km/h increment does not make sense for each iteration of the scenario execution unless it is only to perform boundary condition testing.

On the other hand, for the road friction, a parameter value with a step size of 0.1 will impact significantly in the braking and vehicle dynamics area. So, for every parameter, their range of values should have a step size defined for each of them as part of scenario description and parameter analysis. This will provide better controllability and management of the scenarios for their generation and execution.

The example explained in this chapter utilizes a separate tool with a graphical user interface in "Qt" to control and manage the scenarios. Qt is a toolkit for developing a graphical user interface and supports cross-platform applications (e.g., Windows, Linux). This interface can select scenarios from the database. Depending upon the selected scenario, applicable parameters will be listed from a reference lookup table. This lookup table was created during the scenario analysis phase where the applicable parameters are classified and ranked and value ranges defined based on the feature under consideration and the capability of the simulation tool used to implement them (Figure 4.9).

This tool aims to control, monitor, and manage scenario execution and evaluation of the behaviors. The tool has a graphical user interface and APIs (Application Programming Interface) to integrate it into any validation framework of the organization. The control and management part helps in selecting the feature, applicable scenarios, selecting parameters, and controlling their values for the scenario execution. This also provides controlling the interface with the validation framework, initiating the simulation tool and other associated modules, initiating and shutting down the sequence of the

FIGURE 4.9 Interface tool used for scenario selection, parametrization, and execution.

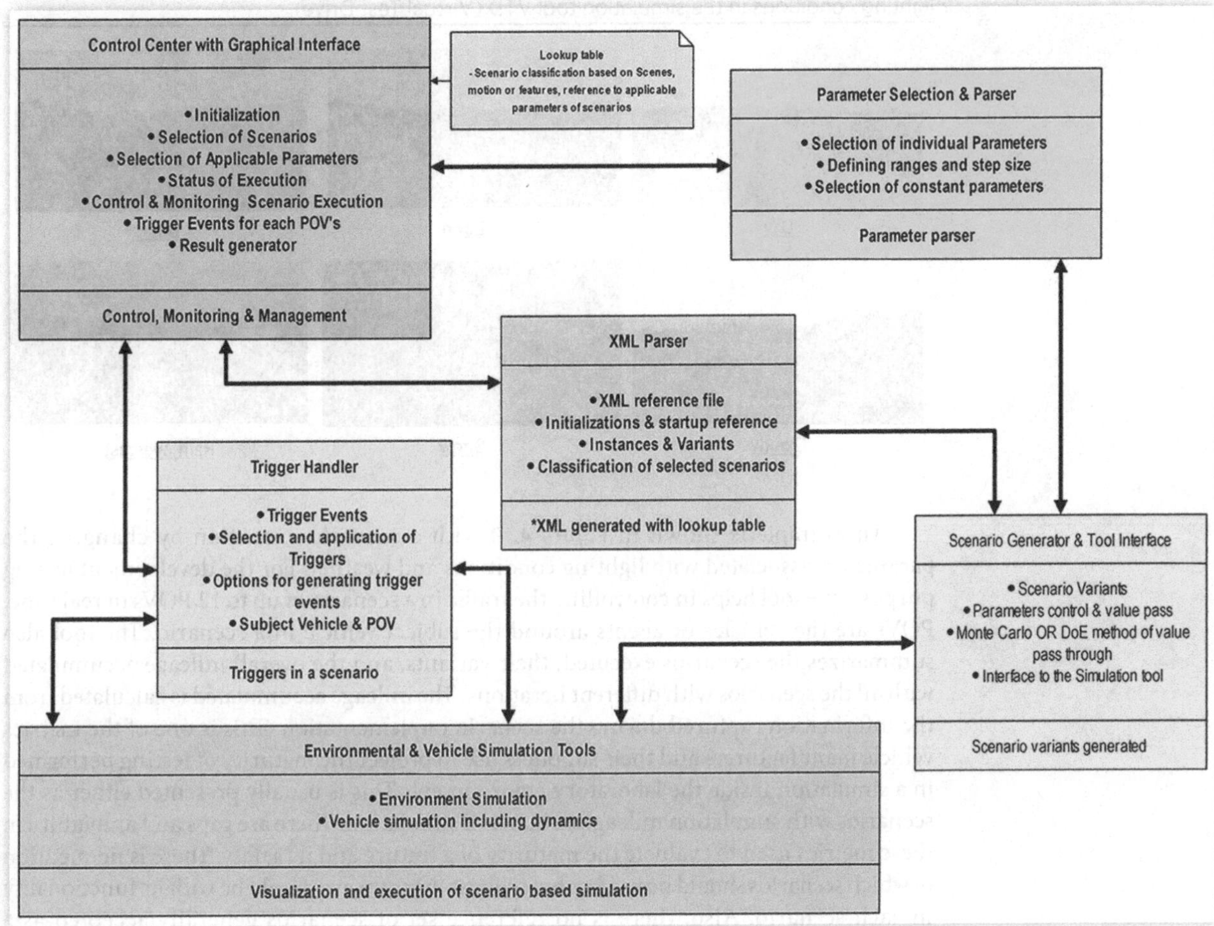

toolchain, and selecting the latest version of the model or software from the integration platform. The monitoring part of the interface tool provides the health status of the validation framework, the status of the simulation tool, the runtime status of scenarios, and its results. It also supports extracting a report and status with details of the performance of the software or model in a particular scenario or a group of scenarios with multiple variants of it.

The interface tool, which was performing control and management for the scenario-based testing as discussed in this chapter, does not generate the applicable values for parameters of a scenario in runtime. Here it generates the values for each parameter with selected step size for values in offline mode considering either Monte Carlo simulation or DoEs and stores those in a *.xml parameter file. A *.xml file is an extensible markup language file, and it is used to structure data for storage and transport. The generation of the values associated with parameters is completed as part of the preprocessing phase before the scenarios are initiated and used for the testing. With this approach, any new batch of scenarios (specific to a scene or a new feature to be tested) requires time to generate parameter files, depending on the number of scenarios and parameters selected for simulation.

FIGURE 4.10 Multiple variants of a scenario generated by different parameter values for lighting conditions in the simulation tool VTD (Virtual Test Drive).

An example is shown in Figure 4.10 with a scenario execution by changing the parameters associated with lighting conditions and weather. For the development testing purpose, the tool helps in controlling the traffic in a scenario of up to 12 POVs in real time. POVs are the vehicles or agents around the subject vehicle in a scenario. The tool also summarizes the scenarios executed, their variants, and the overall mileage accumulated with all the scenarios with different iterations. The mileage accumulated is calculated from the information captured during the scenario implementation. This is one of the metrics vehicle manufacturers and their suppliers use to project the maturity of testing performed in a simulation inside the laboratory environment. This is usually presented either as the scenarios with simulation mileage or hours of simulation. There are gaps and ambiguity in these metrics used to evaluate the maturity of a feature and its safety. There is no mention of which scenarios should be used or how robust the software should be with its functionality in each scenario. Also, there is no reference set of scenarios generally accepted as a benchmark to deploy an automated driving feature. This creates more complexity in defining what scenarios and matrices should be considered acceptable for performing software qualification for automated driving.

4.4. Scenario-Based Testing at Different Levels

Scenario-based testing is a vital test type used for testing ADAS or Automated Driving features. Different test techniques are applied to generate several scenario variants by controlling and adapting the applicable parameter values in a scenario. On analysis, it can be found that different test techniques are applied here like equivalence class, boundary condition, and risk-based classification, based on which one could control the depth of testing for scenario-based tests.

The scenarios can be utilized to perform tests at different levels in the test life cycle. Having a global scenario database and classifying those to different test levels would be helpful. The majority of the scenarios in the scenario database will be used during the model-in-the-loop and software-in-the-loop simulations. The number of scenarios used for

FIGURE 4.11 Scale at which the scenario-based testing performed at different test levels.

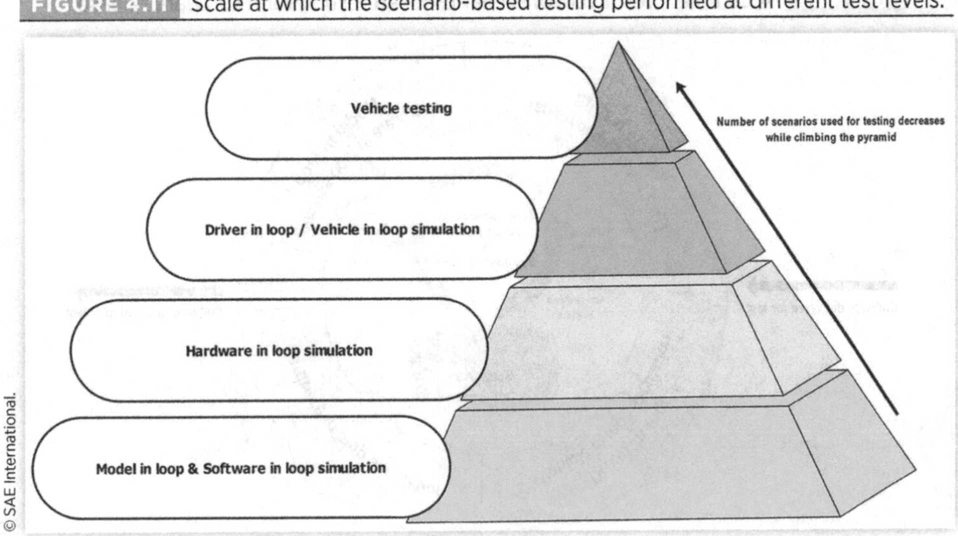

HiL, DiL, and vehicle-in-the-loop simulation comes to a smaller subset from the scenario database that was used for model-in-the-loop or software-in-the-loop simulations. The whole scenario-based simulation utilizing scenario database appears like a pyramid as in Figure 4.11 when moving from the software-in-the-loop simulation at the bottom toward the vehicle-in-the-loop simulation toward the top. Beyond the purpose of simulation, a smaller subset of scenarios is utilized for vehicle testing in controlled environments, test tracks, or on public roads. This also includes scenarios that are tested as part of homologation and approvals.

In the development workflow, whenever a model or software is integrated, it needs to undergo robust testing before being handed over to the testing team. Most of the time, the tests performed after the integration are not robust enough, assuming tests will be vastly carried in the verification and validation. In many cases, the software has been rejected directly after handing it over to the verification team because either the feature is not functional or it is not behaving as required. This can be resolved up to a level if scenario-based testing is also included as part of software integration testing. The scenarios used as part of integration testing can be a subset of scenarios specific to the features under consideration or related to specific bug fixes of different features. The results from the scenario-based tests executed as part of integration testing can be considered as the entry criteria for further detailed verification and validation processes.

Usually, the software releases and test activities are iterative in a production program. The scenario-based testing can also be iterative across different levels. The same scenario database can be used at different levels, as shown in Figure 4.12, for the complete test cycle for a particular software release cycle as part of the project.

One complete iteration of tests covering all test environments for a software release, as represented in Figure 4.12, took about two months. The testing cycle also includes other test types required as part of any software or system release.

In many projects, when speaking with vehicle manufacturers or their suppliers, there is always a question about how many scenarios are to be developed and how many will be used for testing at different levels. There is no specific number to say, and it solely depends on the analysis performed for each feature and the ODD defined for them. For instance, in deploying a Highway Chauffeur feature, there were approximately 1400 pseudo scenarios defined. After the analysis, about 1000 were considered for definition and implementation

FIGURE 4.12 A complete test cycle of scenario-based testing for a software release.

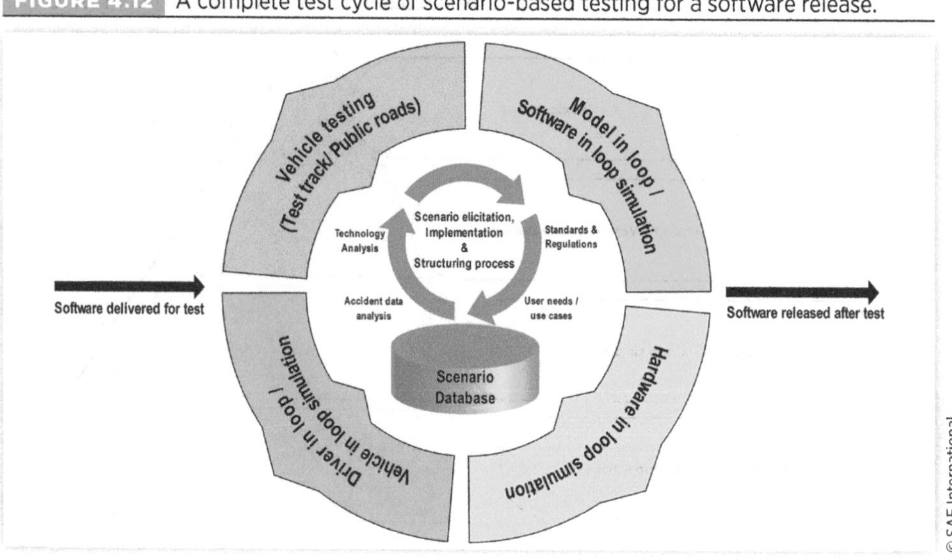

into a scenario database in the first phase. These were further classified based on scenes, features, and applicable test environments as part of scenario-based testing.

The scenario database can be further populated with scenarios proposed and identified from other analysis processes, which are part of functional safety and Safety of the Intended Functionality (SOTIF). For the development testing and feature calibration by tuning the software parameters, the interface tool, which was discussed earlier in this chapter, had a provision to control up to eight immediate POVs around the subject vehicle. This will help to control the movements of the POVs and evaluate the behavior of the subject vehicle in each case.

4.5. Scenario Database Management

A scenario database is the database of all the scenarios implemented, collected, or even purchased from suppliers for scenario-based simulation and testing. A database can be maintained with scenarios implemented using a simulation tool or purchased ones from the tool suppliers [4.17, 4.18, 4.19]. Sometimes purchased scenarios might not be having all the required attributes to apply the above method of scenario parameterization and testing. They might be required to be adapted for use. The approach described here for managing a scenario database is based on a lookup table.

During the analysis and implementation phase of scenarios using a simulation tool, the scenarios are preclassified and grouped into different classes, such as:

- Scenes
- Features
- Geographical location and applicable regulation
- Subject vehicle motion considered in the scenario
- Test environment where scenarios can be executed
- A default risk-based prioritization and classification for test execution

The first five classifications were made during the analysis and implementation phase of each scenario. The scenarios are analyzed and classified for the following: which scene it would belong to, what features can be tested on it, the implementation done for which geographical location (this includes the traffic signs, road marking, etc.), the motion of the subject vehicle whether it is longitudinal or lateral motion considered in that scenario and the test environment where the scenario can be tested. All these should be analyzed and labeled with codes and entered into a lookup table so that each scenario will identify and reference each of these classes.

The sixth class is provided for test execution using the complete scenario database (Figure 4.13). This classification is made from the prioritization matrix created during the scenario analysis phase from the tree diagram. During the analysis phase using the tree diagram, when pseudo scenarios are developed, they are selected and given priorities. They are prioritized first those covering applicable scenarios from regulatory requirements (NCAP, UNECE, etc.), functional safety analysis, SOTIF analysis, and other standards specific to ADAS features, followed by those scenarios which can be utilized for testing features. This correlates to the feature-based classification of scenarios in the database, and the third classification is based on those scenarios which are created explicitly considering

FIGURE 4.13 An overview of scenario grouping in a scenario database.

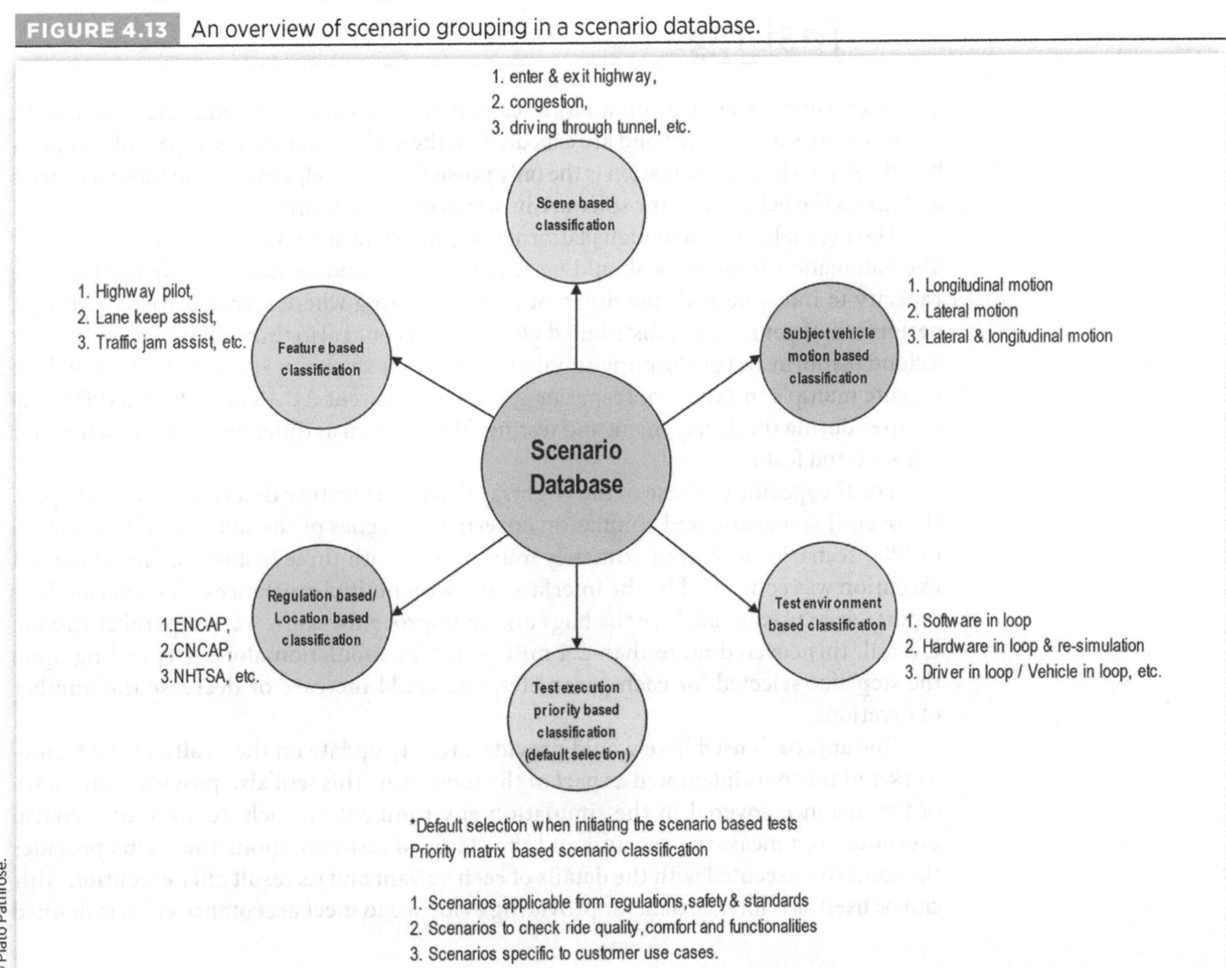

the use cases from the customer [4.9, 4.20, 4.21, 4.22]. The tool mentioned in the above section can list the scenarios from the database referring to the lookup table for any test execution.

The sixth class would be the default scenario list loaded when starting the control and management interface tool and loading the scenarios for scenario-based testing. This corresponds to a list where all the scenarios are listed. It is possible to select other groups or a combination of those groups by using the interface provided in the tool.

Figure 4.13 shows the classifications which were considered in the scenario database. This is more elaborated than the three-layer classification covered during the scenario implementation phase. The tool that interfaces the scenario database plays an important role in selecting and executing those scenarios from each of these classes for each test iteration.

When comparing the scenarios in the database with real-world occurrences, a normal distribution curve can be plotted against the scenarios and their frequency of occurrence. We can find that certain scenarios frequently occur, and certain others rarely occur. This information would help select specific scenarios for testing purposes during regression testing or for evaluating the basic functionalities of the system.

4.6. Automation in Scenario-Based Testing

When executing scenarios on a large scale, it is impossible to handle them manually. Scenarios are implemented and are executed on the scale of thousands as part of scenario-based testing. Hence, automation is the only possibility that helps execute all those scenarios and assess the behavior of the software in scenario-based testing.

Having a reliable automation platform is as important as having a simulation toolchain. The automation framework should be easy to control and manage and should have the capacity to integrate with the different levels of testing where scenario-based testing is performed. If your team is distributed globally, you could also think about deploying it in a cloud platform and performing simulations there, as shown in Figure 4.14. This will help execute multiple instances of scenarios specific to different ADAS or Automated Driving features during the development and testing. This can run as different release pipelines for software and feature.

For the specific use case of the Highway Chauffeur feature described in this chapter, the overall scenario-based simulation covering all scenes of the initial implementation of 1000 scenarios took approximately four months with three instances. The whole test execution was controlled by the interface tool with multiple instances. The intermediate reports were used to analyze the bugs and to improve the software as a parallel stream. Overall, this covered more than 2.4 million km in simulation alone. Depending upon the step size selected for each parameter, one could increase or decrease the number of iterations.

The approach used here would provide a timely update on the health of the framework and the tools integrated as part of the toolchain. This will also provide a summary of the distance covered in the simulation environment on each iteration of scenario execution as a measurement unit and the status of test execution. The status provides the scenarios executed with the details of each variant and its result after execution. This can be used as reference data for providing evidence to meet acceptance criteria defined

FIGURE 4.14 Scaling of scenario-based testing using automation on a cloud platform.

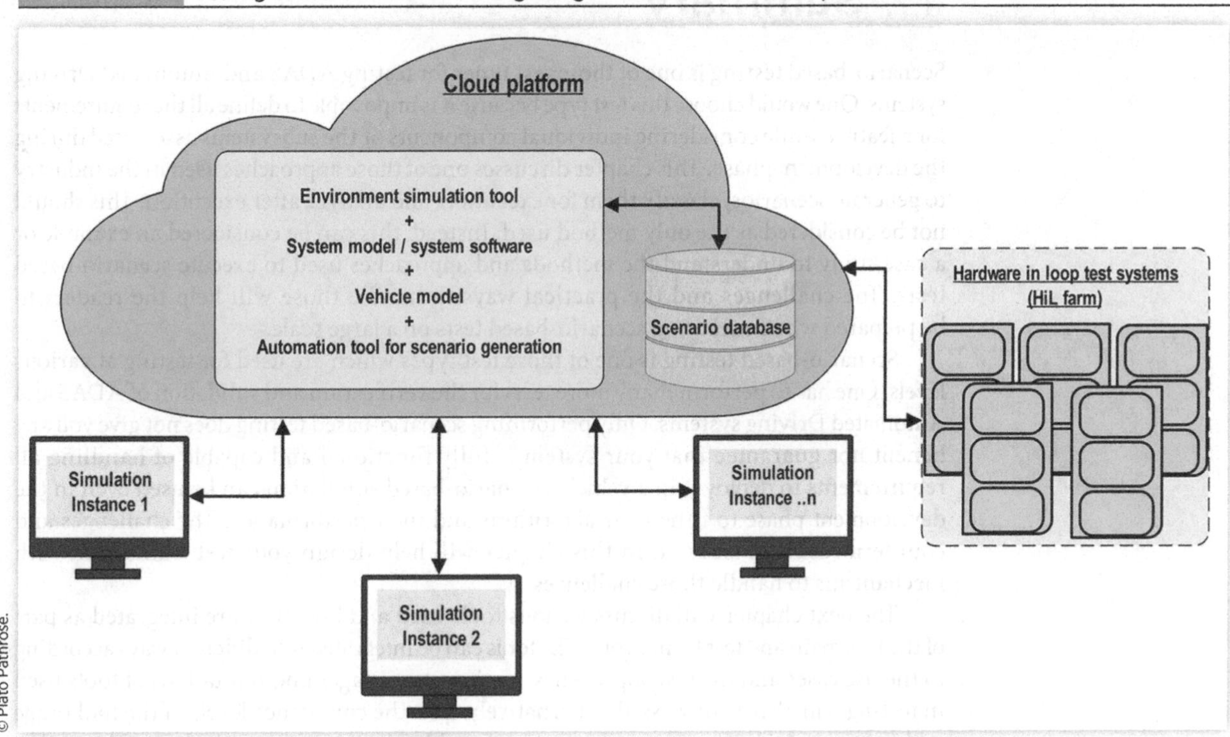

for each feature in virtual testing if it is to be measured based on mileage accumulated in simulation.

The major challenges faced and the countermeasures taken while implementing this approach in a cloud platform were:

- Getting all the required toolchains running in the cloud platform in multiple instances.
- Most of the time, tools become nonresponsive without any reason for which the process needs to be terminated and restarted.
- Simulation tools used for executing scenarios will become nonresponsive or lags after multiple iterations. There should be a mechanism in your monitoring and control of the framework which would detect these.
- Restarting the tool and executing the iteration from where it was stopped. Continuous logging after each scenario iteration is necessary rather than performing the logging after a set of scenarios or iterations.

Optimization of the toolchain and processes and precise monitoring and diagnostics for the validation framework is the critical area to consider whenever thinking about deploying tools for scenario-based testing on large scales.

4.7. Summary

Scenario-based testing is one of those test types for testing ADAS and Automated Driving systems. One would choose this test type because it is impossible to define all the requirements for a feature while considering individual components or the subsystems associated during the development phase. This chapter discusses one of those approaches used in the industry to generate scenarios, classify them for execution, and analyze after execution. This should not be considered as the only method used. Instead, this can be considered an example or a case study to understand the methods and approaches used to execute scenario-based tests. The challenges and the practical ways to handle those will help the readers to be prepared when executing scenario-based tests on a large scale.

Scenario-based testing is one of those test types which are used for testing at various levels. One has to perform many more tests for the verification and validation of ADAS and Automated Driving systems. Only performing scenario-based testing does not give you any benefit nor guarantee that your system is fully functional and capable of handling all requirements to deploy it in a vehicle. Scenario-based simulation can be used even in the development phase to tune your algorithms and their performance. The challenges and countermeasures discussed in this chapter will help design your test framework with mechanisms to handle those challenges.

The next chapter will discuss various tools used and how they are integrated as part of the toolchain and test framework. The tools can be integrated into different ways according to the use cases and the test approaches. Various advantages and limitations of tools used in testing will also be discussed. Alternatively, how the confidence levels of the tool usage and software deployment can be improved will be discussed as critical areas in testing safety-critical systems.

References

4.1. SAE International, “Taxonomy and Definitions for Terms Related to Driving Automation Systems for On-Road Motor Vehicles,” SAE Standard J3016_202104, April 30, 2021.

4.2. Ulbrich, S., Menzel, T., Reschka, A., Schuldt, F. et al., “Defining and Substantiating the Terms Scene, Situation, and Scenario for Automated Driving,” in *2015 IEEE 18th International Conference on Intelligent Transportation Systems*, Las Palmas de Gran Canaria, 2015.

4.3. PEGASUS Project, “Project for the Establishment of Generally Accepted Quality Criteria, Tools and Methods as well as Scenarios and Situations,” accessed July 1, 2021, https://www.pegasusprojekt.de.

4.4. V&V Methoden, “Verifikations- und Validierungsmehtoden Automatisierter Fahrzeuge im Urbanen Umfeld,” accessed July 1, 2021, https://vvm-projekt.de/.

4.5. SET Level, “Simulationsbasiertes Entwickeln und Testen von Automatisiertem Fahren,” accessed July 1, 2021, https://setlevel.de/.

4.6. Kshirsagar, K., Desai, H., Pathrose, P., and Sid, J., “Automated Test Scenario Generation for Testing Highly Automated Driving,” in *VDA Automotive SYS Conference 2019*, Potsdam, Germany, 2019.

4.7. Kramer, B. et al., “Identification and Quantification of Hazardous Scenarios for Automated Driving,” *Lecture Notes in Computer Science (LNCS)* 12297 (2020): 163-178.

4.8. Gitlow, H.S., Levine, D., and Popovich, E.A., *Design for Six Sigma for Green Belts and Champions* (New Jersey: Financial Times Prentice Hall, 2006), ISBN:978-0131855243.

4.9. ISO/PAS 21448:2019, “Road Vehicles—Safety of the Intended Functionality.”

4.10. ISO 31000:2018, "Risk Management—Principles and Guidelines."

4.11. ERTRAC Working Group, "Connectivity and Automated Driving," Connected Automated Driving Roadmap, Version 8, 2019.

4.12. Kuhn, D.S., Kacker, R.N., and Lei, Y., "Combinatorial Testing," NIST Report, June 25, 2012, https://www.nist.gov/publications/combinatorial-testing.

4.13. Liu, S. and Nakajima, S., "A Decompositional Approach to Automatic Test Case Generation Based on Formal Specification," in *Fourth IEEE International Conference on Secure Software Integration and Reliability Improvement*, Singapore, 2010, 147-155.

4.14. Menzel, T., Bagschik, G., and Maurer, M., "Scenarios for Development, Test and Validation of Automated Vehicles," in *2018 IEEE Intelligent Vehicles Symposium*, Changshu, Suzhou, China, 2018, 1821-1827.

4.15. Bock, J. et al., "Data Basis for Scenario-Based Validation of HAD on Highways," in *27th Aachener Kolloquium*, Aachen, Germany, 2018.

4.16. Schuldt, F., Saust, F., Lichte, B., Maurer, M. et al., "Effiziente systematische Testgenerierung für Fahrerassistenzsysteme in virtuellen Umgebungen," in *Automatisierungssysteme, Assistenzsysteme und Eingebettete Systeme Für Transportmittel*, 2013.

4.17. NHTSA, "A Framework for Automated Driving System Testable Cases and Scenarios," 2018.

4.18. Frank, H. and Reinsberg, H., *Leitfaden Fahrbahnmarkierung. Schriftreihe Verkehrssicherheit.* Vol. 17 (Deutscher Verkeharssicherheitsrat e.V., 2014)

4.19. UN.ECE.Transport Division, *Convention on Road Signs and Signals of 1968* (New York: UN.ECE. Transport Division, 2006)

4.20. Weber, H. et al., "A Framework for Definition of Logical Scenarios for Safety Assurance of Automated Driving," *Traffic Injury Prevention* 20 (2019): 65-70.

4.21. Fabris, S., Priddy, J., and Harris, F., "Method for Hazard Severity Assessment for the Case of Undemanded Deceleration," in Presented at *VDA Automotive SYS Conference*, Berlin, June 19-20, 2012.

4.22. Menzel, T. et al., "From Functional to Logical Scenarios: Detailing a Keyword-Based Scenario Description for Execution in a Simulation Environment," in *2019 IEEE Intelligent Vehicles Symposium (IV)*, Paris, France, 2019, 2383-2390.

/Autonomous
/Sensing
/Communication
/Battery
/Navigation
Autopilot mode

5

Simulation Environment for ADAS and Automated Driving Systems

Simulation is an essential step in designing and developing complex systems like ADAS and Automated Driving systems, especially when the product development involves ambiguity with various functions, requirements, and lack of a platform to perform tests. Tools and toolchains for simulation should be carefully selected, especially if it is planned to develop and test safety-critical products. The unfortunate truth in the industry is that, in many organizations, simulation tools are selected primarily based on their cost rather than their capabilities, performance, and quality. Even if the budget has an influence, the tools should not only be selected according to their cost but also based on the analysis results of their capabilities, advantages, and disadvantages. ADAS and Automated Driving system development involve a lot of simulation. Hence, it is very important to understand the tool selection and its integration with an existing toolchain and test framework. Because simulation is the critical element in both development and testing, one must deal with multiple simulation tools across the product development cycle. This varies with different use cases at various phases of the product development. It is most common to use multiple simulation tools as a toolchain for automated driving system development. The tools provide an ecosystem that would resemble the real environment. Using multiple simulation tools to work together is to create a simulation environment that correlates to the real world. This chapter covers different methods and approaches used in the industry for selecting simulation tools and their deployment in the product development life cycle.

5.1. Simulation Tool Selection

Simulation tool selection by any supplier organization has some influence on the vehicle manufacturer's test infrastructure and the tools in many projects. This usually occurs when the vehicle manufacturer wants all its ADAS or automated system suppliers to have the same simulation environment for the development and testing. Usually, for vehicle manufacturers or their suppliers, there might be an existing test infrastructure, including the test framework that is used for multiple projects in the organization. This includes multiple system projects or even different vehicle programs. It is impossible to think about changing the complete test framework for introducing a new tool specific to a project. Thus, every organization tries to have an ecosystem that can accommodate any new tools that are flexible and compatible with their existing test framework.

If a new simulation tool needs to be selected, the decision should not only be based on the cost of the tool. Certain key aspects need to be evaluated on the tool supplier side and the technical capabilities and limitations of the tool before deciding in favor or against it. Along with the cost of the tool, it is essential to understand whether the tool supplier would be flexible and capable of the support and service of the tool even after its deployment. Figure 5.1 summarizes different areas that should be considered for evaluating any new tool to obtain the best value proposition. Depending on the tool and its purpose, the priority of each of these areas can vary and should be considered for the analysis [5.1, 5.2, 5.3, 5.4]. It is quite common in the industry that for certain specific purposes, there is only one tool

FIGURE 5.1 Tool selection criteria.

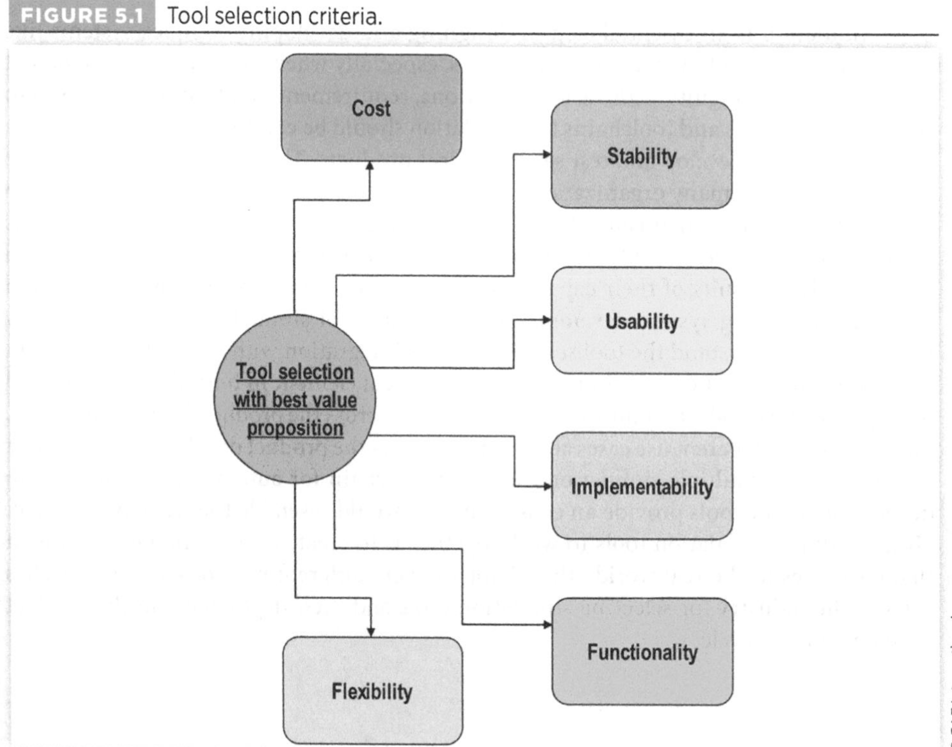

in the market that supports it, and hence, it is not possible to consider this approach in those situations.

1. Cost: The expense associated with the tool purchase, its operational cost, maintenance costs, and training if required should be considered.
2. Stability: The degree to which the tool performs successfully within the technology marketplace and its capability to handle wide operating ranges without failure. (This can be obtained with a market study.)
3. Usability: The degree to which the architecture of the tool under consideration improves system and data usability in the application area while masking system complexities. This also includes the ease with which a user can work with the tool.
4. Implementability: The degree to which the tool is mature, understandable (based on commercial off-the-shelf tools), and support availability. This gives the information if the tool can be trusted with its legacy information or its usage in other industrial areas.
5. Functionality: The degree to which the tool provides support with its functionality and satisfies the need for the project. This checks if the functional purpose is met by the tool or not.
6. Flexibility: The degree to which the tool and its architecture components are scalable and open for further product integrations and compatible with standards. The customization and adaptation of the tool have to be checked here.

With new technologies such as automated driving systems and electrification, many organizations are forced to improve their infrastructure and toolchains for development and testing. Many organizations have procured new simulation tools in the last few years mainly to support scenario-based simulations. One of those is the tools used for performing scenario-based simulation in ADAS and Automated Driving systems. These tools can also be selected considering the selection as mentioned earlier. One important aspect here with the scenario-based simulation tools is that they are used for simulating an environment to qualify a safety-critical system and its features in the vehicle. This brings additional characteristics regarding capability and qualification, which need to be considered in the selection of tools.

Simulation tools have found their way in large scale to develop and test ADAS and Automated Driving systems, especially in the area of scenario-based simulation and testing. They can be classified into three types based on their use cases:

1. Environment simulation tools—These are simulation tools that help simulate the external environment and road traffic in those environments.
2. Vehicle simulation tools—These tools help in recreating a vehicle model with all its characteristics and were used in the industry for a long time and mainly used to simulate vehicle dynamics and behavior in different road conditions. They also found their way to the automated driving domain as vehicle dynamics is an integral part of the ADAS and Automated Driving.
3. Photorealistic environment simulation tools—These tools are used mainly for generating synthetic sensor datasets for training and testing Artificial Neural Networks. They are graphically rich and provide photorealistic images and videos

of the external environment. This can be the major source for data-driven software development and testing. They are widely used because collecting real-world data is expensive and covering certain corner cases in scenarios is impossible or difficult without the simulation tools. These photorealistic simulation tools play a vital role in algorithm development and their performance improvement.

In the current tool market, there is no single tool that would serve all these purposes. Hence, there is a need to use multiple tools, depending upon the use case or specific purpose. There are different tools available in the market for each of these purposes. The safety and reliability of using these tools will be a matter of concern when used in an environment where safety-critical applications are developed or tested. This raises the question: how much can the tools be trusted? Depending upon the usage of each tool, analysis needs to be done on how it impacts the development or qualification of the final product that will help in deciding if these tools should be qualified for safety or not. A general qualification is required for any tool which is used in development or testing. This will fetch information regarding stability, robustness, and accuracy of measurements related to each tool for various applications. It is less likely that most of the tools undergo qualification from the user side for various applications even though that is required. The tool supplier must qualify them as part of their specifications and tool characteristics.

5.2. Co-simulation in Testing

When testing ADAS and Automated Driving systems, it is essential to have the simulation environment match the real world. In many cases, it is impossible to have a one-to-one match of the simulation tool with that of an actual vehicle and its environment. That can be resolved up to an extent with the approach of co-simulation. A co-simulation refers to utilizing more than one simulation tool to evaluate the system and its features by combining strengths from various tools. One such co-simulation environment could be using an environment simulation tool and vehicle simulation tool along with a system model for simulation. Suppose a single tool is planned to be used to test an ADAS or Automated Driving system and its features. In that case, the tool limitations will have an impact on the behavior of the system under simulation and evaluated results. In the current simulation tool market, none of the environment simulation tools are safety qualified, nor do they measure how much it correlates with the real world. This brings in the additional challenge of qualifying these tools whenever they are used for qualifying complex safety-critical systems. There were recent claims from a few simulation tools suppliers regarding the safety qualifications, which have to be further evaluated.

As part of an evaluation project with a vehicle manufacturer, an environment simulation tool was selected and used to create a scenario for evaluating AEB. The same scenario was recreated in a test track with the actual vehicle. Even though the same vehicle parameters were used in the simulation tool as received from the vehicle manufacturer of a passenger vehicle, the accuracy and the data collected in calculating Time to Collision (TTC) and the braking had a deviation of approximately 18% up to a vehicle speed of 100 km/h. This deviation increases above 20% when the vehicle speed is increased beyond 100 km/h and is not linear. This information will be an eye-opener for anyone who blindly trusts simulation tools without evaluating them and identifying how much they match the real world [5.5, 5.6].

The advantage of using a co-simulation environment is that it will help to reduce this gap of staying with the limitations of a single simulation tool. Some tools are good in environment simulation, and some others are good in vehicle dynamics. To get optimum performance and correlation with the real world, one should think about utilizing the strengths from multiple tools together in a test environment.

The simulation environment used for testing ADAS and Automated Driving features usually consists of:

- Environment model: An environment model represents different layers in the scenario implementation, which was discussed in the previous chapter. This includes the road, traffic infrastructure, weather conditions, static and dynamic objects, etc.
- Sensor models: These models would simulate the behavior of active and passive sensors that are integrated into the vehicle. These include sensors such as camera, radar, lidar, and ultrasonic sensors.
- ADAS/Automated Driving system model: This is the model in which different ADAS and Automated Driving functions are implemented. It is usually called the system software model or the model under test.
- Vehicle dynamics model: This is the vehicle model in simulation with the equivalent parameters for chassis, steering, powertrain, and brakes from the actual vehicle. The scope of using a vehicle model in the simulation should match the vehicle behavior with the actual vehicle under test.
- Middleware: This data transmission medium interconnects all these models in a simulation environment. A stable and capable middleware is required for better time-synchronized data transmission between different tools and models, supporting higher bandwidth of data exchange with smaller time delays.

All these different components should be integrated and need to work together to have an efficient co-simulation environment for testing ADAS or Automated Driving systems (Figure 5.2).

FIGURE 5.2 An example of a co-simulation environment.

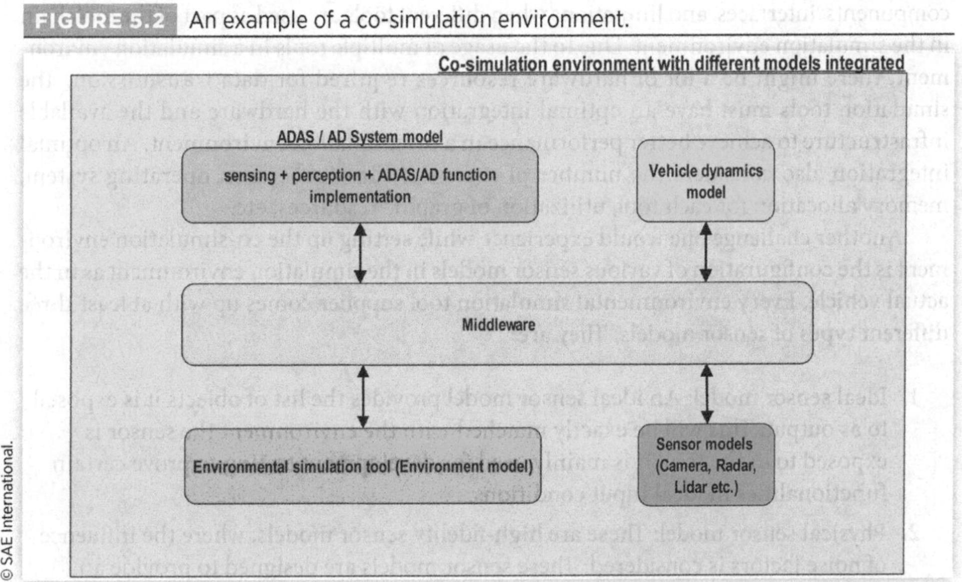

It is not required to utilize all these components together in every co-simulation environment. Depending upon the purpose of simulation, one could decide on those components which should be used in a co-simulation environment. On the other hand, all these components are recommended to be considered for system verification and validation, where the complete system and its features are evaluated. A similar co-simulation environment can be utilized for the development testing but with minimal components in it. The component selection of the co-simulation environment also has a dependency on the organization. For example, a vehicle manufacturer should be concerned about the complete vehicle and its behavior where all the components as part of the co-simulation need to be considered. A tier supplier providing only a sensor or application software or algorithms does not require the whole co-simulation components in their simulation environment as their focus is not the complete vehicle but on a smaller area of the vehicle. Co-simulation is also a prospective way for the homologation and certification for Automated Driving systems. As the regulations for automated driving vehicles are formulated, in the future there will be a need for the evaluation of those systems in virtual environments along with physical testing in the actual vehicle. The co-simulation environment can be considered the test environment for performing virtual testing of the complete vehicle inside a laboratory environment.

One of the significant challenges in setting up a co-simulation environment is the qualification of different tools and components that are part of the co-simulation environment. Suppose there is no measurement available on how much this co-simulation environment correlates with the real world. In that case, it will not help in measuring and trusting the system behavior in this virtual simulation environment even though multiple tools are used. This must be measured by testing it in the real vehicle, calculating the deviations, and measuring the correlation of the components in the simulation environment. One of the methods which were used to perform such a correlation measurement was using the Six Sigma processes for process improvement called Measurement System Analysis, or MSA. Here, one could evaluate the variation in the measurements caused by the process followed while measuring data and the variation induced by the tools used for measurements. The primary source of deviation in simulation environment measurements that can be classified as process variation occurs in the form of data exchange between simulation components, interfaces, and limitations when different tools are used as part of the toolchain in the simulation environment. Due to the usage of multiple tools in a simulation environment, there might be a lot of hardware resources required for data transmission. The simulation tools must have an optimal integration with the hardware and the available infrastructure to achieve better performance in a co-simulation environment. An optimal integration also considers the number of cores used in a processor, operating system, memory allocation for each tool, utilization of graphic resources, etc.

Another challenge one would experience while setting up the co-simulation environment is the configuration of various sensor models in the simulation environment as in the actual vehicle. Every environmental simulation tool supplier comes up with at least three different types of sensor models. They are:

1. Ideal sensor model: An ideal sensor model provides the list of objects it is exposed to as output. This will be exactly matched with the environment the sensor is exposed to in the tool. It is mainly used for development testing to prove certain functionalities in ideal input conditions.
2. Physical sensor model: These are high-fidelity sensor models, where the influence of noise factors is considered. These sensor models are designed to provide an

output that is comparable to the actual sensors used in the vehicle. However, this also has a drawback as these models usually do not match exactly with a sensor that will be used in the vehicle. The output of these sensor models is mostly object lists or processed lists of objects from the environment.

3. Raw sensor model: These are sensor models that translate the sensor environment as raw sensor data that will be provided as its output. These sensors can be used in a simulation environment if the sensor data is processed in the ECU rather than only providing the object lists from the environment data to the control unit.

Depending on the various sensors used in the vehicle and the concept of the system implementation, different sensor models can be considered for setting up a simulation environment.

There is a tendency to utilize sensor models that come along with the simulation tools for all the simulation and test purposes. These sensor models from tool suppliers might not often match with the actual sensors in the vehicle. These days, most vehicle manufacturers use sensors that are new in the market, and the technical specifications related to those sensors are usually not available to the tool suppliers to provide similar sensor models in their simulation tools. Hence, they cannot match their sensor models in the simulation tool with those sensors used by a vehicle manufacturer. This is a major constraint in the tool qualification in many projects when technical specifications are not available. The deviation in the characteristics and measurements from sensor models to that from the real sensors will have an impact on the quality of the data transmitted for processing, and consequently, it will have a direct influence on the measurement and evaluation of ADAS and Automated Driving features in the simulation environment. This will add up some effort for evaluation or benchmarking with an existing sensor model so that these measurements provide the degree to which the measurements from the new sensor models vary.

The sensor suppliers usually develop sensor models as part of their development. It is recommended to procure the sensor models from the sensor supplier rather than use the default sensor model from a simulation tool. It would be more beneficial than using generic sensor models from an environment simulation tool provider without configuring the sensor models with the required characteristics of the actual sensor based on its technical specifications. A radar or camera sensor model from a sensor supplier is expected to be more accurate and can correlate with their real-world sensors. In many cases, even the sensor suppliers might not have a sellable sensor model available for the sensors they produce. This would affect the reliability of sensor models in a co-simulation environment if used without proper benchmarking or measuring the correlation with the actual sensors.

Vehicle models are the simulation models of the vehicle in a simulation environment. These models provide the static and dynamic characteristics of the actual vehicle in a simulation environment during simulation. Vehicle models are usually available from vehicle manufacturers and can be used in a co-simulation environment without much difficulty. Compared to sensor models where the technology and the sensors are changing rapidly, the change in vehicle platforms is not so rapid from the vehicle manufacturers. Hence, these models would be much more reliable to use in a simulation environment. One of the challenges with the usage of vehicle models in a simulation environment is that while working with certain vehicle manufacturers, a special integration effort needs to be considered as certain vehicle manufacturers use proprietary tools for developing their vehicle models. These models might not be having the required interfaces to integrate with a co-simulation environment. For the co-simulation, one could either utilize these vehicle models from a vehicle manufacturer or use a different vehicle model but with the vehicle parameters of the original vehicle model.

5.3. General Qualification of Simulation Environment

Qualification of a simulation environment and the framework for testing must be performed as part of the test preparation. Very few organizations follow a structured approach for qualifying the tools and toolchains used for simulation and testing. When toolchains and test frameworks are established with multiple standard and nonstandard tools for the simulation and evaluation of safety-critical products, it is mandatory to evaluate and qualify the toolchain and test framework. In this section, a general qualification method for the simulation environment is described. The safety-specific qualification of the tools is covered in detail in a different chapter.

One of the common methods used in the industry to qualify a test environment, including simulation tools, is with the help of processes for MSA [5.7, 5.8]. MSA can be defined as an experimental and mathematical method of determining the variation that exists within a measurement system or process associated with measurements. If not considered, the variation in the measurement system can directly influence the overall measured system performance and behavior. Hence, the methods from MSA help in assessing the ability to measure and collect trustworthy and reliable data. There are different ways in which one could apply the methods from MSA and evaluate the simulation environment. MSA methods can be compared with the Plan-Do-Check-Act (PDCA) cycle, or Deming cycle, for continuous improvement of processes and products [5.8] (Figure 5.3).

The "Plan" phase focuses on defining the plan for the complete cycle. This starts with defining the data collection strategy for the simulation environment and the real world, defining the components and parameters which should be evaluated, the evaluation strategy of measured data, and the assessment criteria to check how they vary. This can be performed for the complete simulation environment as part of the toolchain or to a small portion of it. Whether MSA should be applied for the complete simulation environment or a part of it is made based on various factors associated with the toolchain, such as the components that are prequalified, the influencing factor of the measured outputs, and its use cases. This also includes the legacy data and the status of the existing test environment.

The "Do" phase is where all the planned activities in the "Plan" phase are executed. In this phase, the trials are performed, and the samples are collected both from the simulation environment and the real world. Depending upon the area of interest and the organization level in the industry as a supplier or a vehicle manufacturer, one could decide to focus specifically on specific models or the complete models used in a simulation environment.

The "Check" phase will analyze the data collected during various trials in the "Do" phase. Various methods of data representation and analysis are used here. This is the phase where a lot of mathematical and statistical evaluations on data distributions are considered to identify the behavior of the simulation environment in comparison with the real world. Data analysis tools are used to plot the data in different charts and tables for better analysis. This analysis helps in identifying the data distribution and deriving inferences.

The decision-making phase, or the "Act" phase, is where the analysis from the "Check" phase will be used to derive actions. With the collected measurements, the errors and deviations associated with the simulation environment at different input levels are identified. Analysis results of these data are used to derive conclusions and details regarding the

FIGURE 5.3 PDCA cycle.

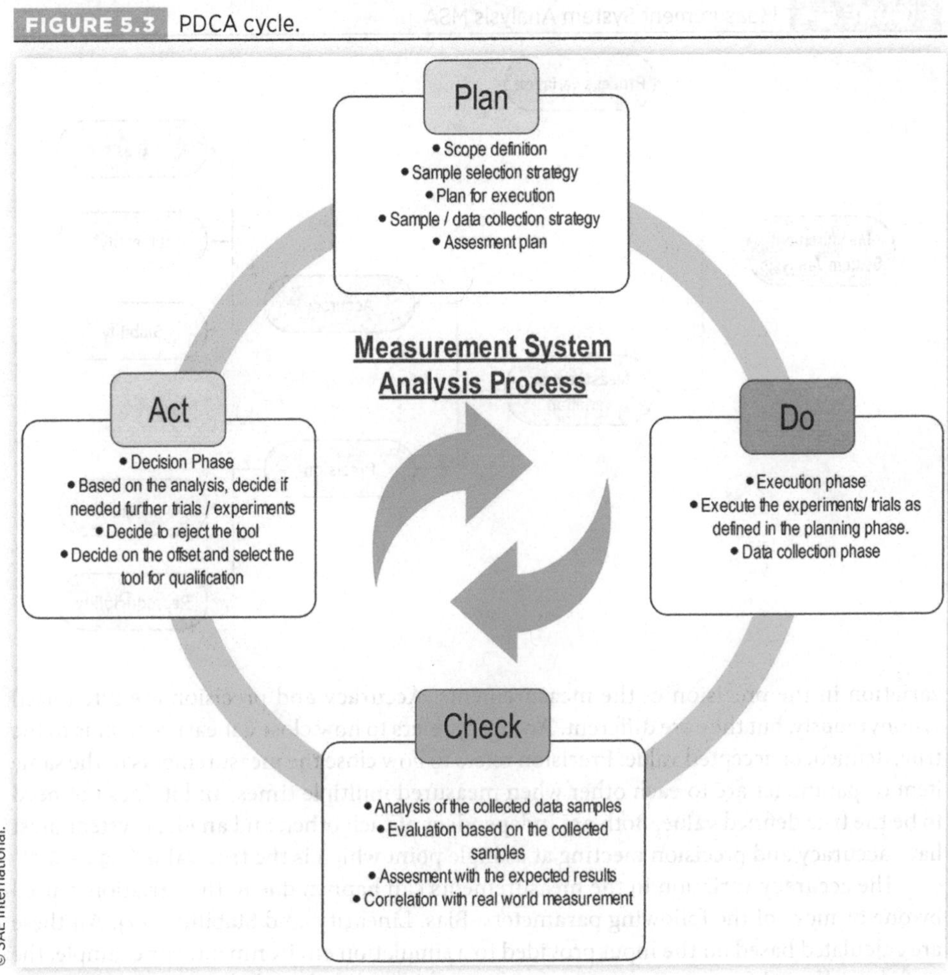

correlation of the simulation environment with the real world. Data from various components in the co-simulation environment are analyzed and evaluated with the help of a test framework. In almost all cases, a test framework acts as an interface to the simulation environment, which would then be used for integrating different test levels from model-in-the-loop simulation to vehicle-in-the-loop simulation. The test framework can be used only as a sample data collector, or it can also be used to perform the data analysis.

MSA covers the measurement from the process variation and the component variation in a process. Here, the process variation evaluation covers the data flow and the steps followed while the measurements are made, or the data is collected while using a co-simulation environment for testing purposes. The main area of focus here is to evaluate the measurement variation of the components of the co-simulation environment rather than process variation. As part of that, different areas where the variation can occur should be identified, such as the data flow, bandwidth availability, methods followed for the measurement from each component of simulation, data collection interfaces, and control and monitoring components of the simulation.

Measurement variations are caused by errors [5.7, 5.8]. These errors can be further decomposed to the error caused by variation in the accuracy of measurements or the

FIGURE 5.4 Measurement System Analysis MSA.

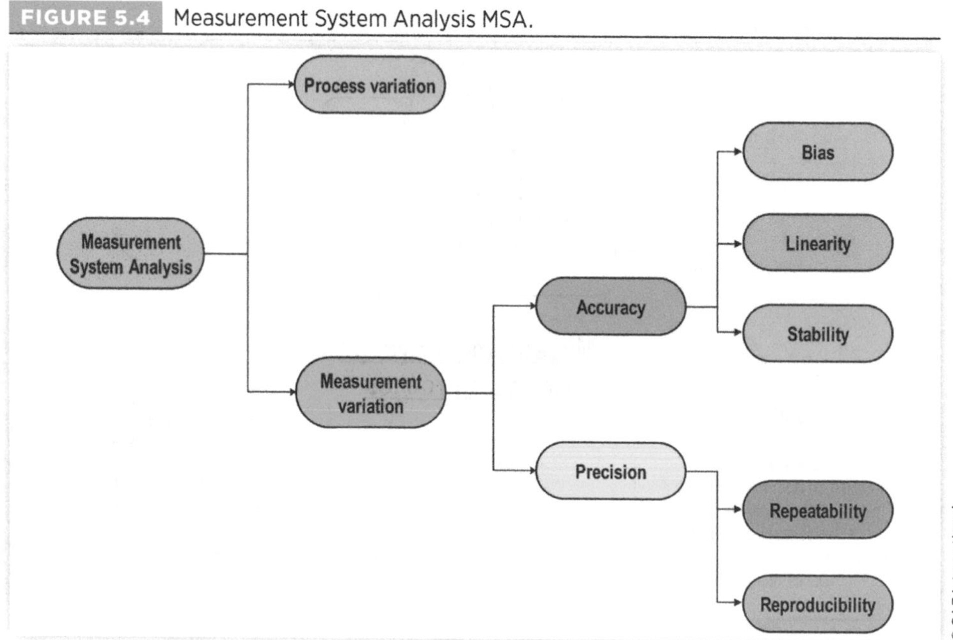

variation in the precision of the measurements. Accuracy and precision are often used synonymously, but they are different. Accuracy refers to how close a measurement is to the true, defined, or accepted value. Precision refers to how close the measurements of the same item or parameter are to each other when measured multiple times, and it does not need to be the true defined value. Both are independent of each other, and an ideal system must have accuracy and precision meeting at a single point which is the true value (Figure 5.4).

The accuracy variation in the measurements can happen due to the variation caused by one or more of the following parameters: Bias, Linearity, and Stability [5.7]. All these are calculated based on the input provided to a simulation environment. For example, the samples collected for comparing the correlation with the real world have parameters such as vehicle speed, ranging from 0 to 150 km/h. This means that the simulation environment must be checked for the whole range of this input parameter of vehicle speed. The output of the simulation environment must be checked for the lower as well as the upper parameter ranges. This is to check whether the simulation environment output is behaving linearly or not, and no other irregularities are observed. The output should be observed to avoid any misinterpretation of output values when the input parameter varies within specific ranges. Along with checking for linearity, the simulation environment output also needs to be evaluated for its stability and bias to identify if it is stable along with its operating range or whether it shows any bias to certain specific input parameter values or ranges.

Precision variation evaluation covers mainly the repeatability and reproducibility of the output of the simulation environment for certain input parameters [5.7]. Here the focus is to avoid the variation due to repetition of tests and ensure that the output remains the same whenever an evaluation is performed with the same input parameters. Likewise, whenever a variation has occurred, it should be possible to reproduce those even after changing the input parameter ranges and then coming back to the original parameter values where the variation initially occurred.

The strategy and definition of the experimentation and sample data collection, including methods used, scenarios to be considered, different parameters and ranges, output data format, and how it should be collected, are defined during the "Plan" phase of the PDCA cycle. All these items defined in the planning phase are executed in the "Do" phase, where one would perform these experiments or trials to generate sample data. The challenge in the "Do" phase is to execute these experiments or trials by replicating the same scenarios from the real world in the simulation environment. Practically, there are always certain compromises made in these experiments. The scenario replication from the real world to the simulation environment is always limited to the used tools. One-to-one replication is not always possible in the simulation environment due to the limitation and flexibility of the tools. The real-world measurement data of experiments or trials are collected, and the same experiments are reproduced virtually in the simulation environment, which is later compared. The real-world tests are usually executed in proving grounds or test tracks or in any other controlled roads.

The most commonly used method for MSA is the PGA (Practical, Graphical, and Analytical) approach [5.9, 5.10]. The practical part encompasses the sample definition and sample collection; the graphical part involves the analysis of the collected samples or measurement data and the generation of graphs for better evaluation. The analysis phase mainly identifies the discrepancy in the output data collected from the real-world experiments or trials to that from the simulation environment. Two of the standard graphical charts generated in MSA are multi-vari charts that visually represent the variability and the Xbar-R behavior chart that shows if the measurements are stable and predictable.

Based on the data analysis and graph generation, the standard deviation can be calculated to establish control limits for output parameter values depending on measured data. For a simulation environment evaluation, multiple output parameters need to be evaluated, and control limits should be established for each of those output parameters differently. A control limit, as per statistics, can be described as the horizontal line above and below the central line over a normal distribution curve within which the data points are controllable. The need to define control limits for the output parameters is to have acceptable upper and lower values within which the measurements are controllable and reliable. This will be easy if the parameters provide data points that are normally distributed. Certain output parameters will not provide normally distributed data. In such cases, these data points must be converted to a normal distribution using statistical approaches like Box-Cox transformation and further processing it like normally distributed data [5.7, 5.8, 5.9, 5.10].

There can be situations when a single tool needs to be newly integrated into an existing toolchain. When there is a need to introduce a new tool to an existing toolchain or a test framework, the new tool is usually integrated with an existing simulation environment rather than replacing the whole toolchain. Suppose an existing toolchain is qualified or stable and robust due to legacy reasons. In that case, one should only consider the qualification for the newly integrated tool rather than qualifying the complete toolchain. Here only a delta evaluation should be performed to evaluate the impact of the newly integrated tool on the overall toolchain. There might also be situations where a complete replacement of the toolchain or a simulation environment is required. Here MSA needs to be performed on the complete toolchain. This is time consuming and requires a lot of experimentation, data collection, and effort to establish a confidence level on the complete toolchain or simulation environment.

There is another approach used in a few of the organizations for qualifying toolchains, especially simulation environments. This is called risk-based evaluation of the toolchain, where the parameters that have a higher risk of variations toward the specification limits are identified. These parameters and their value ranges are controlled with the concept of guarding. Guarding involves adding a guard band as an additional offset over the data

FIGURE 5.5 Defining a guard band within specification limits.

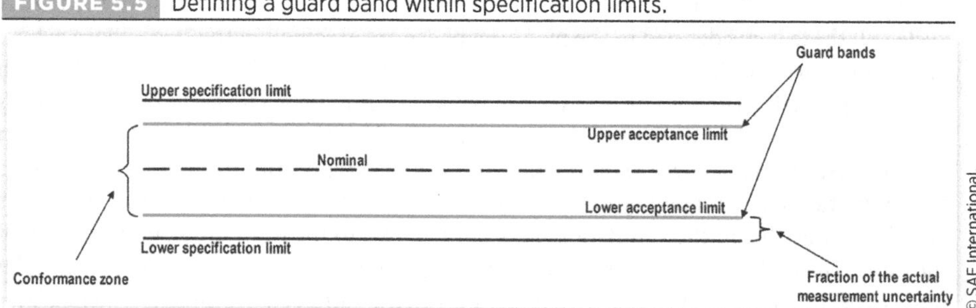

distribution well within the specification limits. This will help establish a control mechanism to adjust the parameters to avoid variations that move out of the defined specification limits [5.10] (Figure 5.5).

MSA for a simulation environment with hundreds of output parameters would be easy if risk-based analysis and evaluation were used. Here the parameters that have a higher impact on the simulation environment are identified, and then the MSA is performed for those. While performing the analysis, the deviation of individual output parameters can be measured and utilized to set the adjusted offset values while calculating those parameters in the output.

5.4. Limitations of Simulation Tools Used in ADAS and Automated Driving

Simulation tools play an important role in the development and testing of ADAS and Automated Driving systems. It is impossible to think about a toolchain used for qualifying automated driving features without a simulation tool. Even though the simulation tools become core components of test infrastructure, there are many challenges one has to deal with while setting up the simulation environment. This section will go through a few challenges faced while establishing a co-simulation environment and using different simulation tools.

In the ADAS and Automated Driving industry, simulation tools are used as part of a toolchain for different purposes. This can start from a simple simulation tool for model testing to a complex simulation environment like a co-simulation environment with multiple simulation tools and models integrated. They are usually used to perform software-in-the-loop, HiL, DiL, or vehicle-in-the-loop simulations. Among these, environment simulation tools have a significant role to play in the ADAS and Automated Driving system simulations. The environmental simulation tools are used for various purposes, but the challenge here is that most of the environment simulation tools are not standardized. They are the main component of scenario-based testing. Even the same tools are used for algorithm developments.

There are different types of simulation tools available in the market which are used for scenario-based testing. This ranges from simulation tools that provide photorealistic environment details to those which focus more on vehicle dynamics and road handling. The resource utilization of these tools also varies from requiring high-end processors and graphical processors to those which are very light and can run with minimal resources. The photorealistic simulation tools are mostly used for development purposes. They are the best source to generate synthetic datasets for artificial neural network training and validation for different sensors and associated scenarios that are otherwise limited to the data from vehicle sensors. These are also used for sensor stimulation in different test environments.

As mentioned earlier in this chapter, it is essential to qualify the tools and toolchains used for testing ADAS and Automated Driving systems. The major challenge one would face with these simulation tools for scenario-based testing is the lack of standardization. The simulation tool from each vendor is different in terms of how the input data is accepted and the scenarios designed and implemented. This is gradually getting improved in recent days, but it will still take time for the market to develop a standardized model. It isn't easy to use different simulation tools for a tier supplier who provides systems or subsystems for Automated Driving systems, including software to different vehicle manufacturers. In most projects, the scenarios designed and implemented as part of scenario-based testing need to be reengineered to make them compatible with a different simulation tool required by the vehicle manufacturer for a particular project. There has been a proactive initiative in the industry to resolve this by the adoption of OpenDRIVE® from the Association for Standardization of Automation and Measuring Systems (ASAM) community [5.11, 5.12]. OpenDRIVE® is an open file format that enables one to specify large and complex road networks that can be used across different tools compatible with OpenDRIVE® format [5.12].

Another effort of standardization that has been initiated is the usage of OpenSCENARIO®. It is a file format with a description of the dynamic content of driving and traffic simulators [5.13]. This will help reuse the dynamic components in a scenario across different simulation tools compatible with the OpenSCENARIO® format. Unfortunately, current simulation tools are far behind in getting these standardized formats integrated into their tools. This will affect the flexibility of using a simulation tool.

Another challenge with the simulation tools is their qualification and confidence level in usage. Currently, the tools used for scenario-based simulation and testing are far from getting qualified while they are being used to develop and evaluate complex safety-critical systems like ADAS and Automated Driving systems. Even the tool vendors developing the tools give more importance to the visual effects and the graphical capabilities than coming up with measurement details on how much the tool parameters correlate with the real world. For example, the tools that provide sensor models are designed based on ideal measurements and calculations, and real-life behavior is introduced by adding noise. While comparing the behavior of an actual sensor to that of a virtual sensor in simulation, there is either significantly less or no correlation between them. So using such an environment without clear correlation measurements to provide input data and evaluating the performance of safety-critical systems only helps in adding additional risk to the quality evaluation of the system and software.

Another challenge while using environment simulation and traffic simulation tools is the usage of open-source tools. In many organizations, the initial prototyping and even the product development use open-source simulation tools, which are not well controlled. There is always a risk associated with those open-source simulation tools if they are not tracked well for their delivery, known defects, and overall performance and quality. If a prototype software is developed and evaluated using open-source simulation tools and later taken into production, it should undergo robust qualification with qualified tools. This is not followed in many organizations, which is risky unless one is completely aware of the errors introduced in measurements from these tools.

Last but not least, the challenge with nonoptimized simulation tools is also important. The simulation tools use a lot of resources like processing power, graphics, and memory. This calls for the need for high-power computing systems with graphic processors for simulations. When these simulation tools are integrated with other models and tools in a co-simulation environment as part of a toolchain, the resource-hungry simulation tools will create lags and unpredictable behaviors when used for long-term operations. Utilizing the available simulation tools and planning long-term scenario-based testing in a co-simulation environment for achieving a million kilometers of mileage without optimizing the tools is impossible. While operating, it is common to experience delays and unsynchronized data flow after a period, mainly because of the nonoptimized tools in the market for their resource utilization. As the optimization of the simulation tools is always considered toward the last part of their development, tool suppliers tend to focus less on optimizing them.

5.5. Summary

This chapter discussed the importance of simulation tools in developing and testing ADAS and Automated Driving systems. It also spoke about how important it is to qualify the tools used to develop and test safety-critical products. The chapter also covered the methods used in the industry for qualifying the tool and the toolchain to evaluate and measure its quality to use for testing safety-critical products. The quality parameters for the tool include confidence level, reliability, and stability and providing trustable measurements.

Various methods discussed in this chapter are from the best practices followed by different organizations in the industry. There have been many initiatives and projects that focus on the use cases and standardization of the tools and the format the tools should support. The initiatives by ASAM like OpenDRIVE® and OpenSCENARIO® are laying the foundation of standardization of tools used explicitly for scenario-based simulation and testing. The last section of the chapter explained the limitations of the tools available in the market and how these limitations would affect the development of a safety-critical system if not appropriately addressed. Establishing a qualified toolchain is challenging in the current market as one must deal with many nonstandard tools for a toolchain. The challenges one would face while using it for simulation in projects will help identify further defects and adaptations required for the tool. One must be ready to handle these challenges while using them for a production project. Qualification of the tool is the biggest challenge as the tools used for the scenario-based testing are also used in the development testing to test and qualify the building blocks of the safety-critical systems.

The next chapter will cover the need for reference data and its importance in the development and testing of ADAS and Automated Driving systems. The chapter will also give an overview of how the reference data is generated, its process, and how one would utilize these reference data to evaluate artificial neural networks.

References

5.1. Morris, A.S. and Langari, R., *Measurement and Instrumentation: Theory and Application*, 2nd ed. (Amsterdam, the Netherlands: Academic Press, 2015), ISBN:978-0128008843.

5.2. ReVelle, J.B., *Manufacturing Handbook of Best Practices: An Innovation, Productivity, and Quality Focus*, 1st ed. (New York: CRC Press, 2016), ISBN:9781420025507.

5.3. ISO 14253-1:2017, "Geometrical Product Specifications (GPS)—Inspection by Measurement of Workpieces and Measuring Equipment—Part 1: Decision Rules for Verifying Conformity or Nonconformity with Specifications," 2017.

5.4. ISO 31000:2018, "Risk Management—Principles and Guidelines."

5.5. ISO 26262:2018, "Road Vehicles—Functional Safety."

5.6. Littlewood, B. and Wright, D., "Some Conservative Stopping Rules for the Operational Testing of Safety-Critical Software," *IEEE Trans. SW Eng.* 23, no. 11 (1997): 673-683.

5.7. Gitlow, H.S., Levine, D., and Popovich, E.A., *Design for Six Sigma for Green Belts and Champions* (New Jersey: Financial Times Prentice Hall, 2006), ISBN:978-0131855243.

5.8. Arthur, J., *Lean Six Sigma Demystified*, 2nd ed. (New York: McGraw Hill, 2007), ISBN:978-0071749091.

5.9. ISO 3951-1, "Sampling Procedures for Inspection by Variables—Part 1," 2006.

5.10. Fearn, T., Fisher, S., Thompson, M., and Ellison, S., "A Decision-Theory Approach to Fitness for Purpose in Analytical Measurement," *Analyst* 127 (2002): 818-824.

5.11. ASAM OSI®, "Open Simulation Interface," accessed November 30, 2021, https://www.asam.net/standards/detail/osi/.

5.12. ASAM OpenDRIVE®, "Open Dynamic Road Information for Vehicle Environment," accessed September 1, 2021, https://www.asam.net/standards/detail/opendrive/.

5.13. ASAM OpenSCENARIO®, "ASAM Open Scenario Version 1.1.1," accessed December 15, 2021, https://www.asam.net/standards/detail/openscenario/.

/Autonomous
/Sensing
/Communication
/Battery
/Navigation
Autopilot mode

6

Ground Truth Generation and Testing Neural Network-Based Detection

Data have a more significant role in developing software these days where the application of Artificial Intelligence (AI) is on a large scale. There are many different ways in which AI is used to design and develop highly automated vehicles. AI-based software components utilize various data for their development, performance improvement, and testing. How is data prepared and used to develop these data-dependent software components? How are they tested? It is a complex topic that is not standardized yet. There are various methods and best practices from other industries that are used in the automotive industry while using AI and these data-dependent software components. This chapter takes you through various methods and processes that are followed in the automotive industry to prepare the data and to use it for the development, performance improvement, and test of these data-driven software components.

6.1. Introduction to Data-Driven Software Development

As artificial neural networks have found their way in many areas of ADAS and Automated Driving such as detection and perception, which are strongly dependent on data. it is difficult and almost impossible to follow the classical V-model product life cycle to develop and deploy these software components. The classical V-model does not consider the new technology areas of software development as neural networks and their testing that are driven by data. Hence, it is necessary to consider a development model that can be applied for these data-driven software components, which strongly depend on data for its development, performance improvement, and testing [6.1]. The data-driven development part of these software components solely depends on the following steps:

- Data collection
- Data preparation
- Software development and performance improvement
- Data-dependent test and qualification

AI is used in different areas of automated driving in different ways. Hence, for the readers to understand better, this chapter will discuss computer vision, a subset of AI and one of the most critical and commonly used fields in ADAS and Automated Driving. Computer vision is about making the machines understand from the visual data [6.2]. This includes detecting objects and classifying them from visual sources like cameras or lidar. As automated driving vehicles utilize many camera sensors, the role of computer vision is substantial in the detection and classification of objects from these camera sources.

How can the machines detect objects and classify them? This is achieved by training those machines like the human brain, feeding in information about those objects to recognize and identify objects with the training they had before. There are algorithms developed that can be trained with the data fed into it, to keep it in memory and detect the same or similar objects when exposed to the algorithm next time. The identification of objects from an image or a video source is achieved by recognizing patterns from the image by mathematical calculations. This is similar to our human brain, which processes the data from the eyes and derives conclusions from it. Few terms are commonly used in this field, such as neural networks, deep learning algorithms, and machine learning. It is good to understand those before moving forward deep into the subject.

Neural networks are a set of algorithms designed to recognize patterns. They can interpret sensor input data by using machine perception, labeling, or clustering. Patterns that they can identify are numerical and in vector format. All the input data, either images or videos, are translated to these numerical values to recognize patterns from them. Deep learning is a part of machine learning in AI. Deep learning is a function in AI that replicates the functionality of the human brain in processing input data and recognizing patterns that support decisions. It can be considered as a subset of machine learning which utilizes specific networks called deep neural networks, which learn without human supervision from a data source that may be not structured or not labeled as reference data. Machine learning is similar to deep learning in its function of mimicking the brain functionality for detecting and classifying the patterns in the input data. It utilizes decision trees and also neural networks similar to deep learning [6.2]. Thus, we could consider it a superset of deep learning that can be trained with or without human supervision; that is not the case with its subset.

The evolution of computer vision in the ADAS and Automated Driving applications was rapid. This brought up the change and pushed in the utilization of more reliable vision-based sensors in the vehicles and using them for multiple purposes for the safety and comfort of the vehicle users. The computer vision components of detection and classification functions using neural networks are strongly dependent on the ability to learn from the data provided to it. Since data is the primary source for developing and testing these software components, they can be integrated into a development framework that covers the full life cycle of the development and usage of those software components, as shown in Figure 6.1.

FIGURE 6.1 A development V-model for data-dependent software components.

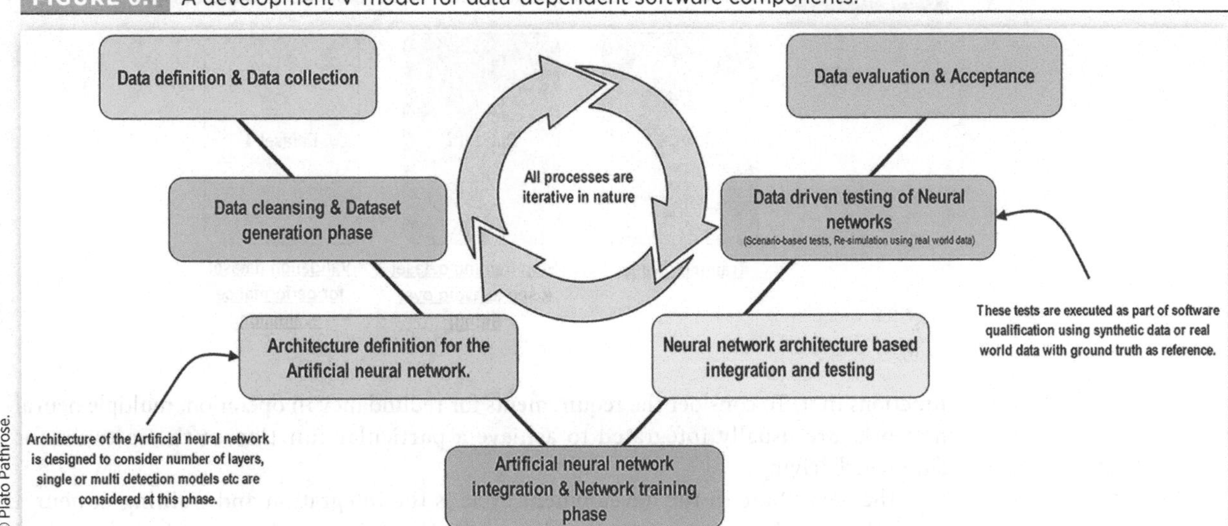

This is a parallel development model similar to the classical V-model for product development. The first phase in the life cycle is the data definition and data collection phase. As a first step, a plan is established regarding the type of data, required resources, and a strategy to collect the required data considering different sources and operating the data collection processes. The second phase is data cleansing and dataset generation from the collected data where, in this phase, the collected data is further analyzed, trimmed, and synthesized. Based on the plan defined in the first phase, datasets are generated according to specific requirements based on various use cases such as the datasets for training, stop training, and validation. These datasets are the reference data that the computer vision algorithms will later use for learning. A human can supervise the learning, or it can be unsupervised.

Data cleansing and dataset generation are the most critical and time-consuming phases in the overall cycle. This phase supplies data for training the algorithms and testing them [6.3, 6.4]. Usually, three different sets of ground truth references or datasets are generated as an outcome of this phase. One set is for the training of algorithms, and this is usually the biggest dataset used as part of development for any neural network application. The second one is the dataset that is used to control and to prevent overfitting of the algorithm from training and to calculate the true error of the network. The third one is for testing the entire neural networks for their functionality. There are different requirements based on which these datasets have to be developed. This includes the accuracy and the type in which objects are represented while labeling, classification of various objects like vehicles and lane marking, and the possible variations which need to be considered based on the architecture of the neural networks (Figure 6.2).

The third phase in data-driven software development is the architecture definition for these software components. Here, in this phase, the architecture definition of the artificial neural networks, various layers in the network that need to be considered, and the design requirements to use these neural networks are defined. Based on the use case, the architecture should consider the number of layers required, and the integration of multiple neural networks is considered. For example, this decides whether multiple neural networks have to be developed to provide the required functions or a single neural network should be developed with multiple

FIGURE 6.2 Classification of different datasets for development and testing.

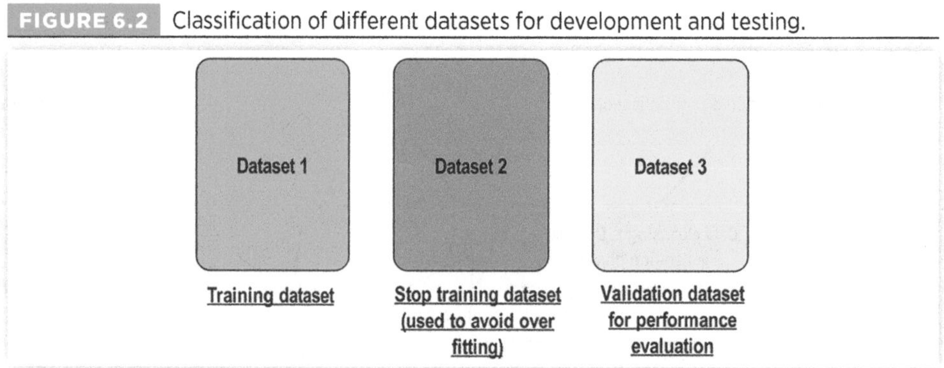

functions in it. To consider the requirements for redundancy in operation, multiple neural networks are usually integrated to achieve a particular function in the software for automated driving.

The next phase in the development cycle is the integration and training of neural networks. Based on the type and quality of the input data used for training, the neural networks learn how to detect and classify objects based on patterns from the dataset previously used for training those algorithms [6.3]. This can be a process that is supervised by a human or completely unsupervised. The size of the dataset used for learning determines the performance of the algorithms in detecting and classifying the objects later, based on patterns from a series of input data to which it is exposed.

The right-hand side of the development cycle depicts the integration and testing of artificial neural networks. The integration phase covers the integration of neural networks based on the defined architecture and the evaluation based on the design. In this phase, the neural networks are evaluated for interdependencies, how the information is processed over different layers and among multiple networks integrated together. This phase primarily checks on how a decision is derived from the neural network. The last two phases in the data-driven development model are testing phases that are executed by means of synthetic or real-world data. Here the neural networks are exposed to the environmental data captured by sensors, such as scenarios from which they are tested if they could successfully detect and classify the objects based on the patterns and accumulated information from previous training epochs. This also includes an evaluation and acceptance phase, which is predefined with Key Performance Indices (KPIs) to measure the quality and performance of these software components.

6.2. Data Annotation and Dataset Generation

Data annotation is the process of annotating or labeling specific data that is of interest from the rest of the data that a machine learning algorithm might use for training or comparison purposes. Feeding data to a machine learning algorithm could be for many purposes like training an algorithm, testing it, or even benchmarking a particular algorithm to another algorithm using the same data.

For a system using camera data, it is images and videos that would be the input data. The images can be individual frames of videos captured from the camera sensors from different positions around the vehicle [6.4]. There are different methods used in the industry to annotate or label the images and videos like bounding boxes and skeletal, polygon, semantic segmentation [6.5]. Out of all these methods, the most commonly used one in the industry is the bounding box method. It is a method in which the objects of interest in an image are enclosed in tightly fitting bounding boxes.

In object detection, there are different ways objects are detected. The most common approach is that the algorithms detect various objects from a particular image frame with a bounding box aligned with a reference axis. This provides the exact position and the size of the objects at a particular instance of time. That is why the reference data or the ground truth dataset using bounding boxes for objects are helpful for comparison [6.6]. Object detection includes a set of tasks done by algorithms that identify the presence of an object, classify it, and then localize it with a reference axis. There can be multiple objects in an image or video source. Various objects are classified by establishing a class for each object type and allocating different colors when the objects are annotated by the bounding boxes.

Dataset generation is one of the most tedious and most demanding markets in the industry these days. Since more and more machine learning algorithms are being used in the automotive industry, the demand for the datasets for training and testing those algorithms is also higher. Organizations meet these demands by different methods. The data source is primarily from the vehicle sensors from which the environment data is captured for generating the reference datasets. Data from the vehicle sensors can be captured using different methods, especially for capturing camera and lidar sensors in the vehicle, the critical data sources where computer vision algorithms are operating. Beyond collecting the sensor data from the vehicle, other methods can be used, such as using an independent Ground Truth System (GTS) to collect and generate ground truths. Another method widely seen in the industry is contracting third-party suppliers to collect and annotate data and even purchase the prepared dataset from any supplier, including required annotated objects. We can broadly classify these data sources as controllable and noncontrollable sources, depending on the end-users' influence on the source of data collection and the requirements of dataset generation.

Datasets sold by different vendors are usually pre-prepared. They are created based on certain specific requirements, and hence, if the end user wants to have a different approach in labeling and with different requirements, they might not be able to change those once purchased. Someone who is collecting, labeling, and generating their dataset can have better control over the requirements for labeling and data ownership and the type of data required. Hence, these days many organizations used to set up a team that focuses on data collection and labeling mainly to address the need for the datasets when using computer vision or any machine learning algorithms. Beyond these, the dataset generation from environment simulation tools is also widely followed for various purposes. It is one of the most flexible and economic approaches used for algorithm development. Photorealistic simulation tools are widely used to generate various scenarios and events, including multiple actors, to create ground truth and train and evaluate the algorithm performance in those conditions.

Algorithm development engineers primarily drive the requirements for labeling and dataset generation. There are certain specific needs for the reference data. These requirements are driven by the method used for algorithm development and how it operates. These needs are translated as requirements for annotation and dataset generation. The goal of having these requirements is to classify objects in a given image source according to their context and texture based on defined object classes and rules associated with those. These requirements act as the basis for labeling or annotating objects in an image through bounding boxes or other methods [6.6]. There will be requirements such as what objects

FIGURE 6.3 An overview of the data annotation process.

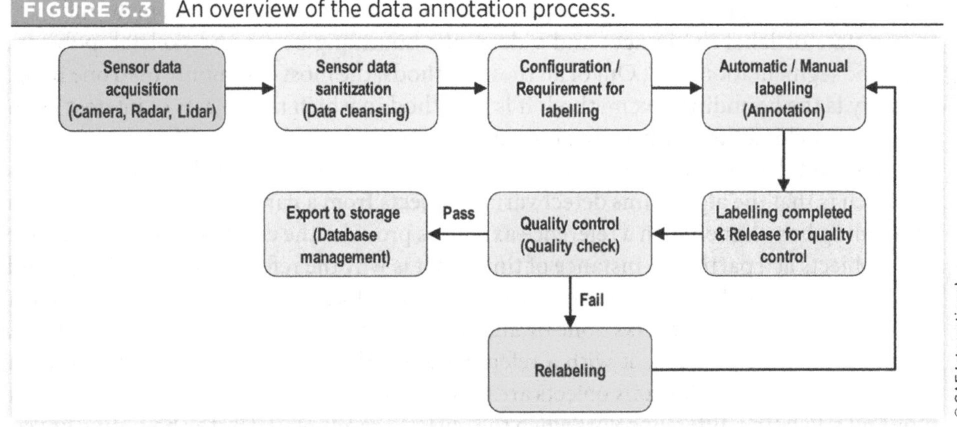

are to be labeled, the minimum size of the label, how accurate it should be in terms of pixels in the image, and what must be done with foreign objects that are not under consideration. Generally, all discernible objects are labeled with 90% or above accuracy in an image with a threshold of single-pixel error. This error threshold can vary if the object is lying in the background of another object.

Figure 6.3 gives an overview of a generic data annotation process in the industry. The images or videos collected from a camera sensor are passed through a data cleansing and selection phase. After the cleaning process of the data, this is passed through a workflow management process where data annotation is performed by employing manual or automated techniques or a mix of both as semiautomated, which is the most common approach.

A semiautomated approach is more precise unless the annotation algorithm is trained enough to provide accurate and precise labels for the objects in an image. The quality evaluation and dataset deployment are the final parts. The quality control is performed to remove the defectively labeled images from the final dataset and send it back for correction.

What should be the dataset size a machine learning algorithm should be trained to use in a production vehicle? This is a challenging question to answer, and one could always expect that the answer is as much as one can. But it is not possible to use an infinite amount of datasets to train an algorithm and improve its performance for a particular project. In the industry, considering the project requirements and the boundaries defined, the suppliers who deliver the computer vision algorithms for object detection always propose a certain number of datasets as the basis for their delivery to a vehicle manufacturer and a performance level measured using specific performance metrics.

The dataset size required for training the algorithm varies with the number of objects that need to be detected, the performance required in detection in accuracy and precision, and the geographical location under consideration. The dataset size also depends on the quality of the dataset used for training the algorithms to achieve a particular quality. Typically, an object class is generated based on the broad division of static and dynamic objects and classified into various subclasses within these groups. In the industry, taking a few examples of production programs with ADAS and Automated Driving features, the dataset used for training the algorithms is in the scale of millions of images and hours of annotated video streams. A supplier providing a detection algorithm for lanes, vehicles, and traffic signs had trained the algorithm with 1.4 million images and about four hours of annotated video data to make it production-ready. Another supplier used about 2 million images as datasets only for lane detection for a European vehicle program. An American supplier who provides algorithms for automated driving features in the vehicle used 3.2 million datasets for training the detection algorithms for vehicles, traffic signs, and lanes over two years. These numbers only speak about the dataset size for the training, and it is assumed that at least 30% of the dataset used in training should be considered for testing as a test dataset. There is no standard definition that states 30% must be considered. Instead, this has been followed in the industry for defining a golden dataset for measuring the quality and performance of the algorithms. Most of this information is not available publicly as they are a technology secret within an organization.

In most of the ADAS and Automated Driving projects in the industry, only two types of datasets are considered, one for training and another golden dataset for testing. The testing followed in the current industry is partial or not robust enough, and there is a gap in following the best practices and methods for data-driven software development. This is mainly due to the time consumed for dataset generation, high costs associated with those processes, or even not having a practically feasible method for evaluating the algorithms.

6.3. Metric for Detection Quality Evaluation

After generating the datasets, how is the detection quality measured? Intersection over Union (IoU) is the most commonly used evaluation metric for measuring the detection quality of an artificial neural network [6.6]. Irrespective of the algorithm used for object detection, this evaluation method can be applied in all cases. The dataset has bounding boxes or any other labeling method used for various objects generated by a manual or semiautomated approach [6.3, 6.7]. The algorithm for object detection predicts bounding boxes based on how it sees the objects based on patterns recognized from the image. The prediction accuracy depends on the object type and the training performed on the algorithm under test for the detection of that particular class of object.

The IoU compares the bounding box from the ground truth image to the bounding box predicted by the algorithm on a particular object in the same image [6.6]. Based on the comparison between two bounding boxes, a ratio is calculated with a numerator carrying the area of intersection of the two bounding boxes and the denominator with the total area encompassed by both bounding boxes from the reference image dataset and the predicted image (Figure 6.4).

A neural network might not only detect a single object from an image. There could be multiple objects which are detected based on how the neural network is designed. The same metric IoU can be used for different objects where the predicted bounding boxes need to be compared with the ground truth data for each detected object. The object classes required for the ground truth are defined, and various types of labels are used while generating the ground truth, including different colors for the bounding boxes based on the class of objects. An object detection algorithm is not evaluated for a single frame or a single image. It will be evaluated with the whole dataset to evaluate its quality. The measurement for a complete dataset can be represented by two matrices, namely, Precision and Recall [6.8].

Generally, the prediction model or the algorithm can be represented based on its outcome using a confusion matrix [6.3, 6.4]. A confusion matrix covers the prediction

FIGURE 6.4 An overview of IoU.

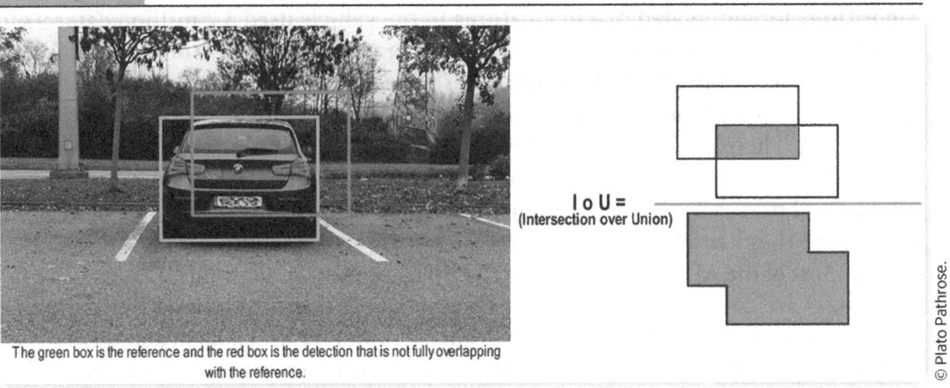

outcome, with the tests covering both positive and negative outcomes. This includes detecting objects correctly, detecting objects wrongly, and possible errors that can occur while considering the whole dataset for prediction quality measurements (Table 6.1).

Precision and Recall matrices can be derived from the confusion matrix. Precision refers to the number of positive class detections from an image dataset that is indeed positive. Recall refers to the number of positive class detections out of all positive detections from the dataset. We should assume that errors are possible, and an algorithm can detect wrongly and provide wrong positive or negative results even if it is trained with a large dataset.

$$\text{Precision} = \text{True positive} / (\text{True positive} + \text{False positive})$$

$$\text{Recall} = \text{True positive} / (\text{True positive} + \text{False negative})$$

Precision and Recall are ratios that can have values from 0 to 1. For improving detection quality, the goal is to improve both precision and recall. Improving the precision will reduce false positives, and improving recall will reduce false negatives. But how much of this can be reduced? In the industry, the focus is mainly given to increasing the precision as it is practically challenging to improve both and have values more than 0.92 for multi-object detection. Reducing the false positives was prioritized when applying the detection algorithms for different ADAS and Automated Driving features.

There is another metric that is used for representing the model performance considering both precision and recall values. The F1-score, or F-measure, is the metric used to represent the detection model performance with a single value [6.8]. It considers the ratios of both precision and recall and derives a harmonic mean of both those ratios. While a 0 value refers to the poor performance of the model, a value of 1 refers to an ideal performance of the detection model.

$$\text{F-measure} = (2 * \text{Precision} * \text{Recall}) / (\text{Precision} + \text{Recall})$$

TABLE 6.1 Confusion matrix for object detection.

	Real true condition/Ground truth	
Predicted positive	True positive	False positive
Predicted negative	False negative	True negative

The use of neural networks for object detection in automotive is not just used for single object detection. It will always be for multi-object detection, and the same approach can be used for evaluating the quality of the detection for multiple objects in the image. The measurement will be performed for each object class in this case.

6.4. Evaluating KPIs for Detection Algorithm

A vehicle with ADAS or Automated Driving features mainly utilizes its front-facing camera and other camera sensors for most of the object detection functions. Lidars are also used in many vehicles, but those are not common. Hence, a front-facing camera is considered to have an important role in the overall sensors on a vehicle for detecting its environment ahead. Detection and classification of static and dynamic objects in a short duration of time as in milliseconds are challenging and critical in decision-making for various ADAS and Automated Driving features, especially when a vehicle is moving at high speed. How can the performance and the detection quality be measured? What are the acceptable detection KPIs one should look for to decide if the performance is satisfactory and production-ready?

One of those integrated detection models for the front-facing camera for a particular project can detect twenty-two classes of objects. These include lanes, vehicles, traffic signs, etc. This section covers a method used in the industry for testing the detection algorithm quality by taking an example of lane detection. Even though these detection algorithms are software components, their evaluation is performed both as a software component and in the overall system. The vehicle response based on the detection is also analyzed to ensure a particular function behaves as intended.

For example, the detection quality is measured from the algorithm level and the vehicle perspective in the lane detection function. The whole system is integrated into the vehicle to provide a feature. In this case, the visual source is the front-facing camera which feeds the detection algorithm. For better understanding, the method used for qualifying lane detection will be explained as an example and will walk through the entire measurement and performance evaluation process. The same method can be applied for the detection evaluation of other object classes. The major challenge one would experience will be to have reliable physical world testing when multiple objects are considered simultaneously.

Most of the testing related to object detection is performed in the laboratory environment, and minimal tests or only robustness tests are performed in the field for most of the ADAS and Automated Driving features. This was not the case earlier when the ADAS features were first introduced. Vehicle manufacturers and their suppliers perform more field testing, and the qualification was purely based on those field tests. While walking through the evaluation process for lane detection, both types of evaluation using reference data for laboratory-based software tests as well as the tests at vehicle level are explained.

The whole evaluation process can be classified as the sample data collection phase and evaluation phase. The most important one is the sample data collection phase, where the quality of data collected determines the accuracy of the detection algorithm. The method described here uses all measurements with reference to the vehicle coordinate system for the data collection and the performance evaluation [6.9]. There are few preconditions to be met while sample data or reference is created to measure and evaluate any function.

The evaluation method used here adopts the Six Sigma method for MSA. As the lane detection algorithm utilizes the camera as the primary input source, certain preconditions for the camera sensor for sample data collection must be satisfied.

6.4.1. Preconditions for Sample Data Collection

Detection of lanes is used by various ADAS and Automated Driving features as an important safety function. The precondition defined here is to make sure to have accurate and qualified reference data. This reference data or ground truth will be used to evaluate the accuracy and correlate the measurement from the software to that of the physical world. The ground truth data preparation for evaluating lanes should ensure the following before collecting the data:

- To have a valid video frame from the camera: The lane detection function is activated after the initialization process, and the required calibration is performed for the camera sensor. There should not be any influence from external factors that affect the field of view of the camera and prevent it from having a stable input. These can be executed using the target sensor with the intended operational configurations like frame rate and resolution, fixed for regular operation. There is also a possibility to use a reference sensor to benchmark.
- Feature availability: The lane detection function is enabled and is available, and the software should be able to detect lanes and patterns from the input source from the camera.
- Range of detection: Valid lanes should be available in the detection range. The area around the vehicle is pre-checked for reliable sensor coverage and range of reliable detection. The area within the range is further separated into zones such as zones A, B, and C, for which the detection accuracies must be measured. For the reference data collection, the lanes and pattern availability should also match the requirements and the geographic location under consideration, such as the lane markings and lane marker width, which varies depending on geographical locations.
- Function state: The lane detection function should be available and active while generating the ground truth. The camera input should be above the threshold value of data quality (SNR) and should not be having any missing image frames, which introduces function unavailability conditions.

6.4.2. Data and Data Types

For the lane detection function, three different data sources have to be considered for reference data collection:

1. GTS input: Data from the reference system
2. State of lane detection functionality: Status of the functionality, whether it is enabled or disabled mode
3. Performance and quality parameters: Predefined performance and quality parameters of the lane detection function

The different data sources that need to be collected for a function under consideration should be identified, and those data types need to be defined. The decomposition and the applicable data types for the lane detection function can be represented as in Figures 6.5 and 6.6.

Based on the input data source, three different data types need to be considered:

1. Integer
2. Continuous data
3. Logical data (1 or 0)

FIGURE 6.5 An example tree diagram of lane detection with its data types.

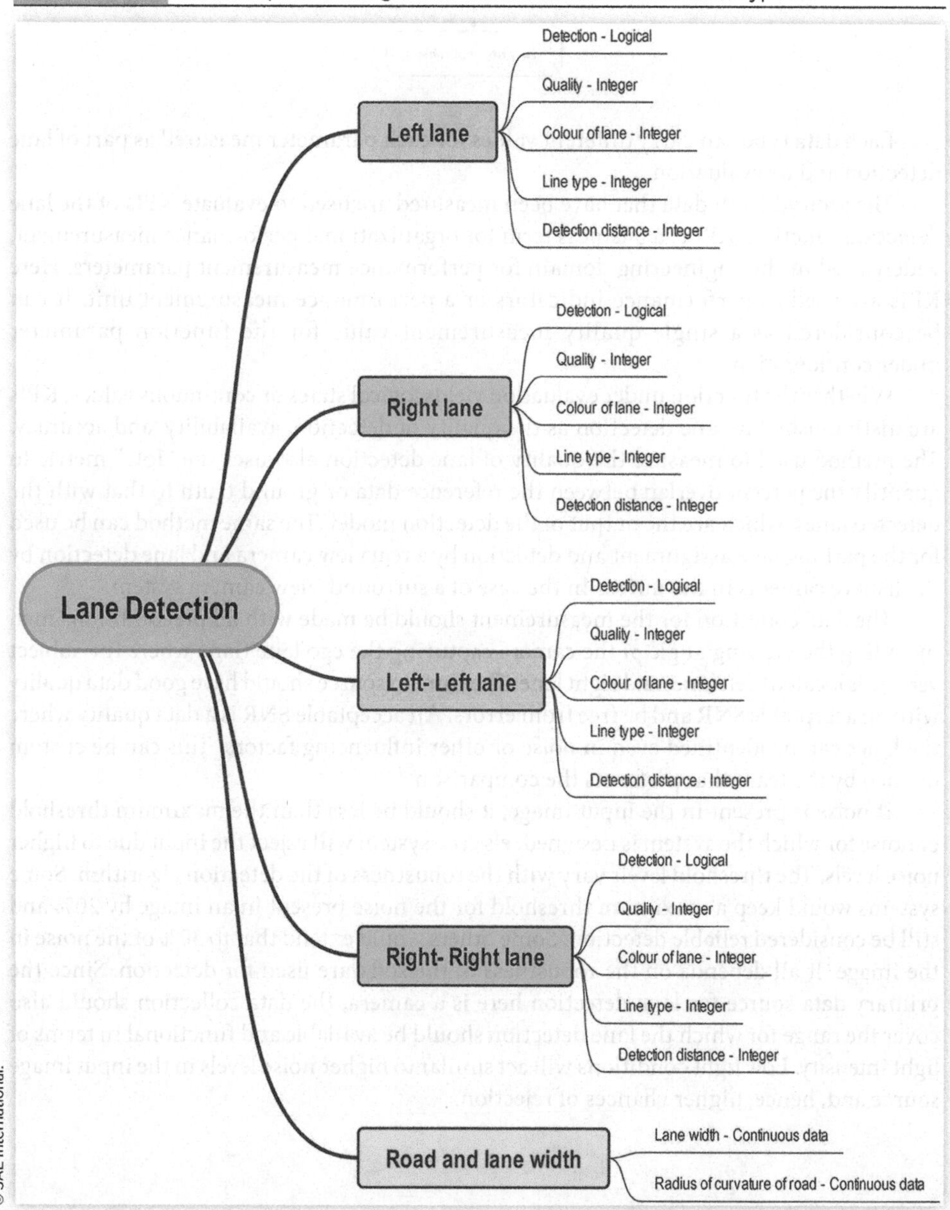

FIGURE 6.6 Data type classification and grouping for lane detection.

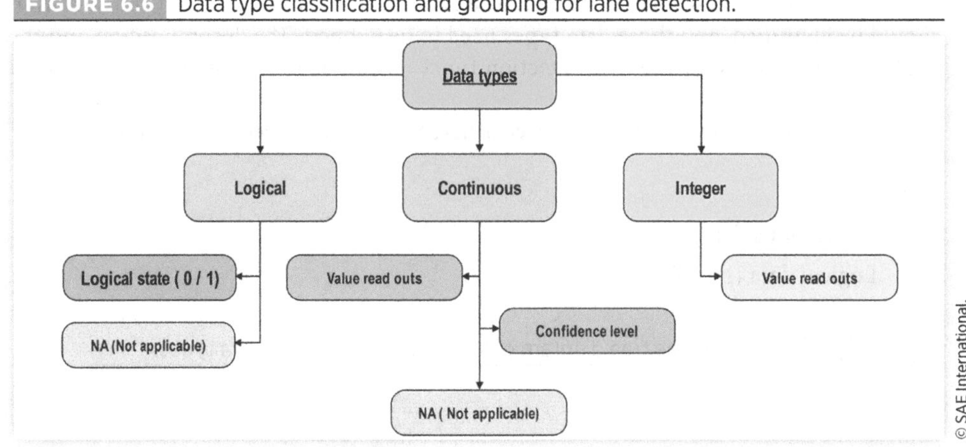

Each data type can carry different values for each parameter measured as part of lane detection and its evaluation.

The ground truth data that have been measured are used to evaluate KPIs of the lane detection function. KPI is a business term for organizational performance measurement, widely used in the engineering domain for performance measurement parameters. Here KPIs are used as performance indicators or a performance measurement unit. It can be considered as a single quality measurement value for the function parameter under consideration.

Whether the function under evaluation yields logical states or continuous values, KPIs are distinguished in lane detection as the quality of detection, availability, and accuracy. The method used to measure the quality of lane detection also uses the "IoU" metric to quantify the percent overlap between the reference data or ground truth to that with the detected lanes, which are the output of the detection model. The same method can be used for the parking lane assignment and detection by a rearview camera and lane detection by the fisheye cameras in the mirror in the case of a surround-view camera system.

The data collection for the measurement should be made with all preconditions met, including the viewing angle of the camera capturing the ego lane (lane where the subject vehicle is located), left lane, and right lane. The camera source should have good data quality with an acceptable SNR and be free from errors. An acceptable SNR is a data quality where the lanes can be identified even in noise or other influencing factors. This can be custom defined by the team that performs the comparison.

If noise is present in the input image, it should be less than the maximum threshold of noise for which the system is designed, else the system will reject the input due to higher noise levels. The threshold levels vary with the robustness of the detection algorithm. Some systems would keep a maximum threshold for the noise present in an image by 20% and still be considered reliable detection. Some others would extend that to 30% of the noise in the image. It all depends on the robustness of the software used for detection. Since the primary data source for lane detection here is a camera, the data collection should also cover the range for which the lane detection should be available and functional in terms of light intensity. Low light conditions will act similar to higher noise levels in the input image source and, hence, higher chances of rejection.

6.4.3. Performance Evaluation (KPI Measurement)

The lane detection performance evaluation is done by comparing the output from the detection algorithm with reference data. Many organizations also evaluate that directly in the physical world, which is equivalent to feature evaluation from the vehicle as part of feature validation.

Comparison of the detection algorithm output with the ground truth data is performed in three steps. One is a frame-by-frame comparison of the detection algorithm output to that of the reference data. The second one is comparing the detection algorithm performance to the overall reference data or ground truth. The third one is comparing the detection algorithm performance at different noise levels with the whole ground truth. The third method is used mainly to evaluate the robustness of the algorithm detection. In addition to these three methods, it is also common with vehicle manufacturers to evaluate the KPIs in the real world at the vehicle level as part of feature evaluation and its robustness in the vehicle.

The region around the vehicle is divided into different zones, as shown in Figure 6.7, and the measurements are made for these zones. The detection quality is measured for each zone separately around the vehicle. The zones are classified according to the sensor range and their coverage. Three different zones are considered from the vehicle for the lane detection function that covers the front-facing camera field of view. The object detection quality is calculated using the Precision-Recall metric for the objects in each zone. In this case, the lane detection accuracy is measured for all zones, and a physical test is performed at the vehicle level, usually for Zones A and B, including other objects.

FIGURE 6.7 Zone classification for camera-based detection algorithm evaluation.

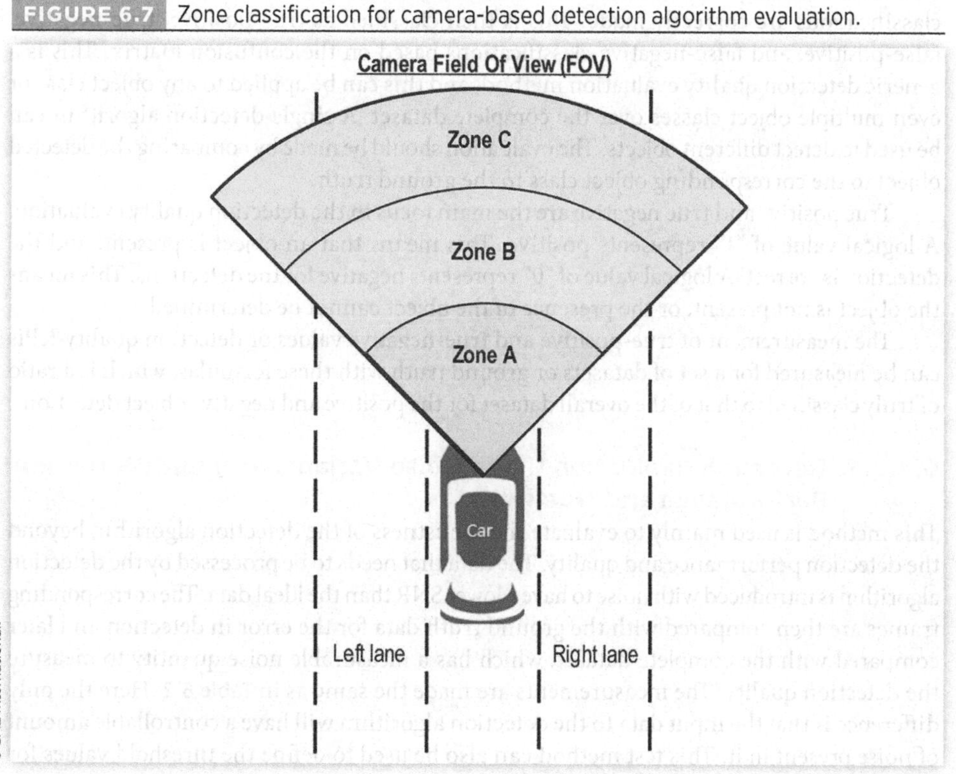

6.4.3.1. Detection Evaluation on a Single Frame (Detection Performance)

The reference dataset or the ground truth data should be free from errors, and the annotation accuracy should be ±1 pixel tolerance. The comparison should be made frame by frame. Here the deviation or the amount of error is calculated for each frame as:

$$\text{Deviation}\left(\text{Error}\right) = \text{Ground truth KPI value} - \text{Object detection software KPI value}$$

Sometimes, it is confusing to understand at what frequency the ground truth data has to be recorded. For example, when the input images from the camera are processed at a frequency of 30 frames per second, the ground truth data should also match this frequency or be closer to it. Usually, the ground truth generated is at the rate of 25 or 20 frames per second or sometimes even less than that to 10 frames per second, where the algorithm processes the images at the rate of 30 frames per second. Tolerance in the number of frames for the ground truth data should be defined in such a way that it will not impact the KPI measurements. There are also semiautomated methods for labeling where the auto labeler will label the sequence of the frames, and manual interaction is required only every 10th or 15th frame in labeling.

6.4.3.2. Detection Evaluation on Complete Ground Truth Dataset (Detection Quality)

The evaluation of the detection algorithm on the complete dataset is to identify the detection quality over a period of time. Here the classification of the object is evaluated to be true or false over the complete dataset, and measurements were made to check whether these classifications are correctly made. This is then classified as true-positive, true-negative, false-positive, and false-negative classifications based on the confusion matrix. This is a generic detection quality evaluation method, and this can be applied to any object class or even multiple object classes over the complete dataset. A single detection algorithm can be used to detect different objects. The evaluation should be made by comparing the detected object to the corresponding object class in the ground truth.

True positive and true negative are the main focus in the detection quality evaluation. A logical value of "1" represents positive. This means that an object is present, and the detection is correct. A logical value of "0" represents negative for the detections. This means the object is not present, or the presence of the object cannot be determined.

The measurement of true-positive and true-negative values or detection quality KPIs can be measured for a set of datasets or ground truth with these formulas, which is a ratio of truly classified to that of the overall dataset for the positive and negative object detections.

6.4.3.3. Detection Evaluation Using Noise Variants as Input (Detection Performance and Quality)

This method is used mainly to evaluate the robustness of the detection algorithm beyond the detection performance and quality. The data that needs to be processed by the detection algorithm is introduced with noise to have a lower SNR than the ideal data. The corresponding frames are then compared with the ground truth data for the error in detection and later compared with the complete dataset, which has a measurable noise quantity to measure the detection quality. The measurements are made the same as in Table 6.2. Here the only difference is that the input data to the detection algorithm will have a controllable amount of noise present in it. This test method can also be used to define the threshold values for the noise a detection model is operational.

TABLE 6.2 KPI evaluation of true positive and true negative.

KPI evaluation	
True-positive quality	= KPIs of positive total/KPIs of total positive expected
True-negative quality	= KPIs of negative total/KPIs of total negative expected
Combined (full dataset)	= KPI of combined (positive and negative) total/KPIs of combined expected

6.4.3.4. Detection Evaluation in the Vehicle (Detection Performance)

The evaluation of the detection algorithm in the vehicle is performed in the open loop in a controlled environment like a laboratory or proving ground or even on public roads. The scope is to have reliable detection of the objects in the vehicle from the physical world for which the detection algorithms are designed. For example, the lanes detected by the detection algorithm are compared to the reference markers and measured for their accuracy in the physical world in a laboratory environment up to a short distance from the vehicle. In proving grounds, sometimes the detected objects are physically verified with measurements from the vehicle to the detected object or compared with the HD maps as reference. HD maps are widely used for comparison when considering lane detection at various road geometry and geographical locations as it is impossible to evaluate the physical lane markings in all areas.

Usually, this method is used to calibrate the object detection using front cameras, side cameras, or even rear cameras. When the algorithm-detected line shifts from the original position of the line in the real world, the correction needs to be considered in the detected line. With the corrected offset, the detected line can be shifted to match the real-world reference. The deviation error can be calculated as:

$$\text{Deviation}\,(\text{Error}) = \text{Ground truth KPI value} - \text{Object detection software}\,(\text{KPI})\,\text{value}$$

The detection performance is measured frame by frame with the reference markers in the physical world and with the ground truth. It is practically difficult to compare all the objects detected within the full sensor range. The checks are usually performed physically within Zone A of the vehicle, as in Figure 6.7, with minimal objects in either a laboratory environment or proving ground. Most of the detection algorithm does not go into a vehicle for testing due to the dependency of vehicle availability or due to cost and time associated with it. Hence this method is not very common with the suppliers and is executed mainly by vehicle manufacturers or even by suppliers with the vehicles provided by vehicle manufacturers.

6.5. Different Acceptance Quality for Detection Algorithms

In Section 6.4, different methods used for the evaluation of the object detection algorithms were explained. All these algorithms have a strong dependency on the data for its performance and evaluation. How is it decided that the algorithm for object detection is mature enough and is of good quality to be deployed in a vehicle?

There are different methods followed in the industry from vehicle manufacturers and their suppliers to quantify the acceptance criteria for neural network-based algorithms like those used for object detection. Since a vehicle program or a project cannot be considered an open-ended contract to improve these neural networks continuously, it tends to be ideal. This has cost and effort associated and is not considered the right approach in most vehicle programs. There are still exceptions where the updates of the detection algorithms are

TABLE 6.3 Sample reference on training dataset used in the industry.

Function	Training dataset
Lane detection	1 million image datasets, 10 h of annotated video data with different types of lane markings. Including wide lanes as in the Netherlands (Training dataset) Location: Europe
Vehicle detection	1.2 million image dataset, 10 h of annotated video data with different traffic conditions (Training dataset) Location: Europe
Traffic sign detection	10,000 images of traffic signs (mainly speed signs) as a training dataset Location: Western Europe (6 countries)

executed in the vehicle to improve performance even after the vehicle deployment by certain vehicle manufacturers.

The vehicle manufacturers and their suppliers take a common approach to quantify the maturity for the detection algorithms for the vehicle deployment based on the size of the dataset used for training and testing the algorithms. This includes the dataset size for training the algorithm and the dataset for verification and validation with the number of defects over a period during testing. In almost all the cases, the acceptance is defined in the dataset size to the defects and the amount of re-simulation tests performed. Re-simulations are performed in a HiL simulation environment where the prerecorded sensor data from the vehicle is fed into the ECU, where the algorithm will process it for detection. The response of the detection function is evaluated for its performance and correctness. Some of the sample training dataset sizes used in the industry are covered in Table 6.3.

As camera-based detection is an integral part of the overall ADAS and Automated Driving features, the quality of detection has a vital role in the perception of the environment and the further response of the vehicle. This will be covered in the next chapter.

6.6. Challenges in Measuring Quality of Object Detection

Even though various methods are applied to qualify the detection algorithm, one can still experience that the vehicle fails to detect other vehicles or traffic speed signs or lanes while driving. This is a common issue with almost all major vehicle manufacturers when they are deployed in the market. As normal humans, we trust our car more and assume the first time that the problem was with the signboard or the color of the vehicle, which went undetected, but if it keeps on repeating, it will disclose that it is a system flaw.

It is challenging to consider all possible inputs a system would experience during its complete lifetime during the development phase. As sensing and detection are an integral part of both ADAS and Automated Driving systems, robust and structured testing methods must be in place for its qualification. Data is an important part of both development and testing. Having an accurate dataset of a considerable size would help in training and qualifying detection algorithms, which must be improved continuously as part of development.

One of the major challenges vehicle manufacturers and their suppliers face is collecting the data and preparing the dataset for training and evaluating these detection algorithms.

As per the current methods used in the industry for sensor data collection, it depends on the sensor characteristics, its calibration status, and its position in the vehicle. Thus, the data usage is limited if the sensor position changes or if a supplier wants to reuse it for a different vehicle program.

Another challenge is having all the variants considered for the detection of objects, and the algorithm is trained and tested for it. When a neural network is designed to detect multiple objects, the detection performance will not be the same for each of those objects when using a particular dataset. Practically, it is impossible due to the costs associated with training and the detection quality and performance for all the objects at the same level. Hence the approach used here is to prioritize the object classes that need to be detected and improve the performance of those objects compared to other objects.

Identification of all scenarios or the environmental conditions the algorithm needs to detect an object is only possible by detailed analysis. For example, ISO 21448 - Safety of Intended Functions [6.10] proposes specific methodologies to identify these scenarios. The algorithm can be trained and tested for those scenarios during its development. Since these analyses are usually executed in parallel with the product development, the outcomes are usually lagging behind the software development. Thus the practical usage of these analysis results during the software unit testing may not be feasible. This can be addressed by spending enough time to analyze and continuously improve the development and tests for these software components.

For homologation purposes or testing against the requirements for vehicle approvals, almost all vehicle manufacturers focus on a smaller subset of the required detection and functionality in the vehicle [6.11]. The tests for homologation are being performed inside a controlled environment such as a proving ground. The detection performance for those specific tests would be better there. On the other hand, their performance would vary on the public road while driving. Considering a robust test strategy that focuses beyond the purpose of vehicle certification is essential in qualifying the detection algorithms and features that use it in the vehicle.

6.7. Summary

Sensing is an important phase for an ADAS and Automated Driving features in the vehicle. Detection algorithms have a vital role to play here. Various algorithms used for object detections these days are based on neural networks and have a strong dependency on data. In this chapter, the data-driven software development and its life cycle as compared to the classical V-model for product development in different phases were discussed.

Artificial neural networks utilize datasets for their training, performance improvement, and evaluation. The quality requirements of those datasets and the process of developing those datasets were covered with examples in this chapter so that the reader will have a first impression about how they are developed. The methods used to evaluate performance and overall detection quality to determine KPIs were explained for production programs. Having reliable reference data and using suitable methods for the development and testing is the key in data-driven software development. The overview of the training dataset used in the industry for realizing certain functions is an informative section. How the KPIs are measured and the challenges in measuring them using different environments highlight the industry's struggle these days while using artificial neural networks.

Perception software being the heart of the vehicles that are highly automated, quality, and performance of the perception software is a concerning factor for every vehicle manufacturer. Challenges associated with the methods used in the industry in delivering a reliable neural network that is designed and tested to cover all the requirements are covered in the last part of the chapter. The proposal to adapt and improve the methods and focus on the safety of the features in the vehicle that uses neural network-based functions should be taken into account with in-depth analysis during the development and testing phase. The output of the detection algorithms, which are part of the sensing in the software stack, is fed to the next layer, which is perception for the generation of the environment model. The next chapter will discuss the perception part of the software stack and various methods used for evaluating the perception software components.

References

6.1. Grigorescu, S., Cocias, T., Trasnea, B., and Margheri, A., "Cloud2Edge Elastic AI Framework for Prototyping and Deployment of AI Inference Engines in Autonomous Vehicles," *Sensors* 20, no. 19 (2020): 5450, doi:https://doi.org/10.3390/s20195450.

6.2. Venables, M., "An Overview of Computer Vision," September 11, 2019, accessed July 2, 2021, https://towardsdatascience.com/an-overview-of-computer-vision-1f75c2ab1b66.

6.3. Geiger, A., Lenz, P., Stiller, C., and Urtasun, R., "Vision Meets Robotics: The KITTI Dataset," *International Journal of Robotics Research (IJRR)* 32 (2013): 1231-1237.

6.4. Geiger, A., Lenz, P., and Urtasun, R., "Are We Ready for Autonomous Driving? The KITTI Vision Benchmark Suite," in *Conference on Computer Vision and Pattern Recognition (CVPR)*, Providence, RI, 2012.

6.5. Meyer, A., Salscheider, N.O., Orzechowski, P.F., and Stiller, C., "Deep Semantic Lane Segmentation for Mapless Driving," in *2018 IEEE/RSJ International Conference on Intelligent Robots and Systems (IROS)*, Madrid, Spain, October 1-5, 2018.

6.6. Rosebrock, A., "Intersection over Union (IoU) for Object Detection," November 7, 2016, accessed July 2, 2021, https://www.pyimagesearch.com/2016/11/07/intersection-over-union-iou-for-object-detection.

6.7. Fernández Hilario, A., García López, S., Galar, M., Prati, R.C. et al., *Learning from Imbalanced Datasets*, 1st ed. (Berlin, Heidelberg: Springer, 2018), ISBN:978-3319980737.

6.8. Powers, D.M.W., "Evaluation: From Precision, Recall and F-Measure to ROC, Informedness, Markedness & Correlation," *Journal of Machine Learning Technologies* 2, no. 1 (2011): 37-63.

6.9. Geiger, A., Lenz, P., Stiller, C., and Urtasun, R., "Road/Lane Detection Evaluation 2013," 2013, accessed July 2, 2021, http://www.cvlibs.net/datasets/kitti/eval_road.php.

6.10. ISO/PAS 21448: 2019, "Road Vehicles—Safety of the Intended Functionality," 2019.

6.11. Huval, B., Wang, T., Tandon, S., Kiske, J. et al., "An Empirical Evaluation of Deep Learning on Highway Driving," arXiv preprint arXiv:1504.01716, 2015.

/Autonomous
/Sensing
/Communication
/Battery
/Navigation
Autopilot mode

7

Testing and Qualification of Perception Software

What is the role of perception in the field of ADAS and Automated Driving systems? "Perception" is the process of understanding the information captured by the sensory organs. Like any of the sense organs in a human body, there are sense organs in the vehicle in the form of different sensors. Perception can be defined as the ability to process and understand the information about the external environment from the data provided by the sensors in the vehicle. In humans, the brain will process the information from different sense organs and understand the environment. All information processing in vehicles will be performed by the software running inside an ECU [7.1]. How can one measure whether the information is processed correctly and if the understanding of the surrounding is correct? In humans, this can be evaluated with different types of tests in which the results are generated based on the inputs provided and by utilizing already known information stored in the memory. The human brain processes the complete information in a fraction of a second with high-speed communication through nerve cells.

This chapter mainly focuses on perception software and its testing. Accuracy and precision in the output of perception software are crucial for the ADAS or Automated Driving features in the vehicle. With the information received from different sensors, the perception software recreates the vehicle environment and events. The response will be planned and executed based on the environment perceived by the software. The complexity of this whole process can be imagined just by thinking of a vehicle moving at a speed of 150 km/h and the software controlling the vehicle based on its interpretation of the environment and the events around the vehicle, which is captured with the help of different sensors in the vehicle. This chapter also discusses how different organizations evaluate the quality of the perception software for their products and the challenges associated with those.

7.1. Overview of Automated Driving Systems

Perception software is one of the software components in the ADAS or Automated Driving software stack. Perception software can be considered as one of the components in the application layer of the ADAS or Automated Driving system software. This accepts input from those software components that process the vehicle sensor data or directly uses the input processed and sent from various sensors as object lists. Object list refers to the processed information by a sensor where the processor within the sensor extracts objects of interest from the environment data.

Figure 7.1 shows a high-level block diagram of the system software with its different layers. The application layer is the topmost layer of the software which is the main component we will focus on here as part of the system software. The lowermost layer of the software is the physical layer, which is purely the hardware part where the Systems on Chip (SoCs), electronic components, and network interfaces are present. The board support package and operating system are the software part which is the lowermost software layer closely bounded with the hardware. This includes the drivers, which are software components to configure the hardware to exchange information and establish communication between hardware and software layers. Middleware comes above that, which is the main communication interface that plays a crucial role in the data and information exchange between different layers. The application layer is the topmost layer where the real ADAS or Automated Driving functions are deployed. This is where all algorithms operate to provide various functions and is trivial to ADAS and Automated Driving features. Middleware also helps exchange information between different software components in the application layer [7.1].

The perception software is the software component in the application layer which accepts data from the sensing part of the application layer or processed information from the vehicle sensors. Vehicle sensors usually provide either raw data or processed data as output from them. Usually, the processed data is in the form of object lists. Object lists are a list of objects identified and classified at a particular instant of time by these sensors. The bandwidth requirement for sending raw sensor data is higher when compared to a sensor providing only object lists as its output.

FIGURE 7.1 An overview of ADAS and Automated Driving system architecture.

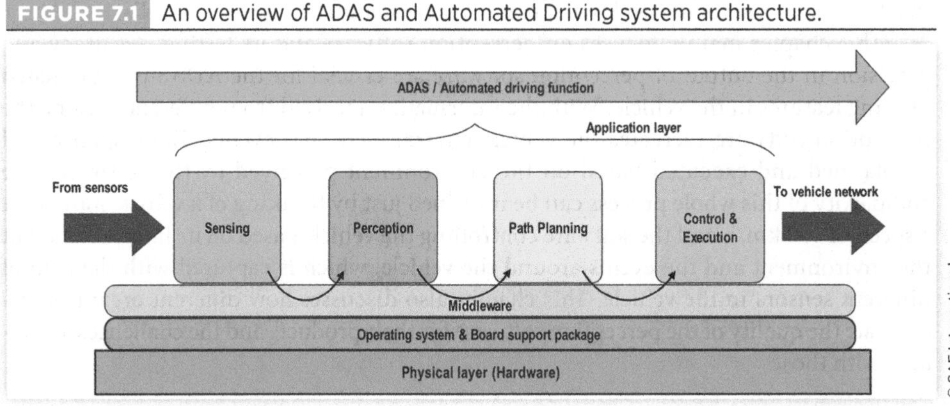

The output of the perception part of the software stack is the environment model. The environment model is the mathematical model of the environment around the vehicle that is modeled based on the details provided by vehicle sensors about the surrounding environment and various events. It is the recreation of the natural environment around the vehicle with the help of mathematical models, including different objects and their behavior. These identified objects can be classified mainly into two categories, Static objects (such as lamp posts, traffic signboards) and Dynamic objects (such as moving vehicles, pedestrians, animals, etc.). There is also a third category of unclassified objects that the sensor detects but fails to classify to either static or dynamic objects. The accuracy and precision of the environment model and its correlation with the real world are the core for implementing any ADAS or Automated Driving features. Any planning and control of the vehicle depend on this environment model. In case the environment model does not correlate with the real world, it will generate the wrong reference of the surrounding environment of the vehicle, and the software will behave based on this wrong reference, which is hazardous, especially when the software controls the vehicle.

The environment model from the perception software is utilized by path planning, which is the next phase of the data flow. Based on the information gained about the surrounding environment and the current state of the vehicle, responses are planned for certain events by this software component. This information is transmitted further to the control and execution part of the software. It is then processed and sent as output signals to the vehicle network that controls the vehicle. The information transmitted to the vehicle network will control the vehicle through various actuators in the vehicle. Once these signals are transmitted to the vehicle network, other ECUs in the vehicle use this information to perform certain functions for controlling the vehicle.

There are many other software components as part of the overall system software stack of an ADAS and Automated Driving system that are responsible for performing essential functions to keep the system available and perform its duties. This includes configuration, diagnostics, calibration, power management, network management, and many more. Since the focus of this chapter is on perception software, the functions and testing of the perception layer are discussed here.

7.2. Perception—An Architecture Overview

In this section, a general overview of the perception software will be covered. It will also explain a few of the main components in the perception software and how the environment model is generated utilizing sensor data from the vehicle. The perception software processes the information provided from various vehicle sensors broadly into two parts: Static and Dynamic objects. Along with the vehicle sensors, HD maps are also used as a source of redundant information to generate an environment model and to position the vehicle accurately in the modeled environment. Figure 7.2 shows a high-level architecture with some of the critical components of perception software.

Perception software has multiple inputs coming from the sensing part of the software where many algorithms detect and classify various Static and Dynamic objects and position them along with the vehicle in a reference map. The main functionalities of the perception software can be briefly classified as:

- Sensor data fusion and tracking
- Localization
- Environment model generation

FIGURE 7.2 A representative architecture of perception software.

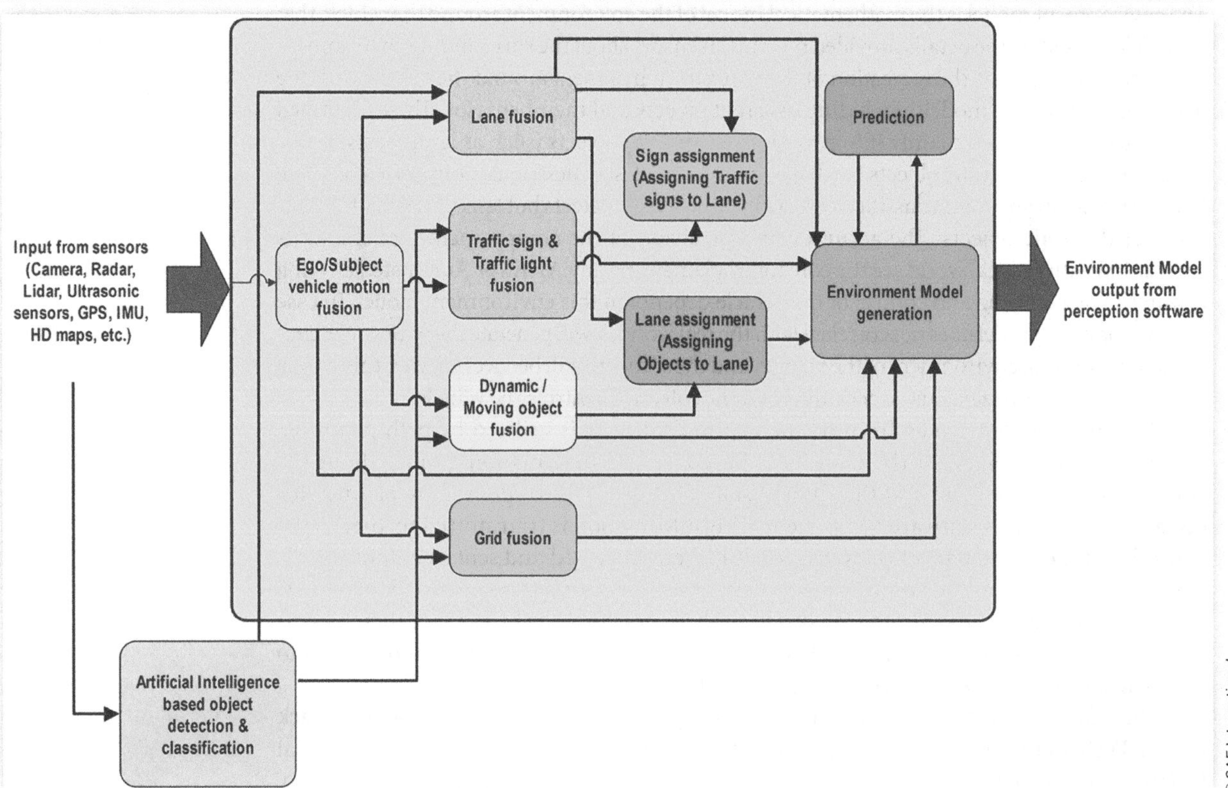

Sensor data fusion is about fusing information from multiple vehicle sensors. The sensor fusion part of the perception accepts processed information from multiple sensors or from the sensing part of the software that processes the raw sensor data and extracts object lists. The primary role of sensor fusion is to combine the information from all the sensors by fusing them to make redundant information about the environment available. For higher levels of automation, redundant information helps in acquiring accurate information than depending on a single sensor type. This is a step to increase confidence in the output by utilizing input from multiple sources about the same environment to derive conclusions.

Sensor data referred to here is the processed data with information about various objects such as Static and Dynamic objects in the environment captured by different sensors. This also includes the position data provided by the HD maps, GPS, and IMU sensors in the vehicle. The sensor fusion and tracking part of the perception covers all the detected objects in an environment and tracks them for their current position and possible motion.

The localization part of the perception is where the positioning of the vehicle takes place. It is mainly about identifying where the vehicle is located in the world. It helps position the vehicle precisely in the environment, whether it is on the road, on which lane it is located, or is in the middle of a lane, etc. The localization utilizes input from various sensors and HD maps, GPS, and IMU in the vehicle and correlates it to the real world, positioning the vehicle exactly in the mathematical model of the environment.

The environment model is the primary output of the perception software utilized by various other software components in the automated driving software stack. Mathematical modeling of the environment around the vehicle based on a reference coordinate system and predicting the future position of the objects are the output from the perception part of the software. Various inputs of different fused data such as sensor fusion, object fusion, grid fusion, and assigning various detected objects to a specific position related to the vehicle position are the key ingredients in developing the environment model. These fusion types and how they are evaluated will be discussed later in this chapter. The environment model has another application in HMI. Different vehicle manufacturers use the environment model information to provide graphical information to the driver as part of driver information systems, such as different instances where the vehicle detects other vehicles, pedestrians, etc., and navigating in different environmental conditions. At the same time, automated driving features are enabled in the vehicle.

7.3. Different Methods for Perception Software Testing

Testing the perception software is complex and is one of the most challenging and demanding tests in the verification and validation in the ADAS and Automated Driving field. Even though perception is one of the software components, it is important in the overall feature behavior. There are different ways in which perception software is tested in the industry. A few of those are discussed in this chapter.

Testing and qualification of perception software are mainly performed in two different ways. One is with the help of using ground truth data or reference data generated using a system like a GTS; The second one is by using a product direct from the market that has a superior perception. Using a superior product as a reference is primarily a method for benchmarking the perception software for its performance and output quality; There is another method, which is not much used these days. It physically compares the perception output with the real world by placing objects and reference lines in a controlled environment. Many vehicle manufacturers perform this with the entire vehicle in a proving ground or inside a laboratory. The first two methods are commonly considered in the industry, and the latter has been in use for a long time back for testing ADAS features even before mature tools and methods are available as like today.

The perception software has different subcomponents. The test and evaluation of the perception software should also cover those subcomponents. Usually, performance and quality evaluation of the perception software focuses only on the environment model without testing various subcomponents of the perception. This shows a gap in reliable testing of the perception subcomponents such as sensor fusion, tracking, and localization. The performance and quality improvements in the generated environment model can only be achieved by knowing which perception component has to be improved. Testing the subcomponents within perception will help in identifying the dependencies in the generation of the environment model and help to improve the overall perception software performance and quality.

The comparison method is the most commonly used method for perception quality evaluation, as shown in Figure 7.3. Input from various vehicle sensors are used to generate output and are compared with a superior reference of the same environment. A superior reference or the reference environment model can be from other qualified reference systems or from a system that has been in the market for a long time and is considered a legacy product that is superior and considered a benchmark in the industry.

FIGURE 7.3 Different references used in industry for qualifying and benchmarking perception software.

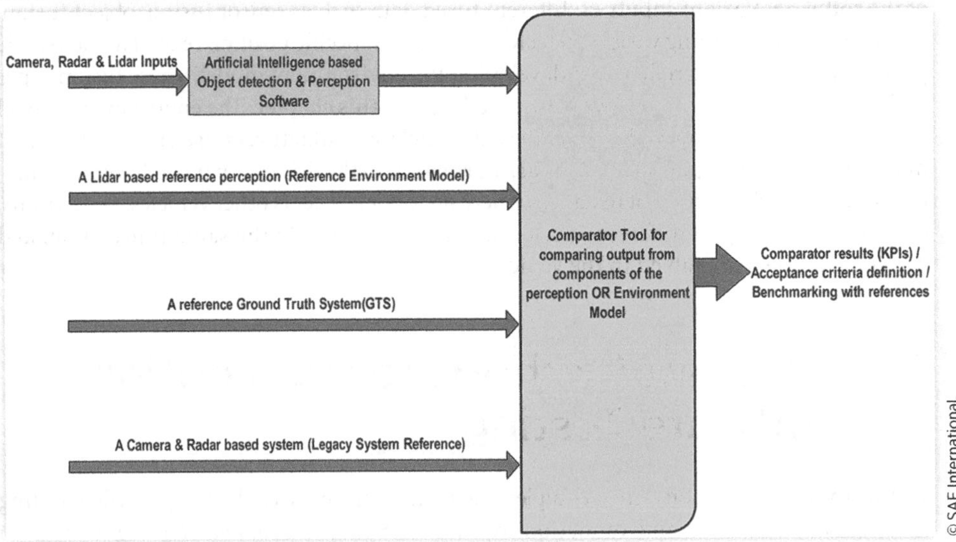

The easiest way to evaluate the perception software during development testing is by using an environment simulation tool. A configuration, as shown in Figure 7.4, can be used to evaluate perception. The scenario input from the environment simulation tool is fed to the ideal sensor and raw sensor models. The perception software utilizes the raw sensor model output as its input. The perception software utilizes the static and dynamic objects and the localization information to generate the environment model [7.2, 7.3, 7.4]. Simultaneously, an ideal output that uses the same input is generated with the help of the environment simulation tool. The output from both is compared as part of the perception evaluation. The reference data or the ground truth is considered the ideal response provided by the built-in reference model of the environment simulation tool. This is an easy check during the development testing rather than a qualified measurement. As the environment simulation tools do not come as qualified simulation tools, there is no guarantee that the

FIGURE 7.4 A configuration of using simulation tool for perception evaluation.

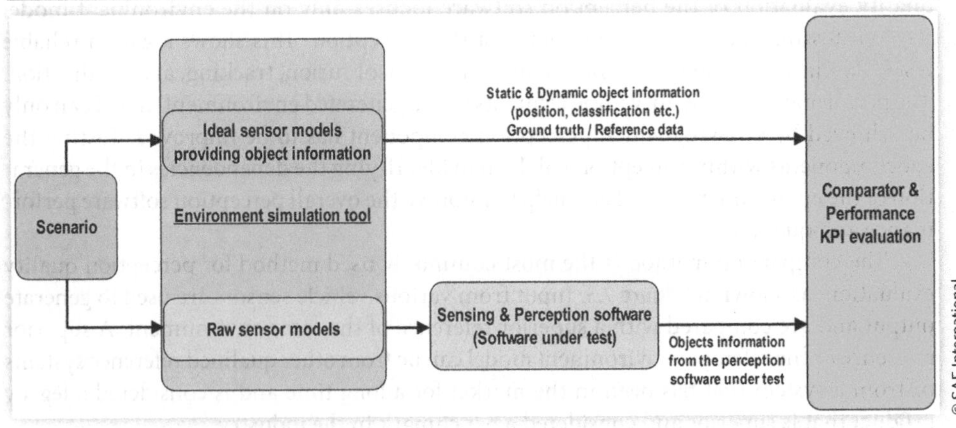

reference system software model provided as a demonstration model along with the tool can be considered a reference. Hence, this method of perception software evaluation is not considered a valid approach for safety-critical product development or testing. On the other hand, the development engineers can use this during the development phase to have a first impression of the quality of the developed perception software. The advantage of using an environment simulation tool to evaluate perception software is that it is possible to design and implement multiple scenarios using the simulation tool and cover the corner-case scenarios, which is impossible to test using other methods. Perception testing using environment simulation tool is still used to cover corner-case scenarios. Still, the quality and performance of the perception measured using only this method is not acceptable for any vehicle manufacturers or their suppliers.

Another method used for perception software evaluation is by using a reference sensor that is more accurate and reliable to generate ground truth or reference data and comparing that with other sensors under the same environmental conditions. For example, a lidar sensor and an HD map can be utilized to generate ground truth or reference data of a particular environment. This can be compared against the camera, radar, and position sensors of the vehicle in the same environment. The use of lidar as the reference sensor can have challenges under certain environmental conditions such as rain and fog, and even in the inner-city conditions where buildings with glass claddings are present. Hence a combination of lidar and HD maps generates better ground truth instead of using lidar alone.

One of the standard methods used these days to generate ground truth to compare the accuracy and precision of the perception software is using a GTS (Figure 7.5). There are many GTS available in the market from various suppliers that can generate ground truth or reference data for any vehicle. These GTS are usually prequalified systems with multiple sensors integrated into them. The quality and the performance measurements are pre-evaluated and calibrated to handle robust environment conditions. These systems usually provide both Static and

FIGURE 7.5 A GTS mounted on top of a vehicle.

Dynamic ground truth references, and it also gives the flexibility to select one or the other based on specific needs. The advantage of these types of GTS is that they require less effort in setting up and generating ground truth data, irrespective of vehicle type and the sensor setup. For a vehicle manufacturer with different vehicle programs and variants with different sensor combinations, sensor position, and wheelbase, it will be challenging to generate the ground truth or reference for each of these variants separately by using only vehicle sensors. With the help of these generic GTS, it is flexible and easy to generate the ground truth data, irrespective of the vehicle type and sensors. The GTS uses a 360-degree coverage with its sensors, including cameras, lidars, GPS, etc., to generate the required qualified ground truth or reference data against which the vehicle software can be tested.

As most of the GTS comes with specification and quality information about precision, accuracy, and quality of detecting various objects at different distances, it helps the users of these systems evaluate their perception software against these as a reference. Usually, ground truth or reference data can be generated either in online or offline mode. In the online mode, the reference data are generated in runtime along with the data capture. This requires the algorithms that are used for ground truth generation to be pretrained and validated. The second method is the generation of ground truth data in offline mode. The data is captured in the initial phase and processed in a laboratory environment to generate ground truth through manual, automated, or semiautomated processes. A hybrid model is usually seen in use in almost all the GTS for better measurable results.

Fourth and the most common method used by various organizations to evaluate the perception software and even sensing part of the software stack is through re-simulation. This is executed as part of the system verification and validation, as shown in Figure 7.6. The vehicle sensor data is recorded and used to generate ground truth employing manual, automated, or hybrid methods. This ground truth is compared against the output from the software under test for the same environment. The comparator compares the processed data from perception software to ground truth to decide the generated output quality and performance.

Perception software is usually tested only for its environment model generation. Testing the subcomponents of the perception software is also required because any defects in those subcomponents could propagate to the overall system and can induce failure. Moreover, qualifying the subcomponents of the perception will help in calibrating the perception of the vehicle and help in the feature calibration.

FIGURE 7.6 Perception evaluation by using the re-simulation method.

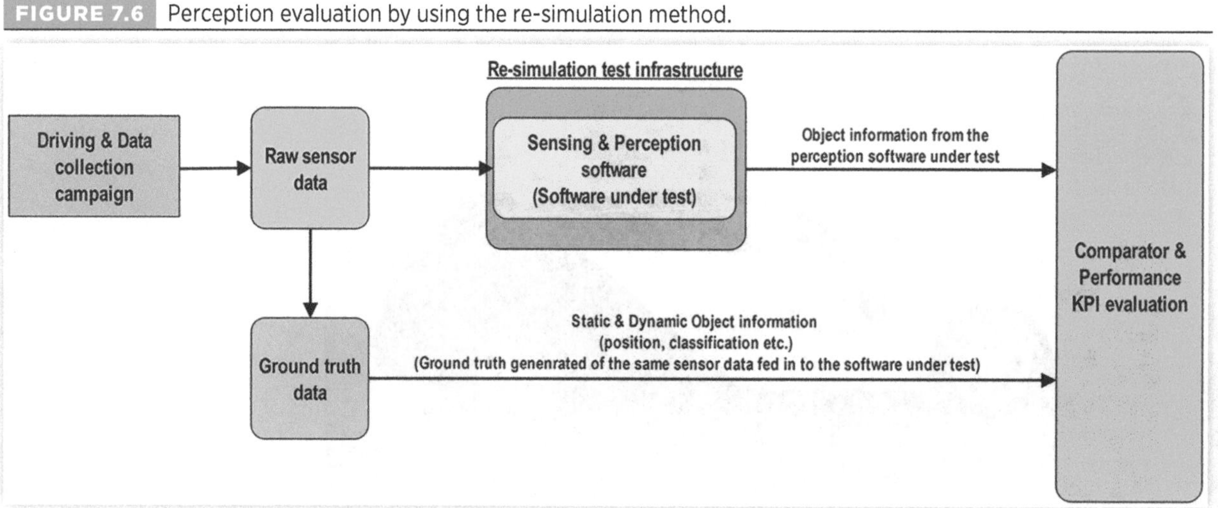

7.4. Methods for Evaluating Perception Software Components

There are various components as part of the perception software, as shown in Figure 7.2. All these components must be tested while evaluating perception software. Various methods that can be used to evaluate different subcomponents of the perception software will be discussed in this section. These methods are applied for evaluating the perception software as part of development testing and are executed using various test infrastructures in the industry.

7.4.1. Evaluation of Static and Dynamic Object Fusion

Object fusion is one of the initial parts of perception. This is a phase where the objects which were detected by various vehicle sensors are fused. These objects are generally classified into two categories, Static and Dynamic objects fusion. Static object fusion involves the fusion of static objects in the environment, such as lane markings, lamp posts, and traffic signs. Dynamic object fusion involves fusing dynamic objects detected by the vehicle sensors such as moving vehicles and pedestrians. It would help to have a brief overview of how the fusion algorithm functions before moving to the testing part of it [7.4]. Even though the entire operation of the algorithm is complex, the explanation here would only provide an overview of the overall operation of the fusion part of perception with an example. Consider an example of lane markings on the road, which are static objects detected mainly using camera, lidar, and HD maps. The vehicle sensors detect and update the position and other characteristic parameters associated with the lane markings. While driving, even though these sensors detect lanes, in most situations, they are occluded or covered by other objects like other vehicles in motion. Sometimes the roads may have no lane markings at all. Even in the presence of these problems, the algorithm keeps the lane availability stable for a few seconds. This can be 2 sec or more, depending on how the algorithm is configured to predict and calculate the lanes when the information is unavailable. For keeping the vehicle in the center of the lane, the centerline within a lane has to be calculated. This is calculated by fusing the borderlines of the lane from various sensors and measuring the centerline from those, which is later used for generating the environment model.

One of the methods used to evaluate fusion algorithm behavior is by the method of using the re-simulation of environmental data from various sensors against the ground truth. The ground truth or reference data for the lanes and the central line is used as a reference and is compared against the calculated lane and central line from the fused lane information from perception software. Physical tests are also performed to evaluate the relative position of the vehicle in the detected lane for a smaller area around the vehicle. How the quality of the calculated lanes is estimated will be discussed later in Section 7.5.

Dynamic object fusion involving the detection and fusion of dynamic objects is performed similarly using ground truth or reference data. Camera, radar, and lidar help in identifying and classifying moving objects which are further fused. All the fused objects are represented with three key calculated parameters. They are the Position, Acceleration, and Velocity parameters of those objects. The tracking part of the perception helps keep track of these objects for their movement and predicts their path for a short duration. This is predicted even if there is no reliable data available from the vehicle sensors or if only one of the sensors is active and no redundancy data from other sensors are available to detect those objects. The lanes that were calculated and assigned to the plane as part of static objects fusion helps dynamic object fusion results in identifying the leading object at a particular instance. Both Static and Dynamic fused objects information are used together for the assignment and placement of these different objects in the mathematical model for calculating their position in the real world. In automated driving features such as Highway Assist or Highway Chauffeur, detection of Static and Dynamic objects are the building blocks of perception.

Along with the objects, lane geometry is also used in controlling the steering movements for the lateral vehicle motion. Leading object information will help in controlling the longitudinal vehicle motion through actuators. Brakes and acceleration can be used to control the longitudinal motion in response to the vehicle in front while driving on a highway.

Dynamic object fusion verification is a significant challenge compared to the verification of static objects. Commonly used methods to evaluate dynamic object detections are using reference vehicles or reference dynamic objects with highly accurate and precise position sensors. These sensors provide position information of the objects and are used to calculate the distance, position, and velocity with reference to the vehicle coordinate system of the subject vehicle. Another method used for testing dynamic object detection and fusion is with the help of environment simulation tools, where the objects are simulated along with the position information. The perception software will detect them and calculate the parameters under test. HD maps and high-precision Global Navigation Satellite System (GNSS) simulators are also used for testing. GNSS refers to a constellation of satellites providing signals from space that transmit positioning and timing data to GNSS receivers. The receivers then use position information to determine the exact location of objects around the subject vehicle.

These tests are also performed in the vehicle, where the test and measurement involve the use of reference objects. These reference objects can be different vehicles or dummies with highly accurate GNSS receivers integrated into them. For usual tests in proving grounds, up to eight objects or POVs are considered at the same time around the subject vehicle with GNSS receivers on them for transmitting exact positions. The information regarding the position of these reference objects or vehicles from GNSS and the vehicle state, including its angular position, force, and orientation from the IMU sensor, is transmitted wirelessly and is used as reference data for comparison. The information exchange from each of these reference objects or vehicles (POVs) can be through the Telematics module or using the Vehicle-to-Everything (V2X) communication protocols in the vehicle. V2X communication protocols in vehicles are used to exchange real-time information like traffic updates or road hazards to other destinations such as a central location or even to specific infrastructures.

Once the ground truth for the Static and Dynamic objects is created, the next phase compares the measured data to the ground truth data for evaluation. Software publishes the fused lane information as four-dimensional vector parameters with magnitude, direction, and uncertainty value. The amount of uncertainty is measured when the fused lanes are compared against its ground truth information. Ground truth will only have pure reference values and not uncertainty. By comparison, the amount of uncertainty in the measured values from the software can be calculated. The software can be improved to reduce the influence of those uncertainties or compensate for those uncertainties in the measurements.

Standard methods used for the measurement of object fusion are by applying the statistical method of calculating the distance between centroids of the distribution and calculating the Mahalanobis distance. The distance of a point of measurement and the deviation from its ideal value can be calculated by measuring the distance between centroids of the measured values from the software under test to the particular point in ground truth data. The centroid of a distribution can be calculated by taking the arithmetic mean of all the points in the distribution. The centroid points are calculated for both measured values and ground truth reference. Distance between those two centroids will provide the number of standard deviations that both readings vary. The amount of uncertainty in measurements can be calculated by measuring Mahalanobis distance. Mahalanobis distance measures the distance between a point "P" from a distribution "D." In other words, if there is a distribution of measured values with mean "D," Mahalanobis distance provides the measurement of how many standard deviations the point "P" is from the mean value "D" of that distribution. The deviation or the difference when comparing the measured values with the ground truth can be calculated with this method. This is graphically shown in Figure 7.7. The improvement of measured values from the software under test can be achieved by calibrating the algorithm or improving the algorithm performance by further training and verification of artificial neural networks. As it is impossible to have an ideal detection and fusion of the objects in any environment, an acceptable threshold of variations can be decided according to the acceptance criteria for performance and quality.

For dynamic object fusion, the same object detection method and comparison with the ground truth are used. Unlike static objects, three parameters—position, velocity, and acceleration—are measured, and corresponding values from the ground truth are calculated. The deviation from the measured value and the ground truth is calculated by measuring a combination of minimal matching distance, the distance between centroids, and the

FIGURE 7.7 Measurement of object fusion by using the distance between centroids and Mahalanobis distance.

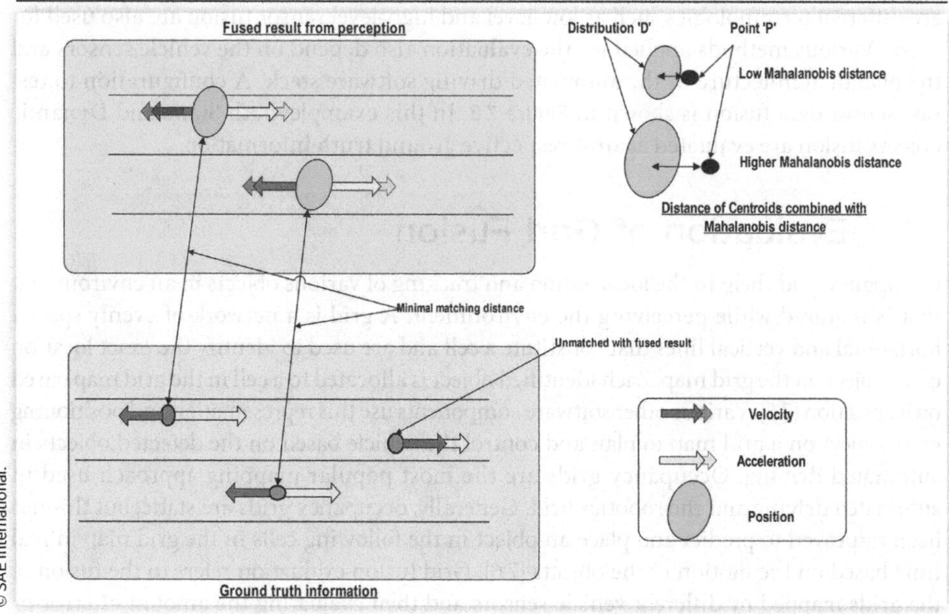

FIGURE 7.8 A comparator framework to test and evaluate raw sensor data fusion.

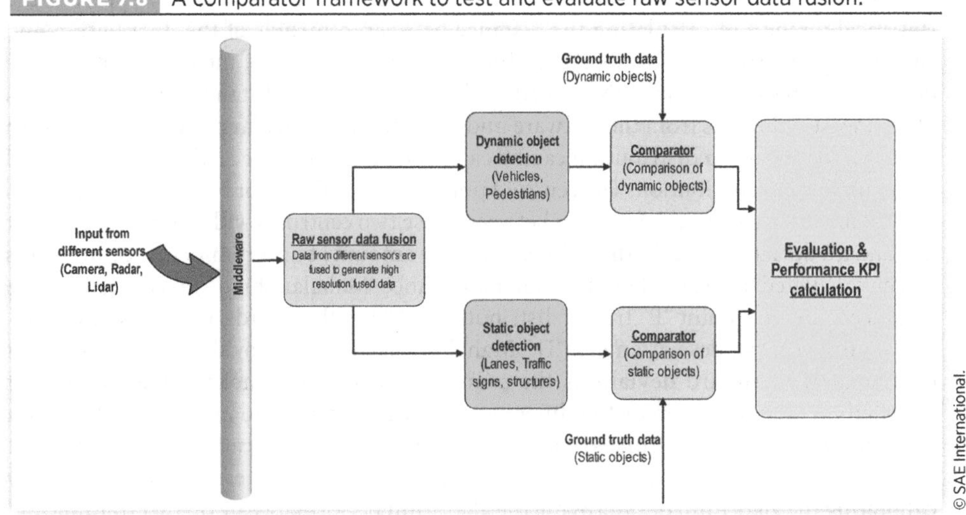

Mahalanobis distance. The minimal matching distance is the distance measured on sets of complex objects that is suitable for defining similarity between two partitioning data clusters that is based on the minimum weight perfect matching of those sets which represents the data. The deviation or the amount of error between them is calculated based on the differences in calculated measurements to the ground truth data. The missing information is classified as an error or defect when the software fails to detect an object or classifies it wrongly, or it may be merged with the detection of another dynamic object.

There are different ways in which the perception software components are implemented. For sensor fusion, there are two different ways of fusion implemented commonly in the industry, one is raw data sensor fusion, and the other one is fusion on object tracking. There are different terminologies such as low-level and high-level sensor fusion are also used for these. Various methods applied for the evaluation also depend on the vehicle sensors and the overall architecture of the automated driving software stack. A configuration to test raw sensor data fusion is shown in Figure 7.8. In this example, both Static and Dynamic objects fusion are evaluated against respective ground truth information.

7.4.2. Evaluation of Grid Fusion

Occupancy grids help in the localization and tracking of various objects in an environment that is required while perceiving the environment. A grid is a network of evenly spaced horizontal and vertical lines that constitute a cell and are used to identify the exact location of an object on the grid map. Each identified object is allocated to a cell in the grid map based on its position [7.5]. Various other software components use this representation and positioning of an object on a grid map to plan and control the vehicle based on the detected objects in automated driving. Occupancy grids are the most popular mapping approach used in automated driving and the robotics field. Generally, occupancy grids are static, but this has been improved to predict and place an object in the following cells in the grid map in real time based on the motion of the objects [7.6]. Grid fusion evaluation refers to the fusion of the grids mapped by different vehicle sensors and then evaluating the amount of error or deviation from the real-world positioning to that from the calculated grid map and positioning (Figure 7.9).

FIGURE 7.9 A dynamic occupancy grid representation for a vehicle in motion.

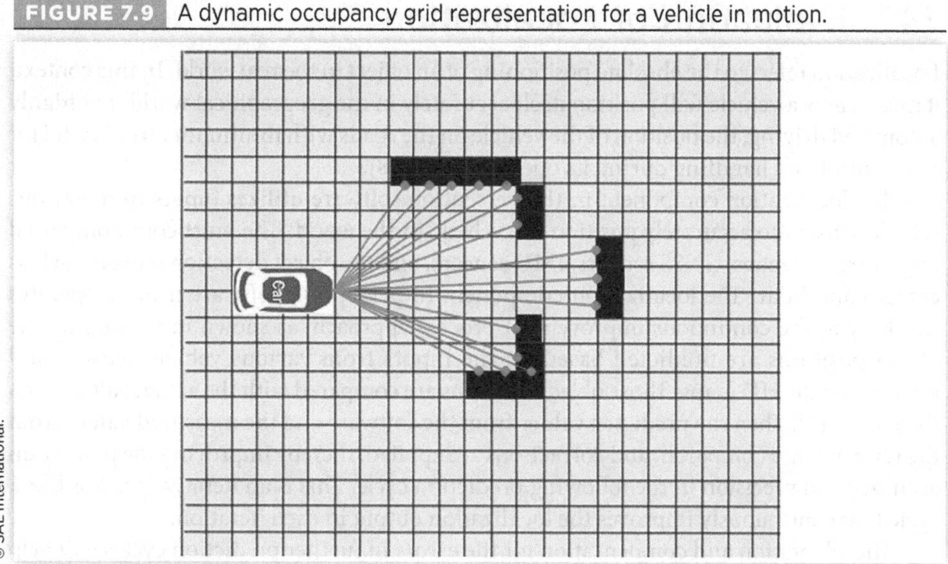

The measurement and evaluation of the grid map are performed against a reference grid map or a ground truth grid map. The reference grid map can be created with a GTS or by employing a highly accurate and precise sensor set within the vehicle. As it is required to consider the motion of the subject vehicle, only the overlapping areas of the grid map at various instances must be considered and evaluated at a particular point in time. The comparison is made between the ground truth and the fused grid map from the perception software under test. The individual points in the grid need to be measured and compared against the ground truth grid map to identify the amount of error or deviation in the fused grid map. The quality of the grid map is always dependent on the quality and performance of the vehicle sensors. Figure 7.10 shows the configuration which can be used to measure and evaluate the grid fusion in perception.

FIGURE 7.10 Comparator test framework for evaluating the grid fusion.

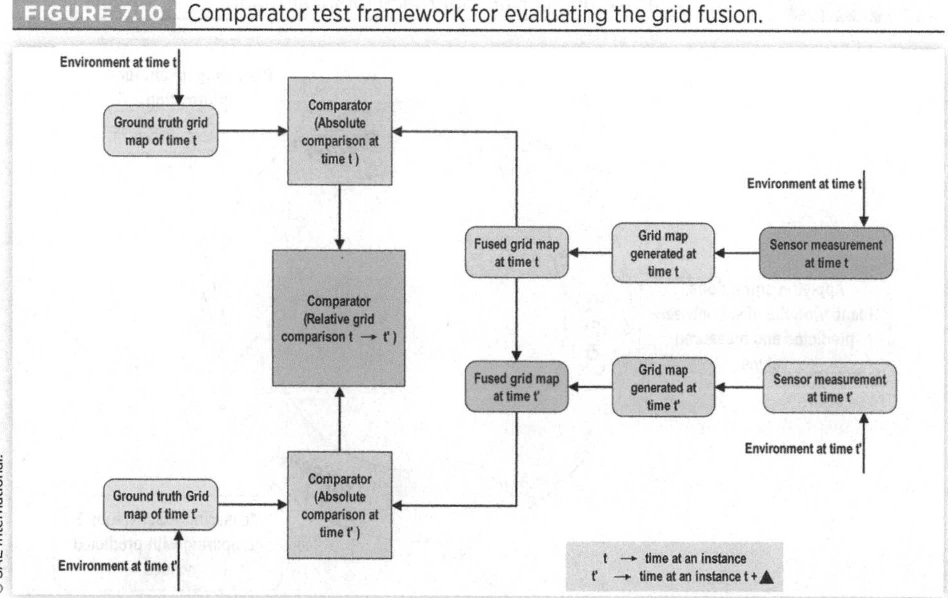

7.4.3. Evaluation of Localization

Localization refers to the absolute positioning of an object in the real world. In this context, it means how a vehicle will position itself accurately in the geographical world. For highly automated driving, the position of the vehicle on the roads with minimum error is vital for safe control and handling during its operation [7.7, 7.8].

The localization component in the perception software utilizes inputs from various vehicle sensors to accurately position the vehicle in the world. The most commonly used inputs are HD maps, GNSS sensors, IMU sensors, and the object detection sensors such as camera and lidar. The localization component in perception software usually operates similarly to the continuous improvement process approach, as shown in Figure 7.11. The object positions are predicted based on the inputs from various vehicle sensors and referencing the HD maps. Those object locations are compared with the actual values from the real world. Then the predicted values from the software and the measured values from the reference are compared, and corrections are applied, thereby improving the prediction accuracy and precision in the following prediction cycle. This is an iterative process like a cycle that continuously improves the localization output in each iteration.

The adaptation and compensation for the errors in further prediction cycles will help to bring more reliable values in the absolute positioning of the vehicle and objects around it to the vehicle position [7.7]. This uses the GPS and IMU sensor information from the vehicle. The motion movements can be predicted by using the course of the road ahead of the vehicle with the help of Kalman filter techniques. Kalman filter is an algorithm that utilizes a series of measured values as input and can predict unknown values over time that are more accurate than those based on a single measurement alone by estimating their joint probability distribution among all the associated variables.

Similarly, visual sensors like camera and lidar can also be used to adapt and predict the relative motion of the vehicle through visual odometry techniques. Visual odometry is the technique of predicting the motion and orientation of an object based on visual data.

FIGURE 7.11 Predict, measure, and correct approach for localization.

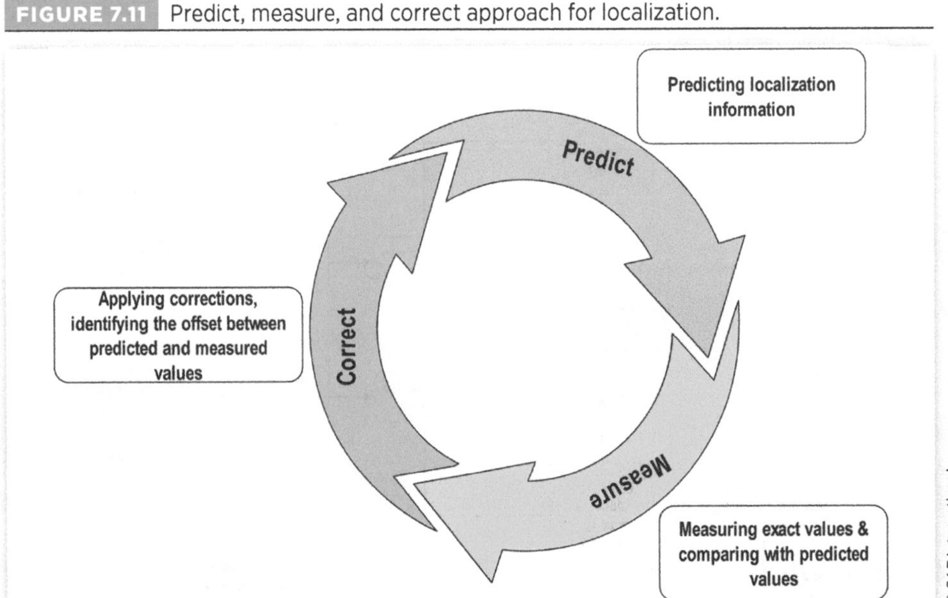

For example, frame-by-frame front camera data can be used to analyze and predict the relative motion of the vehicle, taking into account the camera position in the vehicle and the degree of variation of each frame from the camera.

V2X communication can also act as redundant information with which the absolute positioning and the localization of the vehicle can be improved. V2X communication is a vehicle communication system that communicates with the traffic and other infrastructures in the environment. This communication can exchange information between the vehicle to traffic lights, vehicle-to-vehicle communication, etc. While this is one way to improve localization accuracy, it is heavily dependent on the infrastructure. This technology for localization accuracy determination is still used as part of experimentation and research projects and has not yet been used in large-scale production programs.

Measurement and quality evaluation of localization is performed by comparing the measurements from highly accurate HD maps with localization objects (static) on them. Since there is no standardization available on the localization objects, this is usually available in custom vendor-specific formats. This position information is compared to the localization information from the perception software under test. The deviation in the measurements or the error involved can be measured against the ground truth. Localization accuracy can vary from ±5 up to ±30 cm, depending on the location and the geographical terrain where the vehicle is located. The impact of these geographical conditions on localization error should be considered while designing the perception software and associated functions of automated driving. An acceptance criterion for the error threshold in the position difference from ±10 to ±15 cm of object position accuracy is used in some projects.

7.4.4. Evaluation of Prediction Algorithms

Prediction algorithms are part of the perception software where the future behavior of various objects around the subject vehicle is predicted. Predictive modeling is a statistical method to predict future outcomes or forecast the future based on historical data and the current data about a particular variable [7.1, 7.8, 7.9]. As prediction is a software component using machine learning algorithms, its output is influenced by the past and present data. The reliable output and performance of prediction algorithms solely depend on the amount and quality of data used for training and validation. Prediction models will be better in their performance and quality if the algorithm is trained with vast amounts of relevant training data. Evaluation of prediction algorithm in perception is mainly performed through simulation using environment simulation tools or by re-simulation using prerecorded data from the real world. This includes evaluating the behavior of the subject vehicle, considering certain maneuvers based on predesigned scenarios or events, and tracking the objects around the vehicle.

The standard method used for evaluating prediction modeling or prediction algorithms is recording vehicle sensor data while driving the vehicle and analyzing the prediction algorithms in a laboratory environment based on recorded data from the vehicle. The analysis of the collected vehicle data includes various events captured from the environment from vehicle sensors and comparing those with the predicted values associated with the objects in the environment. The prediction model will also help identify a particular feature within or outside its ODD.

A configuration that can be used to test the prediction modeling part of the perception is through reference data capture and ground truth generation system, as shown in Figure 7.12. Vehicle sensor data recording and ground truth generation are performed in parallel where the predicted values are compared against the actual values from the environment.

FIGURE 7.12 Prediction model evaluation.

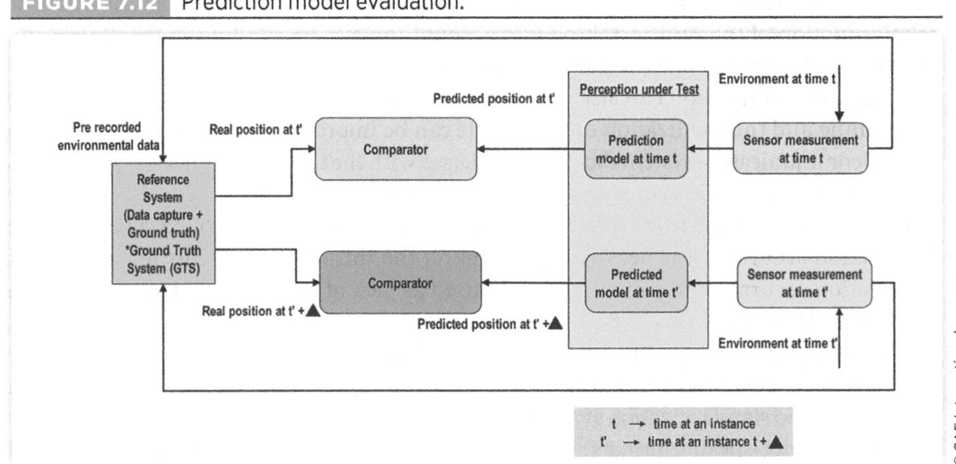

Usually, vehicle sensor data is captured in parallel with the help of data loggers while they are being processed by the ADAS and Automated Driving control unit in the vehicle. The middleware data from this control unit consists of the measured values from the prediction algorithm. The calculated output of the prediction model will be compared against the real-world data while looking forward in time on the recorded sensor data and its ground truth. The ground truth will help in tracking the objects around the subject vehicle in a particular scenario.

Organizations have various acceptance criteria for the prediction model evaluation. An overall prediction accuracy of approximately 0.8 or above within a certain distance from the vehicle is considered an acceptable value for most features. These measurements are based on a set of common scenarios and vehicle maneuvers that are outcomes of the scenario design and analysis phase for different features. The quality requirements vary for different organizations. Prediction is a complex and challenging software component to measure and evaluate. Laboratory-based offline measurement and analysis are commonly used for their measurement and evaluation using a set of scenarios. A structured and standard method and process-oriented approach for evaluating prediction algorithms are not yet used in the industry.

7.5. Measuring Performance and Quality of Perception Software

The previous section discussed various methods used for qualifying different components of the perception software in the industry. This section will discuss how they are evaluated for quality and performance as part of verification and validation. Usually, the perception software quality is evaluated for the software capability to perceive accurately in different regions around the vehicle. The regions of interest around the subject vehicle are divided into various zones that are sensor dependent. The KPIs for perception quality and performance are measured for these zones separately.

The classification of different zones around the subject vehicle is made, taking into account the sensor positioning and the features in the vehicle. The performance and quality

FIGURE 7.13 Zone classification for detection and environment model around the vehicle.

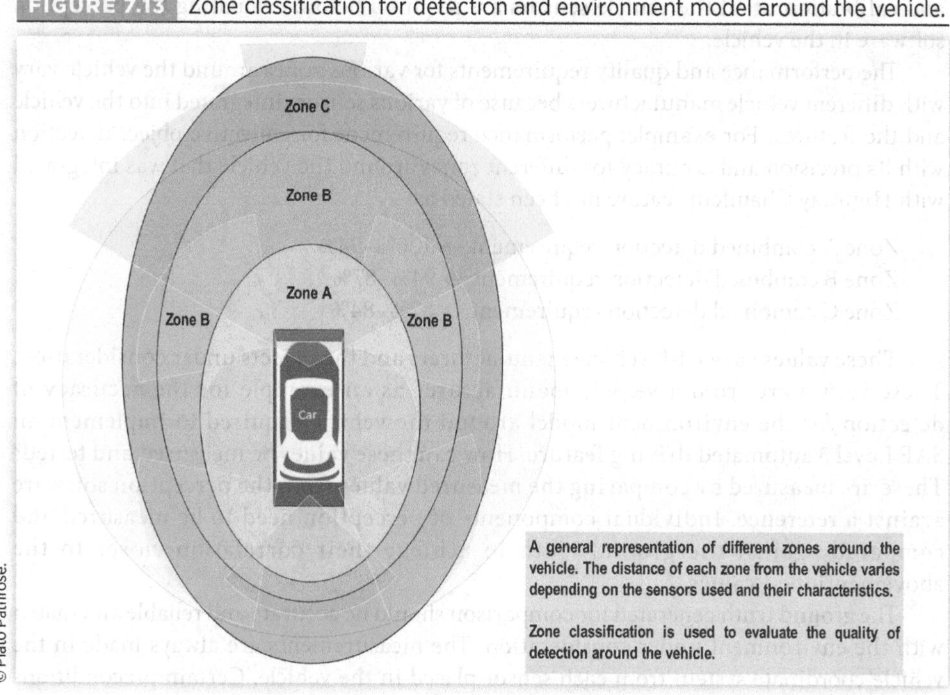

requirements are allocated to these zones. For example, the region around the vehicle can be divided into different zones, as shown in Figure 7.13. The zone width is greater toward the front side of the vehicle, where long-range sensors are located, such as front-facing cameras and long-range radar. In this example, "Zone A" ranges up to 40 m from the center of the vehicle, "Zone B" up to 90 m, and "Zone C" up to 150 m, depending on the sensor capabilities to detect. Classifying the region around the vehicle to different zones will help measure the coverage around the vehicle, accuracy of detection, and precision around the vehicle at various distances. These detection capabilities at different distances influence the quality of the environment model generated from the perception, thereby affecting the vehicle ADAS and Automated Driving features. The distance in the zone classification varies with the sensor range and its capabilities. There are vehicles in testing that can detect up to a distance of 1000 m.

Measurement and quality evaluation of the perception depend on the sensors and their position in the vehicle. Some sensors provide processed information to the perception software of the ECU, and some provide raw sensor data. The example covered in this section considers a vehicle with cameras that provide raw data, radar generates object lists as their output and lidars provides point clouds. Point clouds are a set of points distributed in space with position information and can represent the detected object in a three-dimensional view.

The performance and quality of the environment model are measured by comparing the environment model generated by the software under test against the ground truth data from a GTS. This can be further extended to verify in the real world with real objects. Although the verification in the physical world is complex and limited compared to reference data, it is still performed for the final confirmation and sign-off in many projects. Coverage and calibration of all the vehicle sensors are required before evaluating the perception software in the vehicle.

The performance and quality requirements for various zones around the vehicle vary with different vehicle manufacturers because of various sensors integrated into the vehicle and the features. For example, performance requirement for collective object detection with its precision and accuracy for different zones around the vehicle that was integrated with Highway Chauffeur feature has been stated as:

Zone A combined detection requirement → 100%–94%
Zone B combined detection requirement → 94%–87%
Zone C combined detection requirement → 87%–84%

These values vary with vehicle manufacturers and the objects under consideration. These values are from a vehicle manufacturer as an example for the accuracy of detection for the environment model around the vehicle required to implement an SAE Level 3 automated driving feature. How can these values be measured and tested? These are measured by comparing the measured values from the perception software against a reference. Individual components of perception need to be measured and compared against the ground truth to achieve their correlation closer to the abovementioned values.

The ground truth generated for comparison should be accurate and reliable and match with the environment under consideration. The measurements are always made in the vehicle coordinate system from each sensor placed in the vehicle. Certain preconditions should be met when the samples are collected for measurements and for generating the reference data or ground truth. The methods and processes used for measuring and reliable data collection are derived and adapted from the MSA in Six Sigma.

7.5.1. Preconditions for Measurements

Usually, the performance and quality evaluation is performed at the vehicle level, and certain preconditions are to be met when the measurement and analysis are performed. As the perception evaluation considers more than one sensor at the vehicle level, the preconditions should be considered for all vehicle sensors:

- Valid input and synchronized data: All the sensors in the vehicle are initialized and are calibrated. The output from all vehicle sensors should be time synchronized.
- Availability: The perception software that has to be evaluated is available and ready to be used in the ECU. The software should be configured in such a way that it is capable of accepting and processing the inputs from various vehicle sensors.
- Range of detection: A data collection and analysis strategy must be defined for each zone around the vehicle that detects and identifies various static and dynamic objects. This will play a vital role in measuring data for the evaluation of quality and performance of the perception for various zones around the vehicle.
- Software in the active state: The perception software under test should be in an active state and capable of generating an environment model, accepting inputs from the vehicle sensors.

7.5.2. Data and Data Types

The input data and various data types are defined for the perception evaluation, similar to the KPI evaluation covered in Section 6.4.2. The main difference here is that the complete vehicle is considered along with the input data and their representation. The different types of data that have to be collected and analyzed for calculating performance and quality are:

- GTS reference data (static, dynamic, localization, grid, etc.)
- Perception software availability and operation state (software status)
- Performance and quality parameters (as output)

Various parameters that are analyzed as part of perception software evaluation are of different data types. They can be parameters with only integer values as measurements, parameters that provide continuous data (a sequence of numbers which may be an integer or floating-point data), and parameters providing state of the software that are logical data (0 or 1).

The ground truth data and data types should also be measured and compared against those measured from the software under test. It is essential to have the same data type for the same measured values and the ground truth parameters. The measurements should be made in the same environment with the same scenario and objects. This is one of the most challenging parts, and the test planning phase should cover the requirements for all these aspects. This includes the type of scenario, position, and type of objects in different zones around the vehicle, a static and dynamic condition of the subject vehicle, etc. All these tests are to be executed, and measurements are taken in a controlled environment such as a proving ground.

7.5.3. Performance Evaluation (KPI Measurement)

The performance evaluation of the perception software is made in parts. This starts with the KPI evaluation of the static and dynamic objects calculating the true positives and true negatives along with its combined detection quality [7.10, 7.11, 7.12, 7.13]. Along with the detection of objects like lanes, vehicles, and pedestrians, the classification quality of those detected objects is also considered. Once the detection and classification quality for various objects are measured, the combined detection can be measured from both of those calculated KPIs. This uses the same method as explained in Section 6.4.3.

The measurement should be performed frame by frame to identify the errors present at an instance.

$$\text{Deviation}\left(\text{Error}\right) = \text{Ground truth KPI value-Object detection and classification software KPI value}$$

The overall performance of the perception software in detecting the vehicle environment should be measured considering reference datasets. It includes multiple frames and the errors associated with those for the whole environment under consideration [7.11].

$$\text{True-positive quality}\left(\text{full dataset}\right) = \text{KPIs of positive}\left(\text{detection and classification}\right)\text{total} / \text{KPIs of total positive expected}$$

$$\text{True-negative quality (full dataset)} = \text{KPIs of negative (detection and classification) total} / \text{KPIs of total negative expected}$$

$$\text{Combined (full dataset)} = \text{KPI of combined (positive and negative) total} / \text{KPIs of combined expected}$$

The combined quality represents the quality of the fused static objects in the perception software. Target values are defined as part of acceptance criteria that have to be achieved for static and dynamic object detection and classification. These values vary because of many factors, and hence, a zone-based classification is established to measure these around the vehicle at a different distance, based on various sensor coverages.

Following the same method, KPIs can be calculated for all the subcomponents of the perception software by comparing it with the ground truth data. This will provide us with the combined KPIs for object fusion, localization, and tracking. All these contribute to the environment model generation and its quality. The target requirements for the quality of the environment model can also be defined for each zone around the vehicle. It is not yet common to define the quality requirements for the environment model separately in the industry because of its complexity. Instead, the acceptance criteria cover the accuracy and precision requirements for detection. Measuring the combined KPIs of each perception subcomponent will help identify the amount of error associated with those components and how they influence the perception software. Unfortunately, it is not common in the industry to test and evaluate the perception software in detail because of the complexity and the effort to execute those tests.

Measuring the quality of the environment model components is required because the measured deviations or errors from the subcomponents always vary according to the different controllable and noncontrollable factors that influence the operation. It varies with the input data variation and influences the prediction algorithm while generating an environment model.

Environment model KPI can also be measured using the same method and formulae as explained in this section. The major challenge is to meet the target quality requirements of the environment model for various zones around the vehicle. Even though the quality requirements and KPIs for different zones are defined for the environment model, these are further classified to accuracy and precision of objects, localization, etc. The combined KPIs for all detection and classification will always be less than that has been calculated for the objects or the localization part alone. Usually, for the production programs, the perception software subcomponent KPIs are accepted instead of the combined KPI for the environment model.

7.6. Testing Robustness of the Perception Software

Robustness is the primary requirement for any software that operates in adverse conditions. In ADAS and Automated Driving, the software should be robust enough to handle various factors that influence its output during its operational lifetime [7.14, 7.15, 7.16, 7.17, 7.18, 7.19, 7.20]. The adverse conditions that influence the software during its operation can be either from the external environment or even internal from the system. This section will discuss how essential the robustness of the perception software is and how the robustness of the software can be checked.

It would be nice to have a look at the design part of a system and the software before jumping directly to testing the robustness of the software. The test methods covered in this section were developed based on the methods used for designing robust products. This is one of the easiest ways for designing test cases by making use of the techniques and methods for product design. Taguchi methods were discussed in detail as part of the robust system design in Section 2.6. The same has been used for test design and evaluating the robustness of the perception software. Taguchi methods propose identifying all the factors, including the noise factors that affect the system operation during its operational lifetime. Taguchi proposes designing the system so that it is less affected by those factors or employs a mechanism that reduces the influence of these factors in the system. With the same design concepts, the tests for robustness evaluation can be designed with the help of P-diagrams and SNR measurements. The P-diagram technique can be applied to the perception software to identify all those factors, including noise factors that affect the normal operation and influence the output of the perception software. These analysis results can be used for designing test cases, and the perception software can be evaluated against them.

The perception software accepts inputs from various vehicle sensors and vehicle networks. The robustness evaluation should consider all these inputs and the associated noise generated with each of these inputs. The methods used for evaluating perception robustness are similar to the fault injection testing recommended by the functional safety standard [7.15, 7.19]. Here, in this method, the faults and noise associated with each input data source, such as sensors and vehicle networks, are covered. The test configuration is shown in Figure 7.14 where the noise factors associated with each input source are either generated or injected along with normal data input to the perception software. The software output can be evaluated against those noisy inputs, based on which the perception performance can be measured and evaluated. This will also help improve the software performance and output quality by developing it in a way so that different noises have less influence on the overall operation of the perception software. The perception software can

FIGURE 7.14 Block diagram of a noise and fault injection framework.

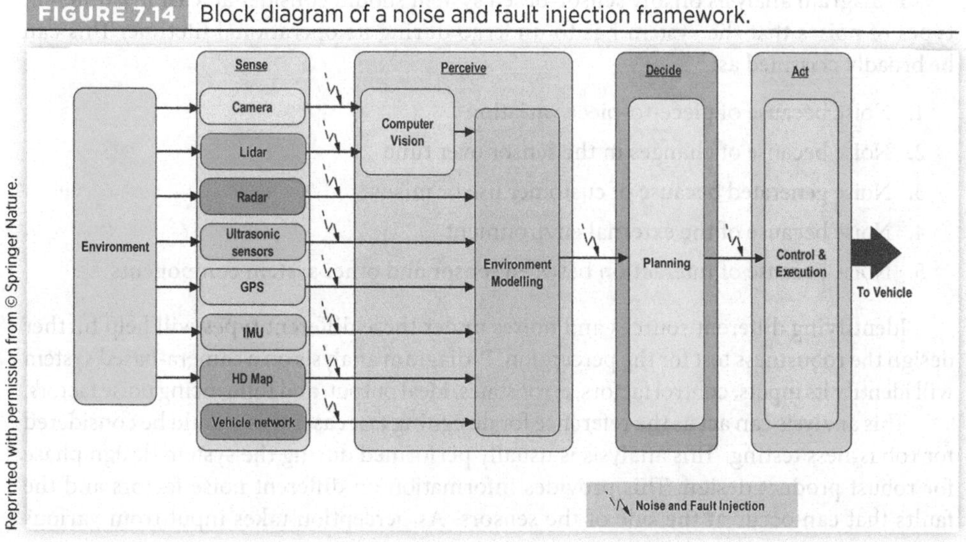

FIGURE 7.15 A sample P-diagram analysis with noise factors specific for camera-based systems.

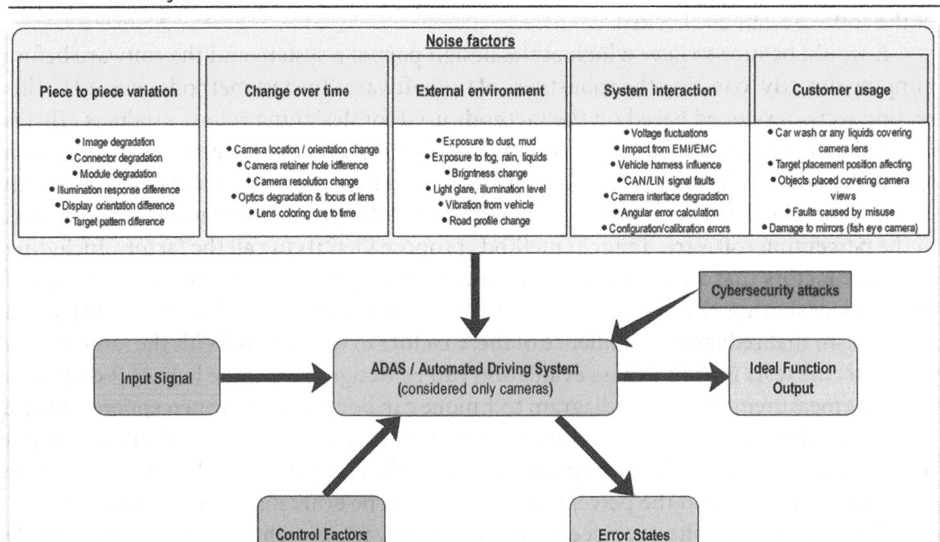

provide reliable output even in a noisy environment or detect noisy inputs and control its output rather than generate wrong results that affect the system functionality.

The automated driving software can be represented as shown in Figure 7.14 with main software components. It accepts input from different sensors and vehicle networks, and the processed output is fed through the vehicle network to various actuators in the vehicle. A P-diagram analysis for each input source is performed to identify different noise factors and other influencing factors on the perception software. A camera is one of the most important sensor types in the vehicle, and a camera-based system will be considered for further detailing of perception robustness testing method. Cameras are used to detect, classify objects, and measure distance in the vehicle. An example of object detection and classification using the camera is considered in the P-diagram shown in Figure 7.15.

P-diagram analysis on any sensor-based system should consider at least five different types of noises that the system has to undergo during its operational lifetime. This can be broadly classified as:

1. Noise because of piece-to-piece variation
2. Noise because of changes in the sensor over time
3. Noise generated because of customer use or misuse
4. Noise because of the external environment
5. Noise because of interaction between sensor and other system components

Identifying different sources and noises under these different types will help further design the robustness test for the perception. P-diagram analysis on a camera-based system will identify its inputs, control factors, error states, ideal output, and influencing noise factors.

This analysis can act as the reference for designing test cases that should be considered for robustness testing. This analysis is usually performed during the system design phase for robust product design. This provides information on different noise factors and the faults that can occur at the side of the sensors. As perception takes input from various sources, such as different vehicle sensors, the noise from each of these sources must be considered. These noises can propagate further into software components and can affect the output of the perception software if it is not designed robust enough to handle them.

The method explained here is from one of the projects where a tool was designed to generate and inject different noises of different magnitudes to various input sources of the perception software. Different noise factors for each input source were identified as part of the P-diagram analysis, including different sensors and vehicle networks. A comparator will compare the output from the perception software to reference data or ground truth. The same tool could inject faults into the data lines of various sensors and vehicle networks when different vehicle sensors were used.

For a camera-based system, commonly occurring faults in the sensor are timing faults and the faults because of a bad connection. Various noises are generated as an outcome of these faults or even directly from other external sources. As there are different ways in which different noises can occur in the system, prioritization and weightage have been provided for different noise factors based on their impact on the software, which was analyzed as part of a P-diagram analysis. Prioritization can be done using a simple FMEA or a dependency diagram technique.

For object detection, the risk associated with noise factors over the communication channel for noises such as Salt and Pepper noises was found to have higher precedence and weightage over other noise factors in the P-diagram analysis. This risk evaluation and analysis were performed by considering various parameters such as frequency of occurrence, detectability, and severity of the influencing factor on a particular ADAS and Automated Driving feature. Most of this information can be collected from the FMEA performed for the system and sensors. Prioritization of all the noise factors based on their influence on the software and chance of occurrence will help select the noise factors that must be considered for testing and designing the software to be less sensitive to those.

Different types of faults and noises are identified using various methods such as P-diagram, FMEA, and even functional safety analysis that help identify and decide specific tests for different noise types and faults in the system. The tool explained here for robustness testing was developed based on these requirements to generate different faults and noises in the system and software. Perception software can be tested with this tool with various faults and noises injected into its input and evaluating its output.

The most common noise types that influence camera-based systems are salt and pepper noise, Gaussian noise, noise because of chromatic aberration, etc. Changing the magnitude of these noises in the camera inputs under laboratory conditions for testing the perception software provides the measurement of the robustness of the perception software [7.20]. Suppose the perception software is able to process input with different noise components to generate an environment model that is accurate and precise reliably. In that case, it will prove that the perception software is robust. It can alternatively be used to design the fail-safe mechanism to identify the level of input noise; the software can process and generate a reliable environment model with which the ODD and feature availability can be determined for the vehicle. Suppose the input noise level is beyond a specific limit and the perception software cannot reliably generate an environment model. In that case, the system can be designed to identify the conditions outside the ODD, and the feature can be made unavailable (Figure 7.16).

Figure 7.17 shows the block diagram of the evaluation method for measuring performance KPIs of the perception software in the presence of different noise types and faults. The KPIs are measured as described in Section 7.5.3. Here the KPIs are measured for different noise types and magnitudes. Usually, these measured KPIs vary for different noise types even if

FIGURE 7.16 Camera images with various noise types—Original image (top left), image with pepper noise 5% degradation (top right), image with salt noise 5% degradation (bottom left), image with salt and pepper noise 5% degradation (bottom right).

FIGURE 7.17 An evaluation method for the robustness of perception of a camera-based system.

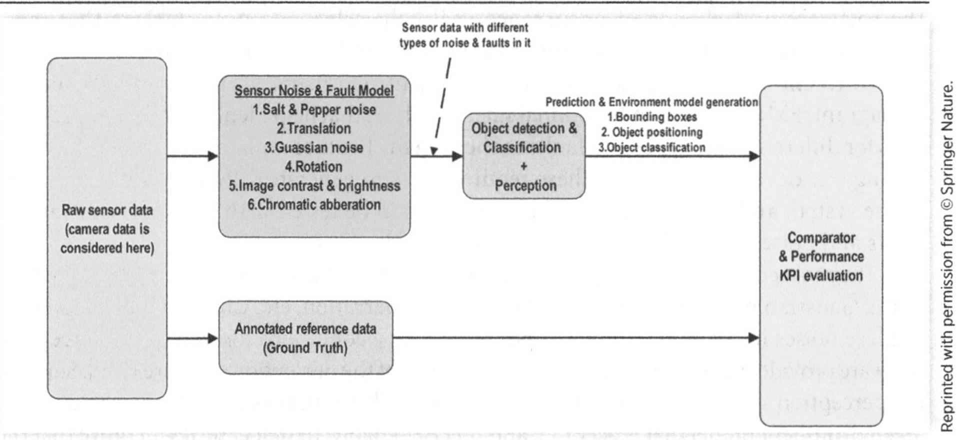

they are of the same magnitude as the perception software will not behave in the same manner in processing the input from different sensors and with different noise types.

Combined KPIs measured with true positive and true negative with the complete dataset for a particular noise type and magnitude or fault type will provide the overall response of the software. The noise and fault generation tool can generate different noise types of different magnitudes depending on the SNR of the data available to the perception software. The tool explained here as an example supports up to four different magnitude levels for different noise types such as 5%, 10%, 15%, and 20% of noise injected into the input data stream of each sensor. Any noisy input data stream beyond 20% to 25% usually does not help generate a reliable environment model.

7.7. Challenges in the Measurement and Evaluation of Perception

Perception is a vital part of an ADAS and Automated Driving system. Even though there were initially different methods proposed for its evaluation, it is a fact that most of them are not practical when considering a vehicle production program [7.21]. This is mainly because of multiple factors such as the involvement of multiple suppliers, not having a qualified ground truth, lack of time spent for evaluation of perception, and higher costs and effort associated with those tests. Perception must be tested and qualified with robust test methods at all levels, which is necessary to guarantee the safe operation of different features in the vehicle.

Perception testing is performed using environment simulation tools with synthetic data as the main source and only minimal real-world data in various projects. There are also exceptions: the tests are only performed in the real world and no simulations are considered. The disadvantage of relying only on the simulation environment is that the simulation tools primarily available in the market are not qualified to test safety-critical systems and software. No information are available regarding the confidence level and the reliability in their data unless they are evaluated. Most of the tool suppliers cannot provide a correlation matrix for the tool compared to real-world behavior.

Using a ground truth reference system and comparing its output with the processed data from the perception software is commonly used in the industry these days. A challenge here is that the GTS are integrated with multiple components from different suppliers, in most cases. There is no qualification data available to show how reliable and accurate is the data coming out from those devices. In most cases, the suppliers only perform a minimum qualification that focuses on one of the components, like static or dynamic objects. As it is challenging to measure the position accuracy of objects across different conditions, only minimal tests would have been performed for qualifying these GTS. When the tests performed for qualifying GTS are limited, there is no guarantee that the GTS can provide accurate measurements at all environmental conditions.

Different vehicles from various manufacturers have different sensor types, positions, vehicle types, etc., and the reuse of ground truth data is always limited for different vehicles. This is one of the major challenges that is seen with different vehicle manufacturers these days. This forces them to create ground truth information for different vehicle programs if any changes are there in sensor type, position, or vehicle types.

The use of legacy sensors as a reference and comparing the new sensors and their performance with those is another approach found in the industry as part of benchmarking. This is mainly performed when a vehicle manufacturer wants to replace a particular sensor type used in the vehicle with a matching sensor from a different supplier. The challenge here is that the checks are minimal, and only a delta evaluation is performed. The impact on the overall software is not usually evaluated as the information internal to the sensors might not be accessible, especially if it is a sensor with the internal processor and provides only object lists as output. This can be an option for specific benchmarking cases but not recommended for a complete qualification of the perception software where a sensor is considered the reference, for which not many details are available either from its supplier or from the field usage.

Even though the technology has advanced, improvements are still required to establish reliable and easy-to-use test methods and cost-effective approaches for qualifying all elements of automated driving systems. Perception testing and its KPIs are still a gray area in many organizations even if those are integrated into the vehicles. This is an area that many vehicle manufacturers and suppliers usually do not discuss even when their products are deployed in the market.

7.8. Summary

Perception software is a complex software component in the automated driving software stack, and its testing is also complex. This chapter provided a high-level overview of the perception software and its main subcomponents. An introduction of each of those subcomponents and different methods used in the industry for testing and qualifying them were discussed. Along with different methods that can be used for testing each of those components, a practical approach to measure the KPIs of the perception software and its subcomponents was also explained with specific examples. The evaluation methods for static and dynamic objects, localization, etc. are covered with methods to compare it with reference data or ground truth.

With perception being the core component in the software stack, the robustness of the software is important. Robustness requirements and a method to evaluate the robustness of the perception software were explained with an example of the camera-based system. Various analyses performed as part of the system and software development were considered for designing the tests related to robustness evaluation. This includes the analysis as part of system engineering, functional safety, and FMEA to identify the possible noise conditions and faults that could influence the system and its features. Testing perception software is still a challenge and is not resolved in most organizations. Various challenges associated with the testing of perception and its components are explained, and the need to improve current practices and methods was discussed. Even though many products are deployed in the market, the testing of perception software is still unsolved for many applications.

References

7.1. Pendleton, S.D., Andersen, H., Du, X., and Shen, X., “Perception, Planning, Control, and Coordination for Autonomous Vehicles,” *Machines* 5, no. 1 (2017): 6, doi:https://doi.org/10.3390/machines5010006.

7.2. Caesar, H., Bankiti, V., Lang, A.H., Vora, S. et al., “NuScenes: A Multimodal Dataset for Autonomous Driving,” arXiv 2019, arXiv:1903.11027, 2019.

7.3. Cityscapes, “Cityscapes Data Collection,” accessed July 27, 2021, https://www.cityscapes-dataset.com.

7.4. Fawcett, T., “Introduction to ROC Analysis,” *Pattern Recogn. Lett.* 27 (2006): 861-874, doi:https://doi.org/10.1016/j.patrec.2005.10.010.

7.5. Bouzouraa, M.E. and Hofmann, U., “Fusion of Occupancy Grid Mapping and Model Based Object Tracking for Driver Assistance Systems Using Laser and Radar Sensors,” in *2010 IEEE Intelligent Vehicles Symposium*, La Jolla, CA, 2010, 294-300, https://doi.org/10.1109/IVS.2010.5548106.

7.6. Milstein, A., “Occupancy Grid Maps for Localization and Mapping,” *Motion Planning* (2008), doi:https://doi.org/10.5772/6003.

7.7. Stachniss, C., *Robotic Mapping and Exploration (Springer Tracts in Advanced Robotics, Band 55)* (Berlin, Heidelberg: Springer, 2010), ISBN:978-3642101687.

7.8. Wen, L., “UA-DETRAC: New Benchmark and Protocol for Multi-Object Detection and Tracking,” arXiv:1511.04136, 2015.

7.9. Sheskin, D.J., *Handbook of Parametric and Nonparametric Statistical Procedures* (Boca Raton, FL: CRC Press, 2011), 109, ISBN:978-1439858011.

7.10. Boncelet, C., *The Essential Guide to Image Processing* (Orlando, FL: Academic Press, Inc., 2009)

7.11. Geiger, A., Lenz, P., and Urtasun, R., "Are We Ready for Autonomous Driving? In: The KITTI Vision Benchmark Suite," in *Conference on Computer Vision and Pattern Recognition (CVPR)*, Providence, RI, 2012.

7.12. Reway, F., Huber, W., and Ribeiro, E.P., "Test Methodology for Vision-Based ADAS/AD Algorithms with an Automotive Camera-in-the-Loop," in *International Conference on Vehicular Electronics and Safety (ICVES)*, Madrid, Spain, 2018, IEEE.

7.13. Mikolajczyk, A. and Grochowski, M., "Data Augmentation for Improving Deep Learning in Image Classification Problem," in *International Interdisciplinary Ph.D. Workshop (IIPhDW)*, Swinoujście, Poland, 2018, 117-122.

7.14. Salay, R., Queiroz, R., and Czarnecki, K., "An Analysis of ISO 26262: Using Machine Learning Safely in Automotive Software," arXiv:1709.02435, 2017.

7.15. Rao, D., Pathrose, P., Huening, F. and Sid, J., "An Approach for Validating Safety of Perception Software in Autonomous Driving Systems," in *Model-Based Safety and Assessment*, vol. 11842 (Cham, Switzerland: Springer, 2019), 303-316.

7.16. Ziade, H., Ayoubi, R., and Velazco, R., "A Survey on Fault Injection Techniques," *Int. Arab J. Inf. Technol.* 1 (2004): 171-186.

7.17. Uriagereka, G.J., Lattarulo, R., Rastelli, J.P., Calonge, E.A. et al., "Fault Injection Method for Safety and Controllability Evaluation of Automated Driving," in *Intelligent Vehicles Symposium (IV)*, Los Angeles, CA, 2017, 1867-1872, IEEE.

7.18. Pintard, L., Fabre, J.-C., Kanoun, K., Leeman, M. et al., "Fault Injection in the Automotive Standard ISO 26262: An Initial Approach," in Vieira, M. and Cunha, J.C. (Eds), *European Workshop on Dependable Computing*, vol. 7869 (Heidelberg: Springer, 2013), 126-133, https://doi.org/10.1007/978-3-642-38789-0_11.

7.19. ISO/PAS 21448, "Road Vehicles—Safety of the Intended Functionality," 2019.

7.20. Takanami, I., Sato, M., and Yang, Y.P., "A Fault-Value Injection Approach for Multiple-Weight-Fault Tolerance of MNNs," in *Proceedings of the IEEE-INNS-ENNS International Joint Conference on Neural Networks*, Como, Italy, vol. 3, 2000, 515-520, IEEE.

7.21. Koopman, P. and Wagner, M., "Challenges in Autonomous Vehicle Testing and Validation," *SAE Int. J. Trans. Safety* 4, no. 1 (2016): 15-24, doi:https://doi.org/10.4271/2016-01-0128.

/Autonomous
/Sensing
/Communication
/Battery
/Navigation
Autopilot mode

8

Calibration of ADAS and Automated Driving Features

During new product development, it is expected that the final product would deviate from the ideal behavior for which it was designed. This usually happens because of many reasons such as the change in design requirements, nonavailability of ideal measurements, and deviation in development and production requirements. Even the development and production processes will also influence a system and induce variations to its ideal behavior. One of the main factors which influence these variations is the environment where the system operates. How can these variations influence the behavior of an Automated Driving system? How dangerous will it be if a system which was designed to perform a particular function behaves differently from its ideal behavior in the real world? These are the usual challenges one would face when a system is out for production.

8.1. Calibration—An Overview Based on Ideality Equation

The world around us is not ideal, and it is constantly changing, so any system being developed and produced is neither an ideal system nor delivers ideal functions. Even though the system is not ideal in its behavior and output, it is possible to improve the behavior and output of such a system to correlate its operations with an ideal system or a reference that is considered standard. An interesting concept from TRIZ developed by the Soviet inventor Genrich Altshuller about Ideality is described here. It is helpful to understand what Ideality means and why calibration is performed in any system so that the functions of a system can be improved to reach its ideal state or the reference state [8.1, 8.2, 8.3]. According to

Altshuller, the ideal state of a system is where all its functions are achieved without any problem. Or it can be simply expressed using a formula:

$$\text{Ideality} = \sum \text{Benefits} / \left(\sum \text{Costs} - \sum \text{Harm}\right)$$

A system is said to be ideal when the overall benefits outrun the harmful functions that it would bring. The Ideality of a system can be measured as a ratio of the sum of benefits to the sum of costs minus the sum of harmful functions. As this is a ratio, it can be improved by continuously increasing the numerator value of useful functions and benefits and reducing the denominator values, which are costs and the harmful functions of the system [8.4]. Why is this relevant to calibration? Calibration is one of those methods with which one could improve the benefits of the system after production.

Calibration means comparing a system and its output measurement to an ideal output or a reference that is considered standard, and the measured output is compensated for errors and deviations to match it with the standard system. The measurement from the system is compared both qualitatively and quantitatively to identify how much it correlates with a standard reference or its ideal output. These measurements will help identify the variation or error associated with the output of the system and its functions from its ideal behavior. The primary purpose of calibration is to identify and compensate for these errors so that the system and its functions can move toward its Ideality. Measurements from the system for its complete operating range are considered for comparison, and statistical methods are used to identify the outliers or the measurements that are lying outside its expected operating conditions and thresholds. The causes for these outliers are analyzed and controlled so that the system output follows that particular pattern for which it was designed. These are typically achieved by controlling specific parameters associated with system operation based on which the variations in the output of the system are minimized. An example is shown in Figure 8.1 with output measurements from a calibrated and non-calibrated system.

For any system produced in the market, the goal is to bring its operation closer to its ideal behavior, which is usually considered during its design phase. For complex systems like Automated Driving systems, it is essential and relevant to have beneficial functions close to ideal behavior and less harmful functions in the system compared with the

FIGURE 8.1 An example of non-calibrated and calibrated system measurements.

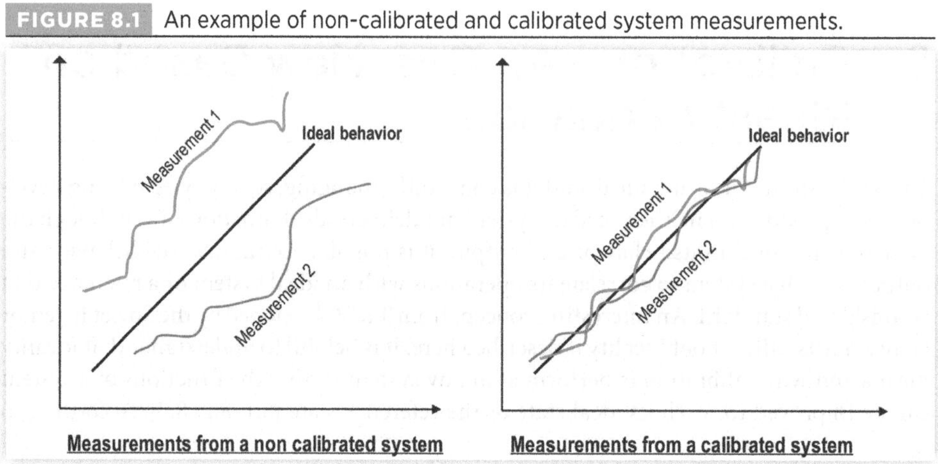

FIGURE 8.2 Methods for improving the Ideality.

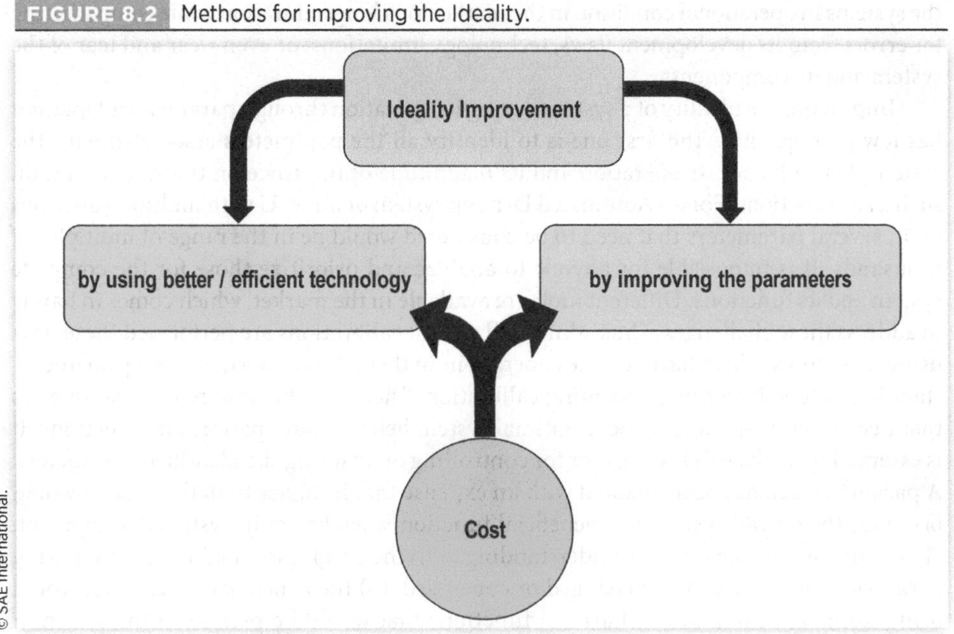

Ideality equation. Functions of a system can be improved mainly utilizing two methods, one is by adapting the parameters associated with the system and the second one is by using better and more efficient technology to compensate for the errors in the system [8.5, 8.6]. Both methods have costs involved, and the decision to choose which method for improving a system behavior is always driven by the cost factor. Figure 8.2 shows the dependency of calibration and Ideality.

Calibration of a product coming out from production mainly focuses on the first method of identifying the parameters associated with a function and adapting those based on its influence on the system output to improve the functional behavior and increase the Ideality of the system during its operation. The second method is about identifying the optimal technology for the system to achieve different functions. This has to be considered during the system design phase as it will influence the architecture, skills, and resources required for implementing new technology. This chapter discusses calibration, which focuses on adapting the system parameters for its output optimization and improvement.

An Automated Driving system accepts input from multiple sensors in the vehicle, and the vehicle is controlled by controlling actuators through the vehicle network. Calibration needs to be performed on both inputs and outputs of the ECU when the system functions influence the vehicle behavior. Vehicle manufacturers mostly use the same set of components or subsystems, such as sensors and actuators, for multiple vehicle programs rather than using different ones for each program. A system behavior while in operational conditions can be optimized by controlling specific parameters acting as input like sensor data or parameters within the system, such as the system timing or operational parameters or both [8.7, 8.8]. These parameters need to be identified and adapted to achieve optimal performance and quality of the system. Calibration refers to the measurement of the deviation from the standard reference or ideal system. The need for these measurements is to adapt the system behavior to compensate for these deviations. Calibration is also a way to keep the systems in operational condition in the market for a longer duration while compensating for errors from its development flaws, technology limitations, or even wear and tear of the system and its components.

Improving the Ideality of a system during its operation through parameter adaptation has few prerequisites. The first one is to identify all the parameters associated with the system that influence its operation and its magnitude of influence on the system output, such as its functions. For an Automated Driving system or an ECU with multiple functions on it, several parameters that need to be considered would be in the range of multiples of thousands. It is impossible for anyone to analyze and prioritize those for the complete system and its functions. Different tools are available in the market, which comes in handy to address these challenges. Thus, almost all system calibrations are performed these days using these tools, which have complex operations in them. It is important to keep an understanding of Ideality while performing calibration. There may be hundreds of parameters that need to be adapted to achieve optimal system behavior and perform its functions. It is essential to analyze the cost factor for controlling or adapting the identified parameters. A parameter that has been adapted with an expense that is higher than the value it would bring for the overall system as a beneficial function is irrelevant in system development. Hence, one should have a clear understanding of (i) the costs associated with the selecting parameter which needs to be changed or controlled, (ii) the beneficial functions it would bring to the system, and (iii) harmful functions that would be present in the system if not compensated.

There are systems present in the market that are developed with drastic unidirectional changes for both numerator and denominator as per the Ideality equation. This can be achieved by either drastically increasing the benefits or reducing the costs and harmful functions. The evolution of centralized vehicle electrical network architecture from the old distributed electrical network architecture, having a Domain Controller rather than multiple ECUs to perform ADAS or Automated Driving functionalities in the vehicle are examples even though the benefits one would have from those systems are higher, the same goes for the cost and the harmful functions that come with it. Still, it is used and considered because the benefits of those systems are more than the harm caused and the cost of those systems.

8.2. Common Types of Calibration in an Automated Driving System

An Automated Driving system can be simply explained as integrating multiple subsystems to provide assisted or automated driving functions or both. The system accepts input from the sensors in the vehicle, capturing the external environment and transferring it to an ECU or a domain controller for processing the information. The processed information generates output for controlling the vehicle, which gets translated as signals in the vehicle network, which are transferred to actuators such as steering and brakes. Calibration of such a system is complex and challenging if the provision for calibration of the system is not considered during the design phase of the system. The system calibration requirements should consider how the system should be calibrated, how it should be evaluated, when it should be initiated and terminated, and how frequently it should be executed. Even though the scope of this book is verification and validation, we would cover a little bit on the design part of the calibration to understand why it should be tested and how important it is in system operation.

FIGURE 8.3 Representation of an Automated Driving system with calibration interface.

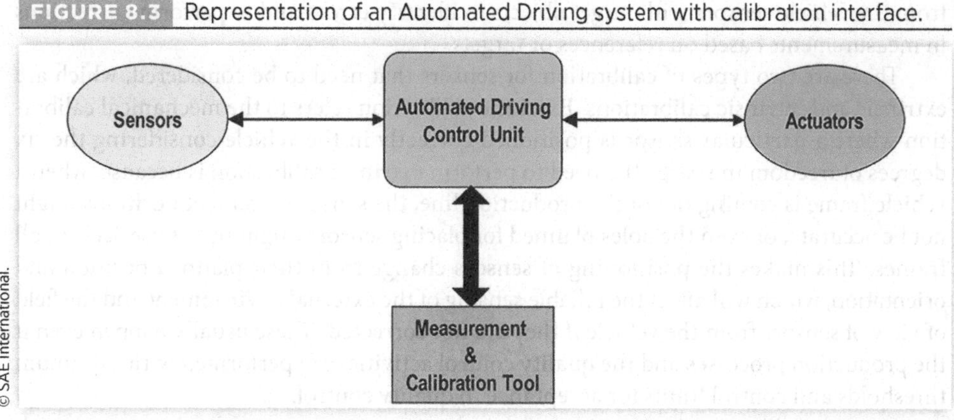

Calibration of Automated Driving systems is performed in two phases. The first one is the calibration of the subsystems involved in the data flow path starting from the sensors to actuators, as shown in Figure 8.3. The ECU can have input data that is calibrated, and the output from it is translated to actuators which are also calibrated and reliable. The second one is the calibration of the functions, mainly the calibration of the software part in the Domain Controller or the ECU. Calibration of all the subsystems and components is the prerequisite for initiating the measurement and calibration of the system functions, which is usually performed considering the complete vehicle.

This section covers only a few of the common calibration types from the system level one would come across in ADAS and Automated Driving systems. Even though the methods used are similar, there is a wide range of tools and infrastructure required for the calibration of different components of the vehicle, like engine calibration, and powertrain calibration. For a complete vehicle, all different domains and components are to be calibrated beyond the ADAS and Automated Driving systems, which is beyond the scope of this book.

Three main calibration types need to be considered for ADAS and Automated Driving systems:

1. End of Line (EoL) Calibration
2. Service Calibration
3. Online Calibration

8.2.1. End of Line (EoL) Calibration

EoL calibration is performed at the end of the line in production. This calibration covers primarily the sensor and actuator calibration associated with the ADAS or Automated Driving ECU in an Automated Driving system. For many vehicle manufacturers, ADAS-specific EoL calibration focuses only on the sensors that capture the external environment and transfer it to the ECU and not actuators. This is because the calibration of actuators is usually performed while performing drive testing or simulating the vehicle to be in a driving condition, such as a vehicle-in-the-loop simulation. For an automated driving vehicle with different types of sensors such as cameras, radars, lidars, and ultrasonic sensors, the output of all these sensors is to be measured and checked for its correctness, such as its accuracy and precision, with a standard reference. Almost all vehicle manufacturers use an EoL calibration facility with controlled environment conditions to capture the measurements from the vehicle sensors with target objects to identify errors and to perform corrections in measurements based on references or targets.

There are two types of calibration for sensors that need to be considered, which are extrinsic and intrinsic calibrations. Extrinsic calibration refers to the mechanical calibration where a particular sensor is positioned correctly in the vehicle considering the six degrees of freedom in space. The need to perform extrinsic calibration is because, when a vehicle frame is coming out of the production line, the sensor placement positions might not be accurate, or even the holes planned for placing sensors might not fit precisely in all frames. This makes the positioning of sensors change from their planned position and orientation, which will affect the reliable sensing of the external environment and the field of view of sensors from the vehicle if they are not corrected. These usually happen even if the production processes and the quality control activities are performed with minimum thresholds and control limits for acceptance in quality control.

Intrinsic calibration is about identifying the correct parameters for the sensors, such as focal length, lens distortion compensation, and so on for cameras and range, azimuth angle, etc. for radars and lidars. As the automated driving vehicles have a heterogeneous sensor setup, the targets used for calibration should also be selected with care. Most vehicle manufacturers use separate targets for the calibration of different sensors rather than a single target. This is because of the complexity and for legacy reasons associated with their production facility. Targets are explicitly selected for the sensors which need to be calibrated. Targets can be both static ones as well as dynamic ones, depending on the sensor which undergoes calibration. With the introduction of lidar, recently in the vehicles, the need for calibration of lidar was also a topic for the vehicle manufacturers. Still, as of today, many vehicle manufacturers do not perform separate intrinsic calibration for lidar at the EoL. Instead, the calibration data from the lidar supplier are used. The environment where the calibration is performed is controlled with accurate and measurable conditions and targets.

The most common calibration facility setup is with controlled lighting of approximately 750–900 Lux (luminous flux per unit area) with checkerboards or lanes with black and white boxes in them as targets mainly for the cameras. Cameras are calibrated by detecting edges and corners of these checkerboards, which are considered as targets. For radars, corner reflectors are used, which are static and dynamic, to adjust the detection and motion with azimuthal angle measurements and its correction. Some organizations also utilize complex single targets with various shapes and reflecting characteristics for heterogeneous sensor calibration, including lidar, but it is not standard across the industry. Once the deviations or the error in measurements from the sensors with reference to the targets are identified, corrections can be applied to use the measured data for further processing and feed the algorithms with corrected values. Even though vehicle manufacturers do not usually consider lidars for intrinsic calibration at the EoL, they usually check for the correlation between the calibrated camera and radar measurements with lidar measurements.

8.2.2. Service Calibration

Service calibration is usually a calibration performed at the service and maintenance centers of a vehicle. Unlike EoL calibration, the infrastructure available at the service and maintenance centers will be limited to calibrating all sensors in the vehicle. Whenever a service job influences the sensors directly or their position, calibration needs to be performed. This is usually required when there is physical damage to the sensors, such as broken cameras or the structure which holds the sensors to vehicle frames, such as windshield damage, side-view mirror damage, and grill and bumper damage of the vehicle. A sensor replacement in the vehicle requires both intrinsic and extrinsic calibration to be performed. Mostly in the service centers, the calibration is performed either with reference to a target or without it (targetless calibration). In target-based calibration, the service center requires an infrastructure similar to that used at the EoL calibration facility at the production centers, which the vehicle sensors can use to measure and adapt for variations. The targetless calibration is usually performed by taking reference measurements and adapting them with reference to another sensor that can be trusted in the vehicle. In this case, the deviations are measured and the offset values are stored in the memory for the new calibration performed at the service center, replacing the previous values which were the offsets from the factory calibration. The sensors will use the new calibration values from here on for their operation.

8.2.3. Online Calibration

For any vehicle with multiple sensors on it, there is always a risk of losing their calibration or measurement errors from these sensors occurring randomly while in operation, which affects the system behavior. These random measurement errors because of calibration loss from sensors can occur because of many factors, of which few of them can be found in the P-diagram analysis in Figure 7.15. One of the most common factors which influence the sensor calibration is the vibrations associated with the vehicle. Vibrations in the vehicle usually displace the sensor from its original position and affect its internal parameters, and thereby errors are induced in their measurements. Wrong measurements from sensors will affect the detection and even the features of the vehicle. For safety reasons, the Automated Driving systems usually have a mechanism to detect these sensor calibration loss conditions and, thereby, shut down the feature availability if it affects the safety of those features with wrongly measured sensor outputs. Online calibration comes to play at this stage, depending upon the operating conditions. Once the system detects that a sensor has lost its calibration and provided wrong measurements, a Diagnostic Trouble Code (DTC) is triggered, and the information is stored in the system memory, which indicates that the calibration loss has occurred. The software calls for its response by initiating certain calibration routines, which runs algorithms that can identify the deviation in measurements and can apply the offsets or corrections while the vehicle is in operation under certain conditions.

The algorithms used for online calibration work mainly using two methods. One method is by considering previously valid calibration data as reference, either from EoL or Service calibration, and transformation of the current measurement data to those while in operation. The second method is by using another reference sensor overlapping with the field of view of the sensor, which lost its calibration. The detected objects from the reference sensor will be considered as references only if they are reliable and have not lost their calibration. Transformation of the measurements from the sensor which lost calibration is adapted to the measurements from the reference sensor with correction values or offsets. Different vehicle manufacturers have different methods and different prerequisites to initiate and complete the online calibration. From the end-user perspective, this online calibration initialization would look like advising the driver to drive the vehicle at a constant speed of 80 km/h or 60 km/h for 20 min, or even driving the vehicle for 50 km or 70 km before the sensors are calibrated. Having online calibration in the Automated Driving system helps keep the ADAS or Automated Driving features available for the user for a longer duration without a need to visit service centers. Calibration requirements will also bring the need for reliability and the system robustness during its development phase.

FIGURE 8.4 An overview of functional calibration in software.

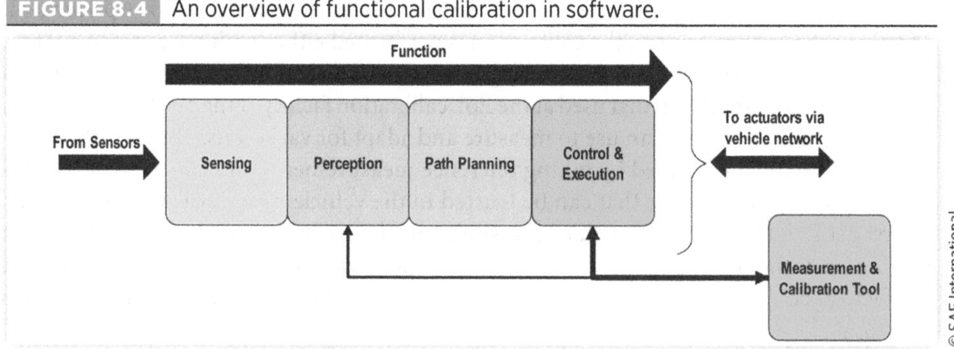

8.2.4. Functional Calibration

Calibration of sensors and actuators are the prerequisite for calibration of ADAS and Automated Driving functions in the vehicle. Unlike sensors like cameras or radar, actuators are always calibrated with the help of simulation environments and calibration infrastructures. Calibration of ADAS and Automated Driving functions are to be performed using structured methods and statistical tools. The parameters that need to be considered for functional calibration from the software part of the ECU are in thousands. The measurements and correction for errors in the software stack are applied mainly to the two main areas as shown in Figure 8.4: the control and execution parts and the perception.

The control and execution part of the software is calibrated mainly to pass on the control signals to the vehicle actuators, which is the main part of the functional calibration. The perception part is calibrated mainly to improve the sensor fusion and the environment model generation. The calibration of perception is performed by tuning the parameters of sensor fusion to have a reliable and accurate positioning of objects from the fused output. Calibration also covers fine-tuning of the mathematical model of the surrounding environment, including adaptation and position of static and dynamic objects with a reference. This does not mean that the other software components will not be calibrated. They are also calibrated, but mostly, calibration of those software components is covered during the software unit development and testing and not during the functional calibration.

8.3. Calibration of ADAS and Automated Driving Features

Calibration of features in the vehicle is initiated after calibrating sensors and actuators of the vehicle. In a vehicle with automated driving features, many subfeatures can be broadly classified as Assist functions, Safety functions, Comfort functions, etc. Calibration of all these functions is performed in the vehicle platform either through simulation or by driving.

When a supplier delivers features to a vehicle manufacturer, there might be less chance that a supplier will get access to the complete vehicle ECUs and the vehicle platform in his flexibility for testing and calibration. There are different platforms or environments used in the industry to calibrate different features starting from a DiL simulation, vehicle-in-the-loop simulation environments to the physical driving the vehicle on a proving ground and public roads for the calibration of different features. Calibration of the functions in an Automated Driving system follows either the bottom-to-top approach or top-to-bottom approach. The most common approach followed in the industry is the bottom-to-top approach where there might be ADAS features already introduced in the previous vehicle models of a vehicle manufacturer, and those are usually taken over as the existing functions which act as building blocks for the automated driving features. Calibration of those building blocks has already been performed in previous vehicle programs (lead program), which is considered as a reference. The calibration parameters are usually taken over from the lead vehicle program where these features are already deployed.

As the ADAS functions act as the building blocks for automated driving features in the vehicle, certain vehicle manufacturers directly move on with the calibration of the automated driving features as a top-down approach and then try to merge it in the middle. The disadvantage of this approach is that the feature interdependencies are not considered for the calibration and will bring additional calibration cycles to deploy the feature optimally. This is also time consuming compared to a systematic method of analyzing all interdependencies. The best approach to perform the calibration would be using the bottom-to-top approach if it is a new development and a hybrid approach of both top-to-bottom and bottom-to-top approach if the functions are being reused from another lead program. Theoretically, the hybrid approach might look weird, but this will help compensate for errors and adaptations for individual features during calibration and testing. The building blocks of the automated driving functions need to be identified, and the interdependencies should be clearly defined to start calibration of these features in the vehicle. Performing the bottom-to-top approach will bring clarity on the interdependencies and the interaction between different functional components, including the triggers from the user and vehicle for fail operation and fail-safe behaviors.

A feature dependency diagram and fault tree analysis are usually used to identify and establish the interdependencies between all the functions in the vehicle. An example of an automated driving feature decomposition is shown in Figure 8.5. The parameters that need to be calibrated for each function must be identified with further detailing of the feature dependency diagram with all the inputs and the associated functions from each block.

FIGURE 8.5 A sample high-level feature decomposition of highway pilot feature.

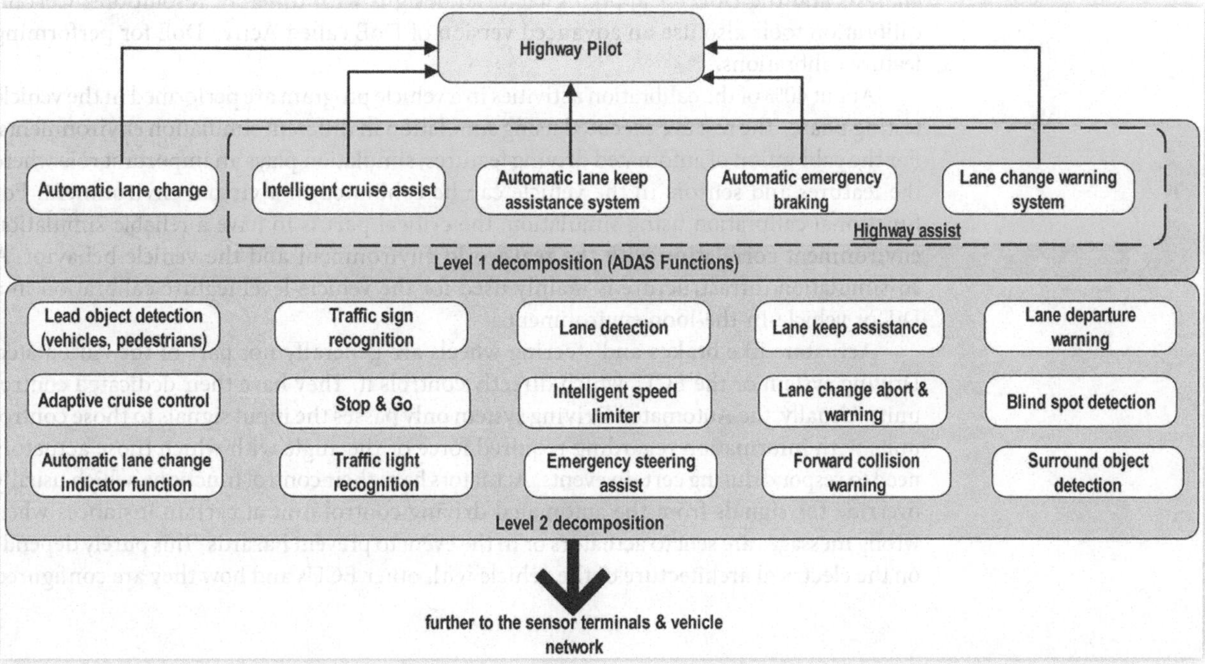

Many organizations used to develop the feature dependency diagram as part of the functional description and functional architecture, which can be used for calibration and testing.

The features can be further decomposed to their lower levels up to the sensor terminals to identify all the inputs. This will help to identify the functions involved and information about the dataflow path with which the calibration can be performed for each of the individual functions to the full feature at the vehicle level. As the parameters that need to be considered for an Automated Driving system in the vehicle are huge in numbers, it is impossible to control each parameter in the dataflow channel. Instead, one of the common approaches used for the calibration in these cases is performing calibration based on parameter prioritization and ranking according to its influence on individual features and their combined effect.

There are different methods used in the industry for the identification, prioritization, and ranking of these parameters associated with each of the features. As of now, in the industry, about 40% of the calibration is performed by considering measurements by driving the vehicle on a proving ground or public roads. The measurements from these drive campaigns are analyzed and adapted for parameter optimization and feature improvements in the vehicle.

The standard statistical methods used to calibrate features in the vehicle are DoE and Active DoE method that uses symbolic regression [8.7, 8.8, 8.9]. This method will help collect data samples and identify the controllable parameters with which the function behavior can be controlled. The main challenge in using DoE is its complexity when the parameters and the levels associated with those are more in number. The complexity will increase with the number of parameters under consideration for each function and the number of samples required as part of the experiments. Fischer's principles are applied for the sample collection, randomization, and blocking whenever the experiments are performed. For reducing the number of experiments to be executed with a wide range of parameters, various factorial designs are used and applied for sample data collection and analysis, depending upon the parameters considered. Statistical tools are used for the analysis and the DoE for applying factorial designs with different resolutions. Certain calibration tools also use an advanced version of DoE called Active DoE for performing feature calibrations.

About 40% of the calibration activities in a vehicle program are performed in the vehicle testing phase. The rest are executed using simulation in different simulation environments. For the calibration of automated driving features, simulation plays an important role where the features and sensors in the vehicle can be calibrated in a virtual environment. For functional calibration using simulation, the critical part is to have a reliable simulation environment correlating with the real-world environment and the vehicle behavior. A co-simulation infrastructure is mainly used for the vehicle-level feature calibration in a DiL or vehicle-in-the-loop environment.

Actuators like brakes and steering wheels are generally not part of the Automated Driving system or the ECU, which directly controls it. They have their dedicated control units. Usually, the Automated Driving system only passes the input signals to those control units with information regarding required force or the angle with which those actuators need to respond during certain events. Actuators have their control functions, which usually override the signals from the automated driving control unit at certain instances when wrong messages are sent to actuators or in the event to prevent hazards. This purely depends on the electrical architecture of the vehicle with other ECUs and how they are configured to operate with priority. The calibration of vehicle actuators is also performed in different environments such as vehicle testbeds and also by physical driving of the vehicle.

Whatever be the configurations in the electrical and electronics architecture of the vehicle, the calibration logic remains the same. The methodical approach for calibration will help in having better control of parameters and helps early in the development and testing phase.

8.4. Calibration Environment for Automated Driving Vehicles

Calibrations inside the laboratory environment or by driving the vehicle require a calibration strategy to be defined first. This includes the plan to collect data and measurements and analyze and optimize parameters. Depending upon the calibration strategy, one could decide how to organize and execute these sample data collection tests and use these to decide on calibration parameters which are used as correction parameters to improve the overall system behavior. Calibration is required to be performed by both the system supplier as well as the vehicle manufacturer.

Two types of test environment or infrastructure for the calibration should be considered, depending on whether it is performed by a system supplier or by the vehicle manufacturer. A system supplier who provides an Automated Driving system is concerned only about the system, sensors, and signals going out from the automated driving ECU to the actuators. On the other hand, the vehicle manufacturer needs to perform vehicle-level calibration, including all the ECUs associated with the feature, which includes the ADAS or Automated Driving ECU and the Actuator ECUs. The boundary considered for calibration by a supplier is a subset of the overall required calibration from the vehicle manufacturer's side, as shown in Figure 8.6.

The calibration tools and infrastructure at the system supplier and vehicle manufacturer depend on the boundary of calibration performed from their side. For example, a co-simulation environment in DiL simulation is the most common test environment used for calibration of the supplier's side for different ADAS and Automated Driving features. On the other hand, a vehicle manufacturer uses a vehicle-in-the-loop simulation or the complete vehicle test environment, commonly termed as Labcar, as a calibration environment. Still, most calibration tests and data collection will be performed by the vehicle manufacturer in the vehicle while driving on a proving ground or public roads. Thus, in a production program, it is imperative for a vehicle manufacturer to provide the vehicles for test and calibration to system suppliers delivering the ADAS or Automated driving features. Otherwise, the system suppliers would mostly stop the testing and the

FIGURE 8.6 A general calibration boundary for a vehicle manufacturer and its supplier.

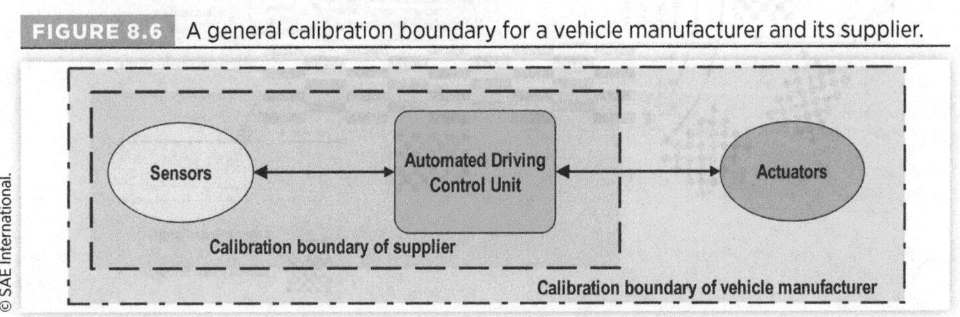

calibration of the function at the boundary where the signals are sent out from the ECU. Thus, the feature would not be complete for the full vehicle. Thus, not testing or calibrating the actuators is part of functional validation.

Calibration starts with the sensor calibration and sensor fusion calibration involving all the sensors in the vehicle. The ADAS/Automated Driving ECU or the domain controller usually initiates this. Calibration of sensors can be performed in the simulation provided that the sensor models which are used in the simulation environment match with the actual sensors used in the vehicle. It is difficult to have the same sensor models used in the vehicle in a simulation environment which is a known challenge. Without having actual sensors in the simulation, the only possibility would be to use a standard sensor available with the simulation and check if those can be calibrated with the calibration algorithm. During the development, this would help check the calibration functionality, but that will not guarantee it would behave the same in the actual vehicle. Unfortunately, there are cases where some of the targetless calibrations for the sensors, like cameras, are tested only in simulation with standard sensor models and have been used in production programs.

Vehicle manufacturers and some of the system suppliers of ADAS and Automated Driving systems commonly use a calibration environment with Labcar. Here, the complete vehicle sensors are positioned like the actual vehicle and simulate the vehicle in motion. Figure 8.7 shows a vehicle-in-the-loop simulation environment that can be used to perform different calibration tests such as EoL, Service calibration, and Online calibration with or without targets [8.10]. A moving vehicle can be simulated in such an environment, and feature calibration can also be performed.

The advantage of using a vehicle-in-the-loop simulation environment for calibration is that the real vehicle sensors and actuators of the vehicle can be used for calibration in a laboratory environment. The driving effect can be simulated through wheel rotation using motors attached to the wheels. This will help calibrate both sensors, actuators, and the ADAS or Automated Driving system in a laboratory environment, thereby reducing the amount of drive testing on roads or proving ground for calibration in the vehicle. Many other use cases for a vehicle-in-the-loop simulation environment were discussed in the second chapter.

FIGURE 8.7 A vehicle-in-the-loop simulation environment used for calibration testing.

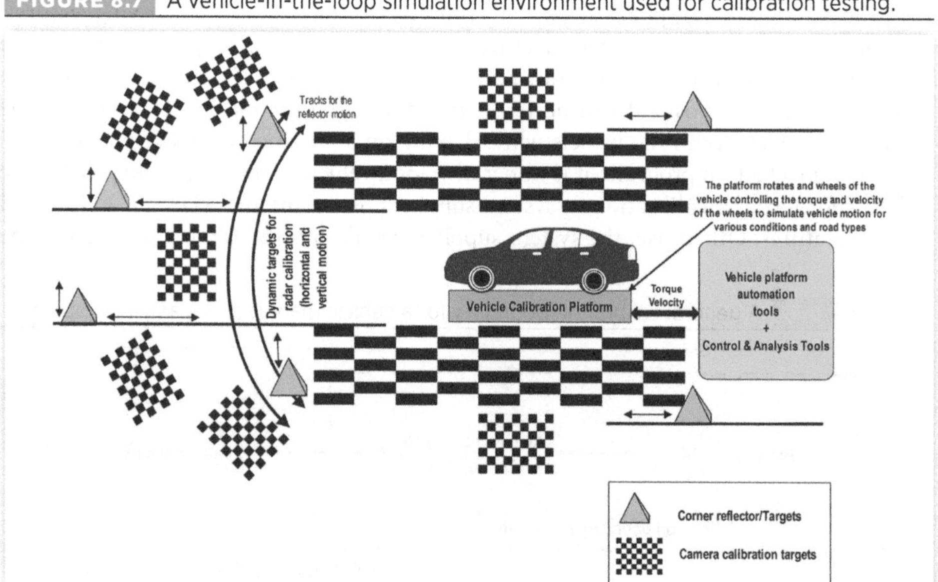

Even though there are different methods used in the industry for calibration of the sensors, actuators, and functions, it is still challenging with the automated driving vehicles and their calibration for large-scale production environments. The challenges are mainly because of the number of sensors integrated into the vehicle and the rapid change in those sensors. Now the production calibration environment of many vehicle manufacturers uses different types of targets focused on each sensor type. Specific sensors like lidars are solely dependent on their supplier calibration information, and the vehicle manufacturers have less control over calibrating those sensors. Above all, the time taken for the calibration of all the sensors and the features in the vehicle is a major concern. With the current methods, the time taken for EoL calibration using targets with multiple sensors consumes hours, depending on the scope of calibration, which is not acceptable in a production line. Feature calibrations are primarily performed with a subset of scenarios and their operational conditions [8.10]. Online calibration of sensors and sensor fusion with many sensors is complex to manage, and cases of algorithm rerunning calibration routines are usually because of measurement glitches. From the end-user perspective, this creates dependency for the vehicles with service centers for recalibrations and they would face limitations in using assisted and automated driving features inside the vehicle because of sensor calibration loss. These are the major complaints the vehicle manufacturers receive these days with their automated driving features on the road.

8.5. Calibration over Diagnostics Interface

Diagnostics refers to the health monitoring and reporting mechanism in a system. This helps identify errors or failures and provides the current and previous status of different systems in the vehicle. All these capabilities of diagnostics are achieved by utilizing certain diagnostic services. In an Automated Driving system, the diagnostic services can follow up on the health of the system and all associated subsystems, the configuration of those, and identify errors and failures, if any, through DTCs. Different diagnostic services are used to calibrate the system (ECU), sensors, and functions. Depending upon the topology of the automated driving control unit in the vehicle network, the interface to an ADAS or Automated Driving ECU can be established by an On-Board Diagnostic (OBD) interface externally, as shown in Figure 8.8. Whenever the calibration routines are initiated, the diagnostic services measure and calculate the correct offset for the measurement from sensors (for sensor calibration) based on specific targets as a reference and can be stored in the memory of the ECU. These offset values are used for calibration of the sensor measurements.

Calibration is usually initiated via a diagnostic interface. This is generally called the calibration routine, where a sequence of reading and diagnostic writing services are executed, and measurements are made regarding a standard reference or a target. These measurements are checked for their correctness, and the offset values are calculated for correction in measurements and stored in the control unit memory. Whenever the readings are available from the sensors, these offset values will be used along with those measurements to improve the accuracy of the measurements and for processing the output data. Offline calibration of sensors and sensor fusion is performed by initiating a calibration routine through a diagnostic interface. This is usually performed in EoL calibration and the Service calibration where the OBD interface access is there for the technician with which they could configure the system in the production environment or service centers.

FIGURE 8.8 Diagnostic topologies of the ECU with OBD module.

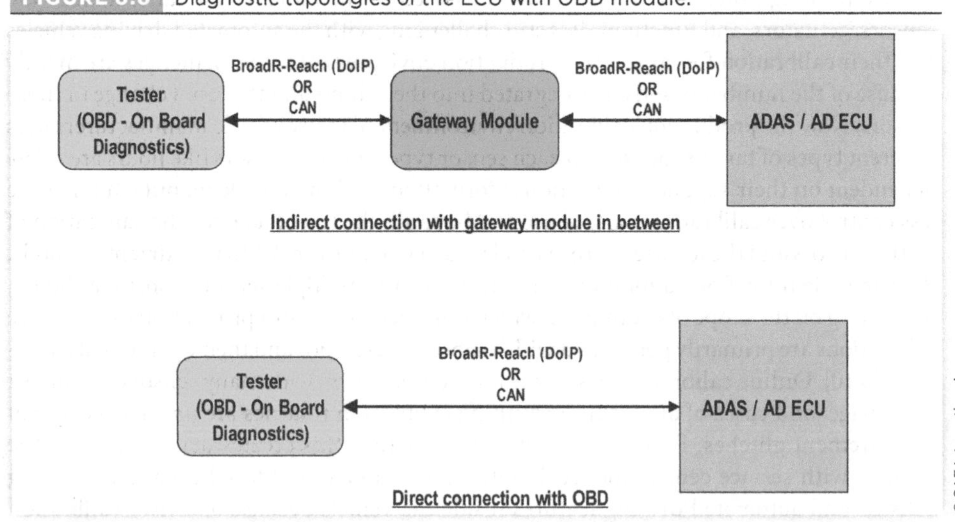

ADAS and Automated Driving systems are designed so that the calibration status of the system and subsystems are checked at regular intervals. Whenever there is a notification of calibration loss, the ADAS/Automated Driving control unit initiates the calibration routines, depending upon the operating condition and the status of the vehicle. Online calibration has a dependency on the environment as well as the state of the vehicle. On every ignition cycle of the vehicle, the control unit usually checks the calibration status of all the sensors and actuators associated with the ADAS or Automated Driving ECU. This is an input for the ECU to decide whether the ADAS or Automated Driving features should be enabled or not for the user and take the necessary steps for executing calibration routines based on vehicle status. When the calibration routines are executed, new offset values are calculated and the memory will be updated with these new values replacing the old ones.

Calibration of the features is usually performed while driving. Some of those include calibrating the oscillations lane-centering function and lane keep functionality on reducing the oscillation of the vehicle in different road geometry, adapting the speed of acceleration to stay within the allowed threshold while engaging cruise control in different road types, adapting the TTC calculated for the AEB and collision warning, etc. Calibration is also required for the systems which are in use for a longer duration in the market. This is to compensate for the wear and tear of the system components and their responses. Because of the wear and tear of the systems, the expected output from those would vary, which needs to be compensated. For example, the wear and tear of brake pads affect the braking effect while applying the brakes. And the calibration would help to compensate for the force applied for braking with worn-off brake pads.

8.6. Summary

This chapter provided an overview of what calibration means and what its benefits are. Since there is no ideal system or ideal environment for operating a system in the vehicle, the features which are integrated into the vehicle need to be adapted or corrected for its optimal operational performance, compensating for all the errors in measurements from different subsystems. This includes identifying the errors in measurements and correcting those in comparison with standard references or targets. Different methods are used to calibrate ADAS and Automated Driving systems in the vehicle, and a few of them are covered in this chapter.

The interdependencies of the system with its subsystems should be identified, and the subsystems of the Automated Driving systems like sensors and actuators are to be calibrated before calibrating the features at the vehicle level. Calibration of the subsystems like sensors would help to achieve optimal input for the automated driving ECU with which it can process the environmental data with more accuracy and precision. The mechanical positioning or the extrinsic calibration of the sensors to the vehicle coordinate system plays a vital role in the coverage around the vehicle and the functions dependent on those sensors. Thus, extrinsic and intrinsic calibration of different sensors and the methods used for calibrating them are important. The importance of having a calibration strategy for the system and the features in the vehicle is the first step in calibration, where different methods are used for measurement and analysis. Depending on the calibration strategy, different methods can be applied for parameter selection, sample collection, and parameter optimization of the system and its functions.

DoE, Active DoE, and methods using Fisher's principles were discussed for the measurements and analysis for calibrating the system and features in the vehicle. For the calibration of automated driving features in a vehicle, the interdependencies must be identified. The functions that act as the building blocks for higher levels of automation need to be calibrated independently before calibrating the high-level functions. The importance of diagnostic interface in calibration and how different topologies for diagnostics are for calibration were discussed.

Autonomous driving systems and the features associated with them make the calibration challenging. The number of parameters which has to be considered for calibration is enormous. Different methods help to control the system by identifying and controlling specific parameters that influence the overall system behavior. The selection of methods and parameters for calibration is challenging unless it is not considered early in the design phase of product development. Calibration also has a secondary responsibility of keeping the system safe by identifying and correcting measurement errors. Thus, the data measured and the correction offset values stored in the memory have a relationship with the safety of the system. During the verification and validation phase, the tests related to calibration are executed as part of safety-relevant testing, which influences the safety of the system and the features of the vehicle.

References

8.1. Altshuller, G.S. and Shapiro, R.B., "Expulsion of Seraphyme with Six-Wings," *Inventor and Rationalizer* 10 (1959): 20-30.

8.2. Altshuller, G.S., *How to Learn to Invent*, 1st ed. (Tambov: Tambov Book Publishers, 1961)

8.3. Altshuller, G.S., "Fundamentals of Invention," Centr. Chernozem Books Edition, Voronezh, 1964.

8.4. Kraev, V. and Kraev, K., "System Ideality-Lesson, Volume 5 Crossref, Vladimir Petrov, Avraam Seredinski, Progress and Ideality," *TRIZ-Journal*, Posted, 2006.

8.5. Petrov, V., "Law of Increase in a Degree of Ideality," Tel Aviv, 2002, accessed August 10, 2021, www.trizland.ru/trizba/pdf-books/zrts-08-ideal.pdf.

8.6. "Ideality," accessed July 4, 2021, https://triz-journal.com/progress-ideality/.

8.7. Stanley, J.C., "The Influence of Fisher's 'The Design of Experiments' on Educational Research Thirty Years Later," *American Educational Research Journal.* 3, no. 3 (1966): 223-229, doi:https://doi.org/10.3102/00028312003003223.JSTOR 1161806.S2CID145725524.

8.8. Yates, F., "Sir Ronald Fisher and the Design of Experiments," *Biometrics* 20, no. 2 (1964): 307-321, doi:https://doi.org/10.2307/2528399.JSTOR2528399.

8.9. Box, J.F., "R. A. Fisher and the Design of Experiments, 1922-1926," *The American Statistician* 34, no. 1 (1980): 1-7, doi:https://doi.org/10.2307/2682986.JSTOR2682986.

8.10. Persic, J., "Calibration of Heterogeneous Sensor Systems," arXiv preprint arXiv:1812.11445, 2018, http://arxiv.org/abs/1812.11445.

/Autonomous
/Sensing
/Communication
/Battery
Autopilot mode

9

Introduction to Functional Safety and Cybersecurity Testing

Functional Safety and Cybersecurity have critical roles in developing and deploying ADAS and Automated Driving systems. As they are complex products that can even influence the lives of their users, they have to be safe and secure for usage. Testing those products should also focus on assuring safety and security while they are in operation. There are a lot of standards in the automotive industry, which focus on the product safety, operational safety, security, etc. of automotive systems. These standards provide exemplary practices and recommendations to follow during the design and for the complete life cycle of the products. Are we considering these recommendations while developing a product, especially when the vehicle programs take only about three years rather than five or six years or even more in the past? This chapter will take you through some of the methods and approaches followed in the industry, highlighting specific challenges associated with those that affect the safety and security of the products developed these days.

This chapter will mainly focus on a smaller set of standards and their use in the verification and validation of Automated Driving systems. Those standards include ISO 26262 "Road vehicles—Functional safety" related to functional safety and SAE J3061 "Cybersecurity guidebook for cyber-physical vehicle systems" and ISO/SAE 21434 "Road vehicles—Cybersecurity engineering" standards for cybersecurity. For the security associated with ADAS and Automated Driving features, ISO/PAS 21448:2019 "Road vehicles—Safety of the intended functionality" focuses on the functions in the vehicle. For ADAS and Automated Driving, functional safety is critical, and so is ISO 21448 SOTIF. It needs to be followed in addition to compliance to established functional safety standards like ISO 26262. While ISO 26262 covers functional safety in the event of system failures, it does not cover safety hazards that result without system failure, which is covered in the ISO 21448 SOTIF standard.

9.1. Functional Safety and Cybersecurity in Automotive

Functional safety and cybersecurity standards recommend its parallel design and development processes which run parallel to the classical product development V-model. It is possible to integrate all these different process cycles with the product development process, even though they are defined differently. In reality, this is not the case. In most automotive projects, which are developing safety-critical products, the safety life cycle is always delayed, and the deliverables from the safety processes are not provided in time as required or recommended by any of those processes. Those who have worked on projects can recognize this better. This is the same for both the functional safety and cybersecurity processes, which delay system development and testing requirements.

ISO 26262 recommends starting safety activities with the safety plan where the safety lifecycle and safety management activities are defined. It is suggested to have an item definition as the starting point for the Functional Safety activities. Item definition plays a critical role in different phases of design, analysis, and development in the automotive ecosystem with multiple levels of suppliers and vehicle manufacturers [9.1]. Projects are even executed without having a defined item definition and instead use reference documents that have similar information for starting the functional safety activities in the project. It is a hidden fact that most suppliers might not have access to the item definition or the safety goals identified and defined by the vehicle manufacturer. Hence, during the verification and validation phase, these suppliers perform tests purely based on the available technical requirements and the available minimum information on safety requirements. Even though it is hard to believe and a pure deviation from the recommended processes by the functional safety standard, this is the reality happening in most of the projects in the industry these days.

Safety Element out of Context (SEooC) is a common term in functional safety for ADAS and Automated Driving systems. This is the method of using components inside the vehicle in safety-critical systems which are not defined initially and designed for the specific application as defined in standard ISO 26262-10. SEooC concept can be considered a vital part of implementing ADAS and Automated Driving functions. These days, a lot of software components are reused from different industries and various products. There are mainly two types of SEooC components. One is those software components that can be used after evaluation based on its field data. The typical term used is the "Proven-in-use" concept to consider the maturity and reliability of the software components because of their long-time usage in a product. The second one is the development of SEooC according to the standard ISO 26262-6. The software can be developed following the standard based on assumptions that need to be proven in the target application. This is a common approach seen across many projects for the software development of ADAS and Automated Driving systems. Even though these software components are adopted from different applications or developed according to the standard, a thorough evaluation must be performed before using them in a safety-critical product. In the industry, the SEooC concept is widely misused. Many open-source software components and software libraries from other industrial products are reused in ADAS and Automated Driving systems, which give an advantage of time, effort, and money during the product development. This is misused and considered a reason for not performing analysis and not collecting relevant data that the target application will be safe enough through testing. This should be avoided, and the SEooC components should also pass through the test cases specifically designed for the application.

For integrating functional safety and cybersecurity in automotive projects, most of the analyses are happening at the initial design phase, including requirement generation [9.1, 9.2, 9.3, 9.4]. On the other hand, the verification and validation part of functional safety is usually evaluated only by inspecting if all the requirements are going to be traced by test cases, coverage of tests, and if the results are in an acceptable state. Even the acceptance criteria defined for various product release phases will be associated with the number of safety-critical defects open to system release decisions. This is also occurring because of the established Development Interface Agreement the level of information shared between the vehicle manufacturer and the supplier is controlled. The testing for safety and security is usually focused on the requirements generated by different analyses. The industry's most common techniques used for safety-relevant tests are fault injection tests and mutation tests [9.5, 9.6, 9.7]. Fault injection tests focus on the system and software response during the event of a fault, and the mutation tests focus on the testing for failure because of the mutation of the software or the system components by external influences. Both play hand in hand when applied at different levels of testing in product development.

In most of the projects, different test techniques are applied randomly in the test execution phase. The drawback of random application of test techniques is that it will limit identifying all possible failure modes in the system and not performing the tests in the required depth to identify those failure modes [9.8, 9.9]. In almost all cases, the functional safety evaluation of a product during its testing phase checks if the requirements related to the functional safety are tested and if the safety goals were met or not. The methodology to generate the tests or the strategy about different test techniques is not usually evaluated, which adds up risk in testing functional safety requirements.

Testing cybersecurity requirements is also a gray area that is not followed as defined processes for different projects in different organizations. During the design phase, the requirements are generated by different analyses of the system and the possible cybersecurity threats. This is used to test the system, which would risk performing minimal testing in the product [9.8]. The methodology used to analyze the threats and the attacks are always dependent on its environment. Suppose a tier supplier defines the cybersecurity requirements for a critical product in the vehicle without knowing the vehicle platform or its interfaces. In that case, it is irrelevant and not applicable as the threat analysis from the supplier will be very much limited to what the supplier knows about the overall product and not considered for the complete vehicle.

Similar to functional safety, the cybersecurity process starts with an item definition of its own. In some cases, vehicle manufacturers create a global item definition, including functional safety and cybersecurity. As certain factors overlap between these domains, both these approaches can be found in the industry [9.2, 9.9, 9.10]. If the overlapping factors between functional safety and cybersecurity were not analyzed during the design phase, there would be flaws in the design for the safety-related system failures triggered by cybersecurity attacks. One should consider these during the analysis phase so not to miss them. This should be one of the major activities covered during the Hazard Analysis and Risk Assessment (HARA) and Threat Analysis and Risk Assessment (TARA), the analysis phases in functional safety and cybersecurity.

In the last few years, cybersecurity has gained traction in the automotive industry as vehicles have changed from classical automotive to software-driven systems with a lot of connectivity and digital products. As the connectivity inside the vehicle is growing like a network of computers running complex software, so does the threats from which these systems need to be secured. Because of rapid growth in technology and different connectivity inside the vehicle, the ways in which the vehicle needs to be secured to prevent any data loss or unauthorized access are also complex. The vulnerabilities associated while using new technology are also there with the current products inside the vehicle. There are many ways in which the vulnerabilities of these systems can be exposed while testing. Correct methods, processes, and tools for the development and testing are the first step in building a secure system. The next step is to maintain the security continuously with updates and fixes.

The verification and validation process for functional safety and cybersecurity are challenging. Usually, this is one of the most complex parts to meet the required specifications from the standards for any production program. Mostly the tests for functional safety and cybersecurity are only executed based on the requirements. As these requirements are the outcome of functional safety and security analysis, the quality and depth of available requirements also depend on the team and the information available to them about the vehicle ecosystem. Beyond the product development processes and the testing for functional safety and cybersecurity requirements, management of functional safety and cybersecurity within an organization influences the products developed [9.1, 9.2, 9.8, 9.10]. The functional safety and cybersecurity management processes, methods, and the team's skillset in an organization give an understanding of how the development processes are followed and what would be the outcome in terms of product quality.

This chapter aims to take you through different test methods and approaches followed in the industry for testing functional safety and cybersecurity areas on safety-critical systems such as ADAS and Automated Driving systems.

9.2. Safety Qualification of Tools and Toolchain

It is usual for a test engineer to depend on multiple tools for testing. The tools and their complexity also increase when complex systems are to be tested. Generally, the tool consists of software and, if needed, additional hardware along it. As per definition, tools support product development in all relevant phases, even though they are not part of the product developed, involved in the manufacturing process, or used as manufacturing equipment. For complex systems such as ADAS and Automated Driving systems, the number of tools one must consider for verification and validation is enormous and mainly used as a toolchain. A set of tools that are collectively used to achieve the purpose of developing a product is called a toolchain. In the toolchain, specific tools are used to serve specific purposes like testing the safety and security of the system, which is the main focus of this chapter.

Unfortunately, in the development phase of ADAS and Automated Driving systems, it is a fact that most of the new tools used are not safety qualified, such as environment simulation tools, feature modeling tools, etc. Most of those tools are selected without proper analysis or evaluation in many organizations. This is a significant challenge in the industry as these tools are even used without qualifying for testing and evaluating the safety-critical products.

Different test methods and types used these days for testing ADAS and Automated Driving systems require different types of tools to qualify them. The major challenge faced these days is either there are no available tools that could serve the purpose or the available ones may not be safety qualified. On the other hand, many prequalified tools from various suppliers are available in the market that is usually used for software development and low-level software testing. Even the safety qualified tools that various vendors sell have to undergo a self-qualification with the help of tool documentation and based on the application before use to make sure that the planned application of the tool is not violating the scenarios in which the tool is qualified. This gives two options, such as the tools that can be procured, which are qualified, or the user can qualify it for specific use cases following certain conditions. Usually, vehicle manufacturers and their suppliers go with the second approach for building the toolchain and infrastructure for testing ADAS and Automated Driving systems.

A method used in the industry for qualifying simulation tools was discussed in Section 5.3 in the previous chapter. Those are mostly the methods used for the general qualification of the tools and toolchains. For the safety qualification of the tool, there is a recommended method from standard ISO 26262 "Road vehicles—Functional safety." It refers to specific parameters to be considered for qualifying the tools and their reliability while using them for qualifying safety-critical systems. A comparison of the tools and toolchains used for a specific application is shown in **Figure 9.1**.

According to ISO 26262-8, three parameters are to be considered to qualify the tools for use in developing safety-critical systems, they are:

1. Tool error detection (TD)
2. Tool impact (TI)
3. Tool confidence level (TCL)

These three parameters can be considered measurable parameters that qualify a tool for its use in a safety-critical environment in product development [9.1, 9.10]. TD is the parameter that provides the probability with which the tool can detect or avoid errors in its defined usage process. This is mainly the parameter that provides the tool capability by evaluating the usage process and the mechanism in the tool, which monitors its health and provides a warning or even ceases operation whenever there is an error in the tool. This will

FIGURE 9.1 A comparison of using a tool and toolchain for testing.

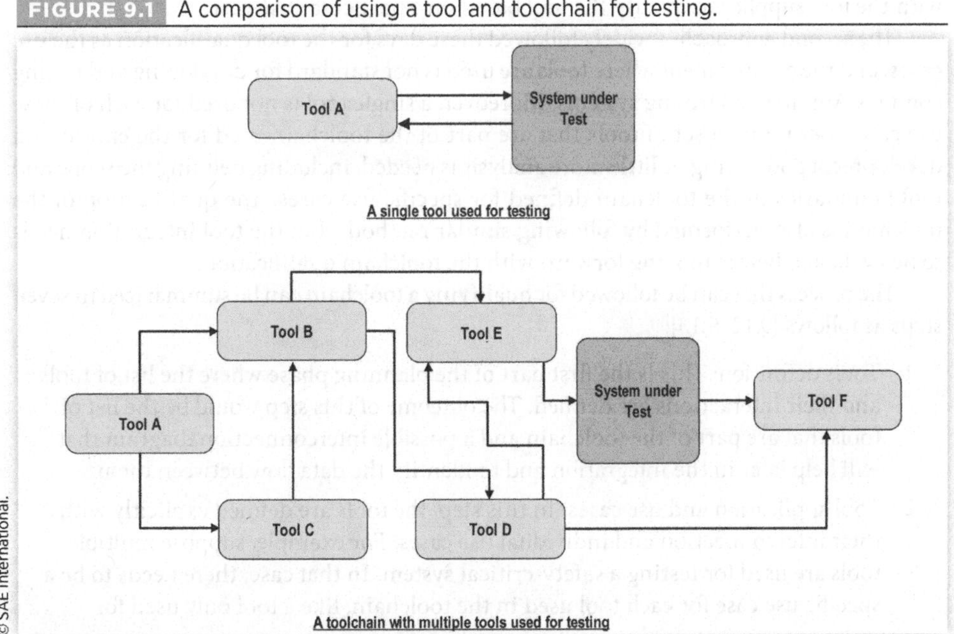

prevent the tool from passing wrong information as input to the system under test, and it prevents providing wrong measurements as output to the user.

Depending upon the measurements and capability of the tools, there are three levels of TD possible for the tool. This will be based on running certain experiments with the tool of a particular usage and defined workflow [9.1, 9.11]. TD1, TD2, and TD3, where TD1 is high detection probability (>0.7), TD2 medium probability (0.4-0.7), and TD3 low or unknown probability (<0.4). These ranges are also flexible and are usually defined based on the tool in use and the maturity of the process followed for development and testing. The measurements for calculating the probability of TD are made with a standard prequalified system following a defined use case as per the tool qualification plan. The input and output measurements are collected for different ranges of inputs, including error conditions to analyze and classify to different probability ranges. There can be a situation where the same tool can have different error detection probabilities based on its operating ranges.

TI is the parameter that is defined as the possibility that a malfunction of a particular software tool can introduce or fail to detect errors in a safety-related item or element being developed. This parameter provides the measurement of how a tool will influence a safety-critical system. This parameter holds two values, TI1 for the tool with influence and TI2 for those that do not have any influence on the system. With these two parameters, TD and TI, with their possible values based on different classification, a 3×2 matrix can be used to calculate the TCL for particular tool usage. A tool can have different confidence levels allocated as TCL1 for a low confidence rate, TCL2 with medium confidence, and TCL3 for high confidence based on specific purpose and usage. A particular tool can be rated with a different confidence level in different use cases. The change in TCL depends on the use case of the tool and its impact on the system under consideration; hence, the scope and purpose of the tool on specific use cases play an essential role in allocating the TCL.

There are two approaches seen in the industry for the safety qualification of the tool. The tool supplier qualifies the tools and provides the report, and a delta evaluation and qualification are performed based on its application, which is the old method. The second one is the tool supplier provides the tool, which is not qualified, and the user needs to work with the tool supplier to qualify the tool based on a particular use case.

The second approach is widely followed these days for the tool qualification as the use cases, and the environment where tools are used is not standard for developing and testing complex Automated Driving systems. Moreover, a single tool is not used for each of these use cases, but rather a set of tools that are part of the toolchain used for the end-to-end development and testing. A little more analysis is needed, including defining the scope and tool boundaries in the toolchain defined for specific use cases. The qualification of the toolchain is also performed by following similar methods, but the tool integration needs to be evaluated before moving forward with the toolchain qualification.

The process that can be followed for qualifying a toolchain can be summarized in seven steps as follows [9.12, 9.13]:

1. Tools definition: This is the first part of the planning phase where the list of tools and their interactions are defined. The outcome of this step would be the list of tools that are part of the toolchain and a possible interconnection diagram that will help later in the integration and to identify the data flow between them.
2. Tool application and use cases: In this step, the tools are defined explicitly with their interconnection and individual use cases. For example, suppose multiple tools are used for testing a safety-critical system. In that case, there needs to be a specific use case for each tool used in the toolchain, like a tool only used for

providing parametrization, another tool for analysis of the results by different statistical methods, etc. Those functions of the tools must be listed, and any part of the tool reused from an existing tool infrastructure must be mentioned. Usually, all the tools get integrated with a test framework that has been used in the organization as a test platform. This might bring specific conditions to consider as certain tools are reused and few tools are newly introduced to this framework. The user guide for the tool, along with the list of limitations, errors, and possibly other constraints, would be input during this phase.

3. TI calculation: All the tools used in the toolchain must be evaluated separately with their interfaces to measure TI on the system under consideration. The evaluation is performed to identify if any of those tools in the toolchain injects any error into the development or testing of the system or if the tool misses in detecting any errors while using it for system testing. The ratings for each tool need to be allocated based on analysis and even by executing sample tests or experiments for data collection and analysis.
4. Error identification: The error identification of the tools has to be evaluated separately after performing a logical validation of the tool and its use from the interconnection diagram. There are two components here that need to be evaluated. One is the error that can occur from the tool and the error that can be introduced by an operator using the tool. The error from the operator should be evaluated as part of the TI phase where it would provide information if the tool has a mechanism to prevent the error introduced by an operator of the tool or not. Each tool must be evaluated for its accuracy with different data types, precision, stability, and linearity over its complete operating range. Statistical analysis, including the error calculation for repeatability and reproducibility, must be performed and analyzed. The results are used to prove that the error exists and the tool capability to identify those overall operating areas is valid. Depending upon the different use cases, mainly the file type compatibility between tools, parsing the decimal points, converting the data to different formats, etc., are a few of the common errors one could identify in a toolchain during its evaluation. Once the error analysis is completed with the proof from available data, the operating limits and functionality limits of each of the tools in the toolchain can be defined.
5. Define checks and restrictions: Once when the analysis for the error conditions in the tools are completed. The results will provide information on the operational conditions and their limitations, and specific use cases for which the tool would not be suitable are identified. This is the step that will help in deciding the TD capability. Based on the analytical data, the probability of the occurrence and the chance of detecting each of those identified errors are allocated. The probability values are assigned based on the capability of the tool to detect it and evaluate its impact on the specific use case which it should serve. The whole process of probability allocation is similar to the risk evaluation method or using the FMEA tool. It is not a single-person activity to establish the probability values for different tools in a toolchain.
6. TCL calculation: TCL is calculated for each of the tools in the toolchain. An overall confidence level is estimated for the complete toolchain. Practically, TCL has a strong relationship with TD. The TCL is mostly allocated based on assumption rather than using a data-driven approach. If the error identification process

involves quantitative measurements, the TCL must be allocated based on available experimental data rather than assumptions.

7. Evaluation results: The whole process needs to be captured, and the evaluation methods and the proof of evaluation must be recorded. The results must be kept as the reference and a basis for allocating any confidence level to the tool and the toolchain. The evaluation results might get checked as part of specific safety assessments for the project as well as this can be a reference in the future for any reuse of particular tools and toolchain for any new product development or testing.

These seven steps describe the qualification of the tools in a toolchain for specific use cases [9.1]. It includes both qualitative and quantitative approaches recommended from functional safety standards (ISO 26262) as well as the MSA as part of Six Sigma methods [9.14, 9.15]. The acceptance criteria for the errors concerning different data types as recommended by Six Sigma processes are as in Table 9.1. These values were considered according to the recommendation by the AIAG (2002) for the manufacturing environments. This was taken as a reference for the tool qualification, and some required adaptations can be considered for simulation tool qualifications if their integration tolerances with the test framework are known. The evaluation methods also vary depending on the type of data each tool processes and the sample data collection mechanism used for each tool in the toolchain evaluation.

The development process and the qualification of the tools before their deployment to use in the testing or development of the safety-critical system are necessary and affect the system itself. A typical scenario seen in many new product developments is usually an introduction or replacement of an existing tool with a new tool in a qualified test framework, as shown in Figure 9.2. In this case, the delta qualification of the new tool is to be considered depending on the use case and the application. Tool qualification is a rigorous process, and it is not expected to be performed frequently for the software updates in the tools and the framework. But an inspection on the updates for the changes and trial runs are to be considered [9.13, 9.14].

TABLE 9.1 Criteria for acceptance on tool measurements as per MSA.

			Criteria		
Data type	Error classification	Evaluated KPI	Accept	Accept with caution	Reject
Continuous data type	Accuracy	% Bias % Linearity	Below 5%	Between 5% and 10%	Greater than 10%
	Precision	% Gage R&R	Below 10%	Between 10% and 30%	Greater than 30%
	Resolution	Number of distinct categories	Greater than 10	Between 4 and 10	Less than 4
Discrete data type	Accuracy	% Appraisers w.r.t standard reference	Greater than 90%	Between 70% and 90%	Less than 70%
	Precision	% Within or between the measurements	Greater than 90%	Between 70% and 90%	Less than 70%

FIGURE 9.2 Boundary definition for a new tool in a toolchain for qualification.

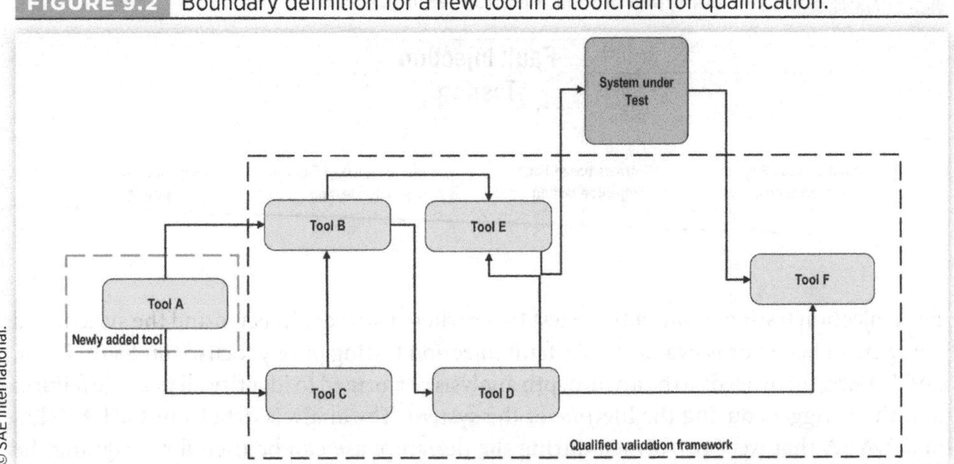

9.3. An Overview of Functional Safety Testing

Having qualified tools and toolchains for testing helps achieve reliable results and quality output while evaluating a system. This section will take you through different methods used for safety testing in ADAS and Automated Driving systems. The main goal of testing for safety is to prove that the functional safety concept and the technical safety requirements are met and tested in the system and its functions [9.1, 9.10]. Rather than considering the safety testing as individual test cases for requirement-based testing and mapping those test cases to technical safety requirements, the concept of test design must be applied. Testing for safety has to be considered in different levels of testing, starting from the unit software testing and hardware testing until the system verification and validation. Different test environments, including various tools and test methods, have been used at different levels for safety-related testing [9.16].

There is a misunderstanding that fault injection and mutation testing are the only test types used for safety testing. Testing for the safety of the electrical and electronic components and the software starts even from the model-based development and testing. The functional safety standard ISO 26262-4, 5, and 6 recommends multiple test methods and approaches for qualifying the system, hardware, and software for meeting the test requirements for different Automotive Safety Integrity Levels (ASILs) [9.1, 9.10]. Testing for functional safety covers both a process-oriented approach and the application of different test methods. This includes the development processes, reviews, and different test levels, including verification and validation processes. The processes followed for product development mostly depend on the organizational process maturity.

The complete functional safety spins mainly around three terms: Fault, Error, and Failure [9.1, 9.10]. Fault refers to an abnormal condition that may cause a reduction, loss, or capability of the functional unit to perform a function. "Error" refers to the deviation from the measured or computed value to its original or true value. Failure refers to the termination of the functional unit or a system to perform its intended function. The main goal of testing for functional safety is to identify if a system can handle most of the faults and still be operationally safe without causing any hazards to its users. As mentioned above,

FIGURE 9.3 Different types of fault injection testing.

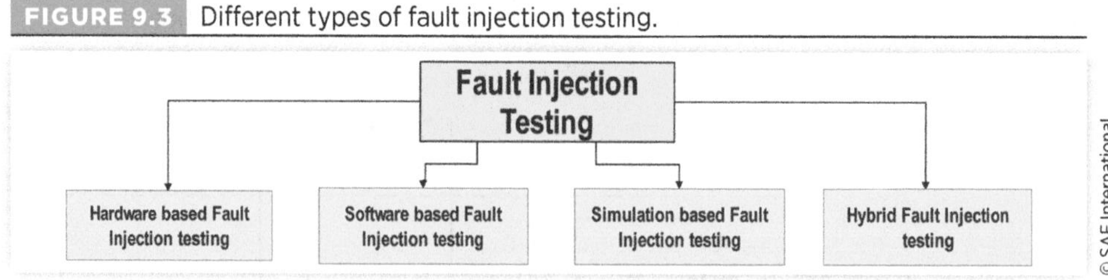

fault injection testing is one of those test types where faults are injected and the system and component behavior is evaluated. As fault injection testing purely focuses on Fault, Error, and Failure, there needs to be an in-depth analysis performed to identify all possible failures and their triggers during the lifetime of the system. The analysis data from FMEA, FTA, and HARA that were performed during the design phase can be used for designing the fault injection tests. Different types of DFMEA can be used (System, Software, and Hardware) as input and references for the test design.

Fault injection tests can be broadly classified into four types, depending upon the application area of fault injection tests, as shown in Figure 9.3. They are:

1. Hardware-based fault injection
2. Software-based fault injection
3. Simulation-based fault injection
4. Hybrid fault injection

Hardware-based fault injection focuses on the physical layer of the system, such as the printed circuit board and the components, and interfaces including the processors [9.1, 9.10, 9.17]. The hardware FMEA and FTA act as input in designing the tests for hardware-based fault injection. Beyond FMEA and FTA, another primary source for identifying multipoint fault is the Dependent Failure Analysis (DFA) that is performed during the safety analysis of the system. DFA helps identify failures that can hamper the required independence or freedom from interference between various components, including hardware, software, or firmware, which may ultimately lead to the violation of safety requirements or safety goals defined for the system. Usually, the hardware-based fault injection also tests the lower layers of the software, such as the operating system and the board support package in the system. From the HARA, different hazardous failures would have been identified because of single-point and multipoint hardware failures. Single-point faults are faults in an element that are not covered by a safety mechanism and that lead directly to the violation of a safety goal. Multipoint faults are faults that can lead to violation of safety goals in combination and without any response within a particular time duration.

Unfortunately, those are usually not considered for mitigation or providing redundancy to keep the system operational. Fault injection tests evaluate whether the risk mitigation mechanism is functional or not by injecting faults by external or internal means. Even though there are a lot of possibilities to have the failure in different areas of the hardware, very minimal sensors are utilized in the system as hardware components to detect those inside an ECU. This is done mainly because of higher costs for the production and qualification of those ECUs with expensive additional sensor components. Another major challenge the automotive industry faces is the use of new processors in automated driving ECUs, which may not be qualified or may not have any field data to check their operational behavior. Many new systems developed in ADAS and Automated Driving are using new processors in the market, mainly to achieve the required processing power for these systems. Different tools and devices are used in the hardware testbed for these. The faults can be forced or injected with or without contact. This can be injected using external hardware or even low-level software that closely operates with various components such as watchdog timers and processors. There are even methods where the faults are injected using electromagnetic signal bursts. Because of the higher costs and complex infrastructure required for hardware-based fault injection tests, most fault injection tests are performed by utilizing the low-level software interface to SoCs. The test strategy should define what techniques are to be considered and how much testing is enough to declare the hardware-based fault injection to be declared as acceptable.

Software-based fault injection tests are performed at different levels in the software stack. These tests can be executed in the operating system, base software, middleware, or the application layer of the software. Many software components are open source or reused from existing products during software development, and part of the software comes from different suppliers. Integrating various software components and testing them to guarantee the safety of the product is challenging. For the test design of software-based fault injection, one of the main inputs is the software FMEA, DFA, and Event Tree Analysis (ETA) [9.8, 9.9]. ETA is similar to FTA with a difference that the primary focus of the ETA is the failure and its consequences. Even though software FMEA is an essential tool used during safe and quality software development, it is rarely considered during software development and design. The faults in the software-based fault injection tests are either triggered by masking specific signals or feeding wrong inputs and then analyzing the behavior of the Safety Monitoring Function (SMF) and its response. Common faults considered for the fault injection tests in the software are compile-time faults and runtime faults, which consist of time-synchronization and interrupt faults, exceptions, code insertion faults, etc. Many tools are available in the market, such as Ferrari, FTAPE, and Doctor, which help generate these faults with the help of fault and error generators to test the overall software.

Mutation tests are executed as part of the fault injection testing where the software characteristics change because of mutation or with the influence of external factors. Mutation tests consider these external influences and associated error or fault, and they are tested in the software [9.17, 9.18]. Standard mutation tests include data corruption tests in the volatile and permanent memory and its interfaces, data manipulation in different software interfaces and in protocols, etc. A typical application of mutation tests is during the data access cycles, where a mutation operator will manipulate the data passed to the destination, which the software will process. The evaluation of those tests is performed by analyzing the software response and the possible error identification by the software. A general concept of mutation testing is shown in Figure 9.4.

Simulation-based fault injection is the most common fault injection test performed in safety-critical product testing of ADAS and Automated Driving systems and their functions. The use of different simulation tools has been advantageous in designing the tests and identifying how faults would impact a system even before the system is produced. The simulation-based fault injection can be executed in different test environments such as model-in-the-loop, software-in-the-loop, and HiL. Simulation tools are used to generate triggers for the faults in the system or its function under test. The fault injection tests are these days commonly covered as part of software-in-the-loop and HiL simulation tests.

FIGURE 9.4 A concept of mutation testing in software.

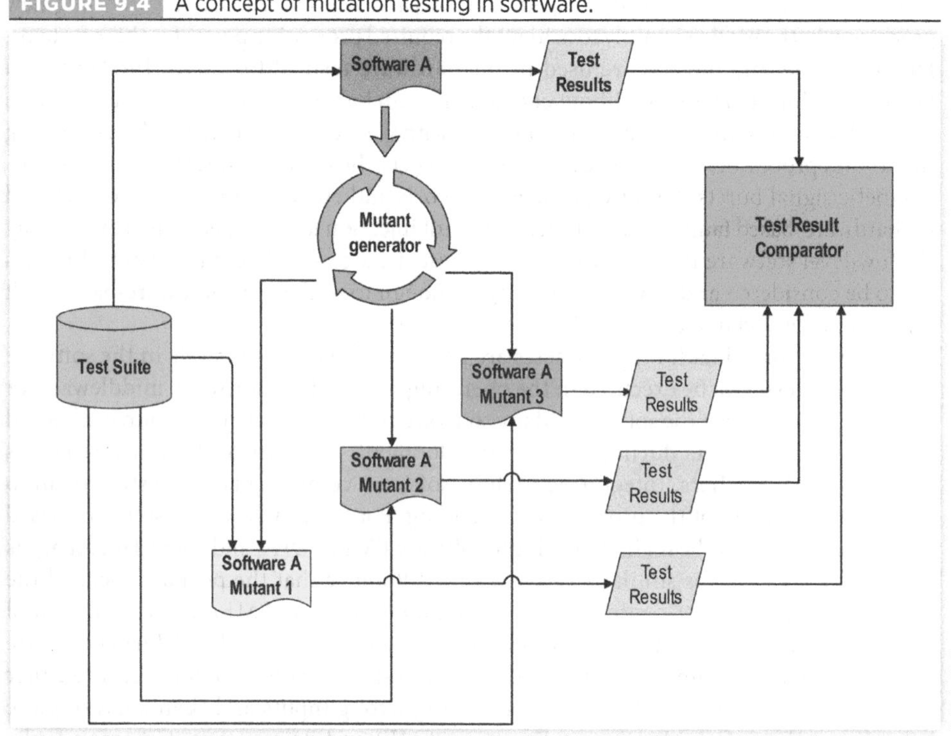

Along with those tests, different test techniques will help extend the fault injection tests to uncover defects in the system. One of those commonly used test techniques for safety-critical system tests is negative testing. Negative testing in the simulation is a test technique that considers the system under invalid inputs, user-introduced errors and faults, and checks its impact on the system. There are many tools available in the market that design and generate test vectors using different algorithms for covering the faulty conditions and the normal ones. Markov chain Monte Carlo simulation is one of the most common methods among these for test vector generation.

Hybrid fault injection tests include one or more types of fault inject test techniques used simultaneously. This is one of the most common approaches used to utilize the advantages of different types of fault injection tests at different levels of testing. The most commonly used hybrid fault injection method is the combination of simulation and software-based fault injection, which goes hand in hand with the software unit tests and the integration tests. Software and hardware fault inject tests are widely used when performing system-level verification and validation in laboratory environments. An example of it is using software-based fault injection and HiL tests, where hardware faults are injected from the HiL test infrastructure. Alternatively, the software-based fault injection can be replaced in the HiL testing with simulation-based fault injection. Here the faults are injected in the sensor models in simulation that accept input from environment simulation tools for processing environmental data for different ADAS and Automated Driving functions.

Scenario-based testing used for testing different software components can also be considered for fault injection tests. This is one of the methods used for the qualification of data-driven software components testing. ISO/PAS 21448 "Road vehicles—Safety of the intended functionality" standard recommends specific test methods which should be considered for the complete functional verification and validation, which mainly focus on the safety of the functions associated with the system [9.1, 9.4, 9.10]. This covers the complete data pipeline from sensors to actuators to perform an ADAS or Automated Driving function.

Safety-related testing also depends on the depth of testing considered at different levels. Test depth is determined by the use of different test techniques and the specific components covered for testing. Fault injection tests are usually designed to cover all possible faults and error conditions that a system would experience during its lifetime. Creating test cases specific to individual requirements will not help while defining test cases for functional safety tests. Still, for estimation and measurement of progress in testing, there are many projects where safety testing KPIs are evaluated based on the number of test cases and requirements for functional safety. For vehicle programs, the maturity and the acceptance defined for gates are also based on the status of functional safety requirement coverage, the presence of safety-critical defects, and the traceability of functional safety requirements with the test cases.

9.4. Fault Injection Testing Using Diagnostics

One of the standard methods used for executing fault injection tests in the system verification and validation phase is through the diagnostics interface. Unified Diagnostic Services (UDS) is a standard diagnostic communication protocol in the ECU. This is widely used by almost all automotive electronics manufacturers as a standard protocol and has its roots in the standard ISO 14229 "Road vehicles—Unified diagnostic services (UDS)" [9.19]. Some of the standard fault injection tests for testing software and hardware monitoring functions in the system are explained here.

Examples of testing SMF in software:

- *Explicit switch off of the software component during external fault events*: The software component under test can be Quality Management (QM) software or with an ASIL rating A to D. ASIL is a risk classification scheme as defined by ISO 26262 "Road vehicles—Functional safety" standard. QM software is not developed according to ISO 26262 ASIL A to D but is developed following well-defined development processes. Usually, QM software is not recommended to be used for realizing safety-related functionalities in an ECU. In some cases, the QM software has also been used to perform some of the safety-relevant functions that are not critical after undergoing in-depth analysis and evaluation.

For the particular architecture in consideration in Figure 9.5, the SMF for the software is expected to take appropriate actions in the event of an external fault condition or in the event of any malfunction of the main software component. Diagnostic services like Write Data Identifiers can be used to disable the operational software intentionally. This approach makes it possible to test the functionality and response time of the monitoring function, which is usually called a Fault-Tolerant Time Interval.

- *Testing by memory trap creation for the trusted and non-trusted partitions of the memory*: This is a method to test the memory trap handling in the software. Memory traps can occur as a result of one of those events such as Non-Maskable Interrupt, in the presence of an instruction exception, in the presence of a memory management exception, or when there is illegal access to memory [9.5]. Traps are always active,

FIGURE 9.5 Diagnostics—Fault injection testing for SMF (software).

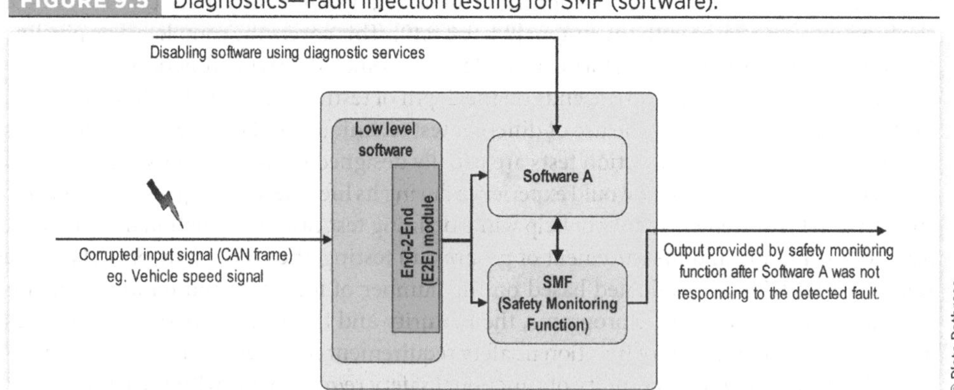

and the software cannot disable them in general. The testing focuses on the system response in the presence of memory traps because of one or more of these events and if an error flag is set along with publishing and storing DTCs. This is usually performed with Diagnostic Read and Write services.

- *Switching off the software component during internal fault events for analyzing software monitoring function:* This example is similar to the first example. Here the difference is the scope of the test is to analyze the internal fault response within the software and not about the output. How the fault is handled internally when different faults are being injected so that the system is operational or moving to the fail-safe and how the SMF handles it is checked here. The software components vary with different safety ratings from ASIL A to D or even QM.

Similarly, the diagnostics interface can also be used for testing the hardware monitoring function response in the event of hardware failures.

- Injection of faults to the internal communication protocols like Inter-Integrated Circuit interface (I^2C), Serial Peripheral Interface (SPI), etc. I^2C and SPI are serial communication interfaces that are commonly used for the communication of microprocessors and controllers with other components in the electronic circuit boards. These fault injection tests in hardware are usually performed to check the watchdog timer that constantly monitors the processor status. A failure or malfunction in the communication bus that connects the processor with other components is a safety goal violation based on the analysis performed as part of functional safety for the system safety goal definition. With the injection of such a fault in the communication interface, as shown in Figure 9.6, the functionality of the watchdog timer and its response can be tested. The response from the system includes the generation of DTCs, which are tested using the Diagnostic interface.
- *Bus protection tests:* The faults can also be injected via a Diagnostic interface to the internals of the system to test the bus protection of different internal communication channels. This includes manipulation and injecting faults in the communication channels of I^2C, SPI, SMBus (System Management Bus), etc. The faults can be triggered over diagnostics or with other means such as using external Bus Testers to test the bus protection functionality of different communication channels. Most of these tests are usually covered as part of the hardware testing. Even the system testing considers similar tests for a CAN Bus using CAN bus testers.

FIGURE 9.6 Diagnostics—Fault injection in hardware for testing the watchdog timer.

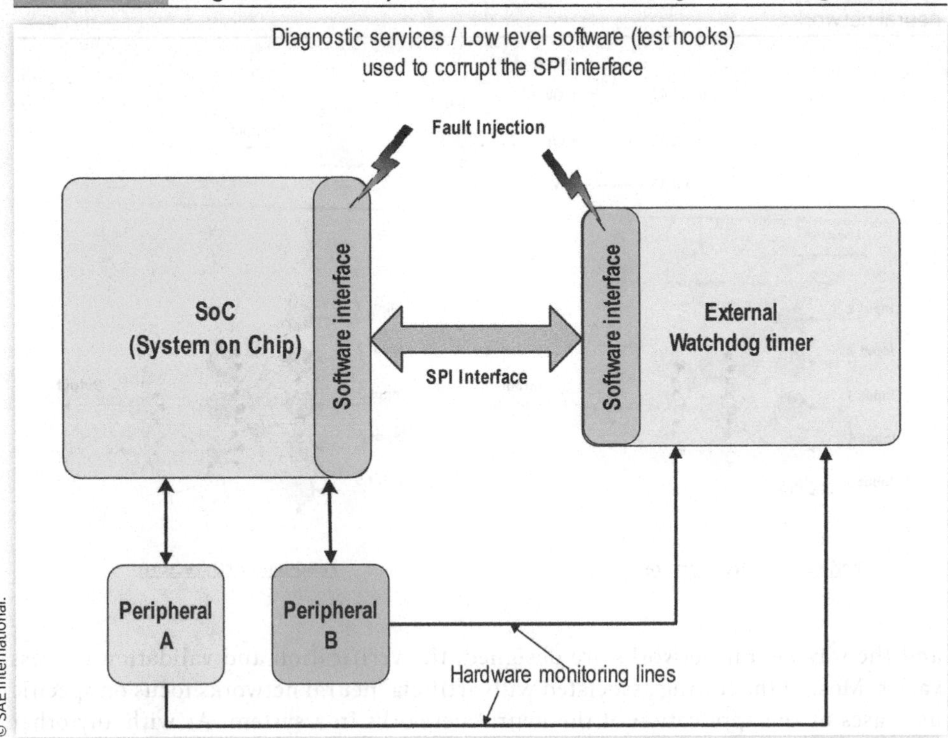

9.5. Safety Testing of Artificial Neural Networks—An Overview

Artificial neural networks attempt to mimic how a human brain operates, rather than cells or tissues, where they work with mathematical models and calculations to store and process different information types [9.20]. Neural networks are used widely in the automotive industry and especially a lot in ADAS and AD systems these days. They have drastically improved the performance of these systems more than any classical old-time methods. The application of neural networks in safety-critical areas also has a lot of risks and chances of failure. These risks and the chances of failure are reduced with rigorous verification and validation of the neural networks in the system. This section will take you through some of those methods used in the industry for qualifying neural networks in the ADAS and Automated Driving and discuss different challenges in getting a neural network qualified to use in a safety-critical system.

Neural networks mainly have two types of architectures: Feedforward architecture and Feedback architecture, as shown in Figure 9.7. Generally, neural networks have three main parts: a weighted input part, a transfer function, and an output part. Weights can be simply considered as the amount of knowledge in the network or the strength of the network. Weights allocated to a neural network depend on the learning a network would have undertaken and how the network is integrated with different layers based on the architecture [9.21].

Artificial neural networks development and testing is part of data-driven software development and testing, explained in Section 6.1. Depending on the scope, use case,

FIGURE 9.7 General structure of a Feedforward and Feedback architecture of artificial neural networks.

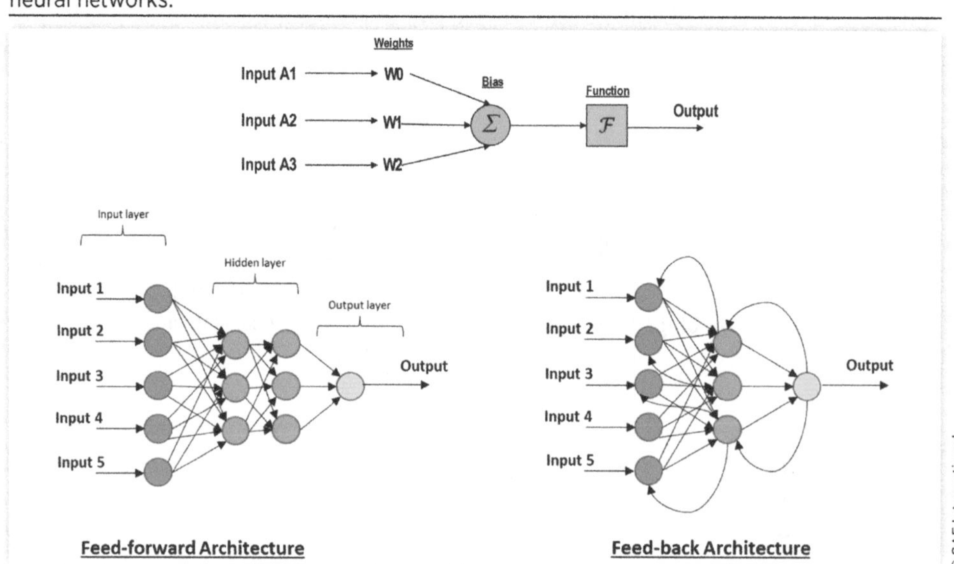

and the way neural networks are designed, the verification and validation process varies. Most of the testing associated with artificial neural networks focus on specific use cases or the application of the neural networks in a system. As with any other software development process, artificial neural networks are also expected to follow the process and to have requirements and architecture specific to a use case [9.21, 9.22]. Unfortunately, it is sporadic to find detailed requirements and architecture for neural networks in many production projects where they are used. It is either because the vehicle manufacturer is not provided with this information claiming it is an intellectual property of the supplier or the supplier sometimes may not even have that information. During the design phase of neural networks, acceptance criteria must be defined, such as required performance during operation, resource utilization in the system, operational conditions, the time required by a neural network to become error free when exposed to an error, etc.

The development of an artificial neural network is an iterative process. The improvements and the adaptations in the neural network must be strictly tracked. As the changes in neural network developments are rapid, a process-oriented approach must be considered, including configuration management. Depending on how the neural network is used in a system, either pretrained or adaptive, configuration management should be followed in development and testing. In most industrial use cases, neural networks are not developed specifically for a particular use case. Instead, most of them are adapted or customized for certain other existing applications [9.22, 9.23].

As with any other software component used in safety-critical systems, neural networks should also need to be used, knowing whether the system uses operational monitors or monitoring functions to identify and prevent the error from passing out of it. One of the most common issues found in the system with operational monitors is that they usually influence the intended function and the architecture of neural networks. Hence, it is crucial to know about the use of any monitoring functions in the system as a prerequisite to designing the neural networks accordingly.

Verification and validation of artificial neural networks are performed employing process-oriented methods and based on specific KPIs. One of the most common applications of artificial neural networks is in front-facing cameras to detect lanes and objects. Pretrained neural networks are used in this case. They are the main components in certain safety features in the vehicle, such as lane keep assistance and AEB. Testing of neural networks evaluates the network accuracy and checks if the network has gained enough knowledge to be precise with its predictions based on its training and if the behavior matches with the requirements defined [9.22, 9.24]. Neural networks use at least two types of datasets, one for their training and another for testing. The quality requirements of these datasets should also be defined as part of the requirements, as they play an important role in deciding the weights of the neural network. To avoid any error being passed on or any influence on the weights of the neural networks, the accuracy, and the quality requirements should be strictly followed. A process-oriented way of controlling the dataset generation and its source would help in having a better-quality dataset. One common error is mixing up various datasets to generate a larger dataset as most organizations do, where the importance is given to the size of the dataset rather than the quality of the dataset. This is a bad practice; as for robust development and qualification of the neural network, one would need both quality and size.

Standard testing approaches [9.23, 9.25] used in the automotive industry for verification and validation of the neural networks can be broadly classified as:

- Testing for the design and architecture of neural networks
- Testing for the functionality of neural networks

The design and architecture-based testing cover the internal components of the neural network, including its structure, number of layers, transfer function, and nodes. The main area of focus for testing here is resource utilization, neural network structure, time constraints, and memory utilization in real-time operation. The implementation of the transfer function and the input and output layers are also checked. Many tools are available to analyze and measure these in a neural network. These are associated with mathematical and statistical analysis on the structure and the contents of the neural network in its static and runtime conditions. One of the major drawbacks in verifying and validating the neural networks is the environment and infrastructure used to measure the parameters associated with the neural network. Measuring and analyzing all these in a high-power PC environment does not add much value and mostly are not helpful. Using the right hardware platform will help identify the memory utilization, performance, and need for further optimization of the neural network for a particular use case. The hardware platform plays a significant role in providing the necessary resources for the optimal performance of the neural network. The knowledge components or the weights in the neural network nodes measure how much the neural network has learned for specific data patterns.

The learning and testing cycle, called epoch, influences the weights of the nodes in a neural network. Planning methodical training and testing epochs will help control the knowledge component of the neural network beneficially. Rather than using a learning dataset with a mixture of data characteristics for learning, the datasets should be organized for certain specific characteristics, and the training and testing should be performed in structured epochs, as shown in Figure 9.8. This structured method will help understand the knowledge stored in the neural network, like the completeness of specific characteristics knowledge and what is not available or their limitations. With the limitations in the knowledge available in the neural network, it is possible to evaluate its influence on the overall system behavior. This can be measured with the test dataset as well as the use of

FIGURE 9.8 An iterative approach of training and testing epochs for artificial neural networks.

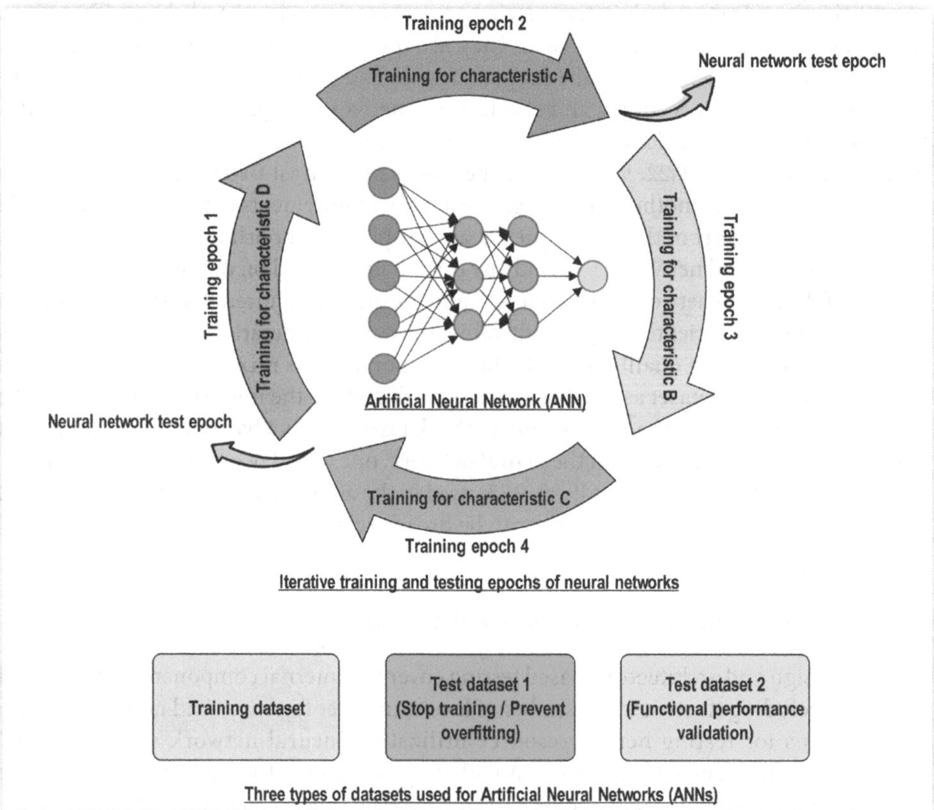

scenario-based testing. This information can be used for calculating the residual risk and the performance KPIs of the neural network for a particular use case [9.23].

The measurement of the knowledge accumulated by a neural network using different test types and methods will provide critical information on the capabilities and reliability. Measurement of precision and accuracy of the neural network for its intended functions like lane detection or vehicle detection will indicate the training performed and the maturity of the neural network. The residual risk information will provide information on whether there needs to be a fault management mechanism to be considered or not for the general application in the system. These will also provide the confidence level of the functions of the neural network for that particular application.

Functional verification of the neural network almost follows the same methods as discussed before. The only difference is that the functional verification also checks how well the neural network performs a particular function. The test methods used for testing neural networks are almost the same as any other software component. The only difference is the area of focus is different for qualifying artificial neural networks [9.23]. The measurement parameters focus on the functionality of the neural network, including its stability, robustness, and how well the neural network is performing in delivering a function.

Functional verification of the neural networks also uses the datasets for training and testing as a minimum to evaluate the quality of the neural network performing a particular function (e.g., object detection). The datasets are classified to evaluate the neural network for its stability and calculation of the accuracy in its detection, mean average precision, etc. The measurements are made for individual frames or images as well as for the complete dataset. The details on how the functional verification is performed are covered in detail under Sections 6.3 and 6.4. Standard methods used for testing the functionality of the neural networks are performed either at integrated software-level or system-level testing using scenario-based testing or utilizing re-simulation methods in a software-in-the-loop or HiL test environment. A combination of synthetic data as well as real-world data is used. It is very common that the acceptance criteria for neural networks are only defined based on the functional verification of the neural networks. Scenario-based testing and its parametrization approaches are widely used to expose the neural network to all possible scenarios and evaluate its performance.

For the training of artificial neural networks, there is a massive demand for data and datasets. This demand is now met in the industry through the use of synthetic data along with real-world data from the vehicle [9.23, 9.24, 9.25]. Recently there has been large-scale usage of photorealistic environment simulation tools for dataset creation for training and testing artificial neural networks. The learning and knowledge of the neural network is the key to its functionality on any particular application. The quality and the quantity of the datasets, along with the methods applied for the development and evaluation, determine the overall quality of the neural network and its fitness to use in a safety-critical application.

9.6. An Overview of Cybersecurity Testing

Cyber-physical systems have become standard components in the automotive industry. The evolution of technology and the need for higher processing power and data transfer speeds improved the electrical and electronics architecture inside the vehicle. The evolution of connectivity outside the vehicle, along with the advancements of internal connectivity, brought many advantages, so did the risks and threats on cyber-physical systems in vehicles. ADAS and Automated Driving systems, which are safety-critical systems that provide safety features in the vehicle and take control of the vehicle in certain specific conditions, will increase additional risk and hazards if they are under cybersecurity attacks.

Developing cyber-physical systems as a closed system without external influence is impossible. Most of these systems accept input from various sources, including sensors, ECUs, and other systems in the vehicle, and provide output to control the vehicle through various actuators such as steering and brakes. The only possible way is to develop the system to be secure and prevent any possible attacks because of system vulnerabilities. This is possible with a process-oriented development approach and rigorous security testing [9.2, 9.3]. The process-oriented development, including tools used for development, testing, threat analysis, and risk assessments, is available as guidelines in the standard SAE J3061 "Cybersecurity Guidebook for Cyber-Physical Vehicle Systems." Most of the organizations used this standard as a reference for their development and compliance to cybersecurity until recently; there is a migration to the newly introduced standard ISO/SAE 21434 "Road vehicles—Cybersecurity engineering" and the UNECE regulations.

Similar to functional safety processes in system development, cybersecurity processes for product development and testing are also classified as different subsections. This extends from the concept development to the production and rollout of a system in the market. They are starting from item definition to continuous monitoring and updates of the system in the vehicle. The main scope of verification and validation of the system concerning cybersecurity is to prove that the system meets cybersecurity goals defined during the design phase. Cybersecurity concepts, goals, and requirements are generated like processes similar to functional safety processes. Similar to HARA in functional safety, cybersecurity has TARA, which will act as the foundation for requirement generation for cybersecurity in different domains [9.26]. The testing of those requirements is performed to prove the system meets the required cybersecurity requirements.

Cybersecurity testing is executed in different ways, and it varies according to each organization and the product. Cybersecurity threat analysis and the risk mitigation actions which are part of the requirements are not usually disclosed to a broader audience beyond the development team. Hence, the requirements-based testing is usually performed as development testing by software and hardware engineers. Cybersecurity testing typically seems like software-only testing as the risk mitigation actions for security threats are implemented mainly in the software than hardware. The system-level tests are usually performed considering the vehicle interfaces.

The tools used for security testing include those used for attacking, investigating, or defending attacks on a system. In many organizations, rather than qualifying any tools specifically for the usage, they use the tool classification, which is used by offensive security for Kali Linux [9.27], which is one of the well-accepted security testing environments in the industry. Cybersecurity testing processes are yet to gain traction in the industry as it is not yet standardized across different organizations. Even though the security risks are higher, the processes related to cybersecurity are established differently in each organization. The new standard ISO 21434 aims to bring all those processes as standard across different organizations in the industry.

Three different types of testing are considered to evaluate the cybersecurity compliance of a product. This includes the development testing, which focuses on the cybersecurity requirements and evaluating those as part of the functionality checks once it is implemented. This is the functional testing part of cybersecurity, which focuses on verifying cybersecurity requirements in different domains. Different test techniques are applied as part of software testing to check on the functionalities of implemented risk mitigation functions of the security mechanism. This is executed as software testing and low-level hardware tests; some extend to hardware and software integration testing.

Another type of testing performed in cybersecurity testing is Fuzz testing. Fuzz testing is a type of automated or semiautomated test technique used to identify the defects which cannot be identified using functional testing methods or testing specific functional requirements. Fuzz testing is software testing that can help to identify security vulnerabilities in software, networks, interfaces, etc. The main area where the fuzz testing is performed is the interface between different software components [9.28]. This could be on any software layer and even in the operating system where the interfaces are checked for any possible vulnerabilities. Fuzz testing is executed with specific tools with a fuzzer component that injects different types of input data and analyzes the output of the system or software, as shown in Figure 9.9. Similar to mutation testing, fuzz testing focuses on all types of input data. It will make sure the software component or the system can handle the memory leaks, invalid input conditions, corrupted database access issues, and unhandled exceptions in operations, which are typical defects in the lower layers of the software in a system. The application of this test technique can also be extended to system integration and even vehicle integration tests. These tests can be extended to the system network interfaces like CAN, LIN, and Ethernet, within the vehicle and the interface to the external world like GPS, Wi-Fi, and mobile network communications to the external world.

FIGURE 9.9 A generic configuration for fuzz testing.

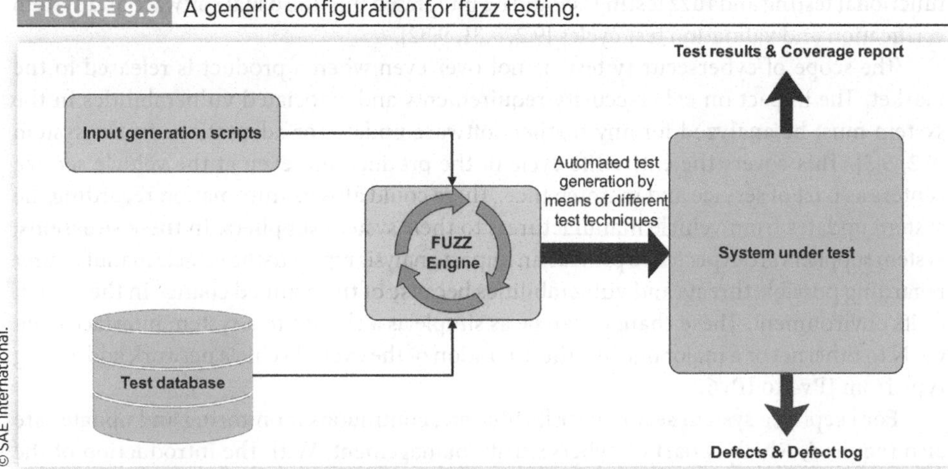

A P-diagram during the system concept, FMEA results during the system design phase, and detailed interface diagrams of software, hardware, and the system can help design the fuzz tests. The TARA should consider software, hardware, and system separately to identify all possible risks for which mitigation actions are to be planned. Fuzz testing should be integrated with the standard verification and validation processes, and the strategy should define when these tests should be executed.

The third type of testing in cybersecurity verification and validation is penetration testing. In almost all organizations, penetration testing is executed by an independent team with the mindset and perspective of an attacker. This is executed as part of cybersecurity validation where different attacks are planned and executed to evaluate whether the system is secure as well as to identify whether the attackers can utilize any existing vulnerabilities to gain access to the system and manipulate it. Penetration testing is executed as an independent test activity. Sometimes it is even executed by a different organization.

Cybersecurity organizations have grown in recent times, mainly in these areas where they execute projects of only performing independent penetration tests in different products. The validation strategy established for the penetration test determines the depth of the tests and the different attack types that should be considered. This can be with minimal attack types or with a whole range of different attack types over some time. A smaller set of different attack types used for performing penetration tests can be found from the automotive attack database established by experts [9.29, 9.30, 9.31].

Sticking to a predefined set of attack types, especially for penetration testing, is a limitation on the depth of testing. The technology evolution can bring additional attack types, and it varies with different applications in different systems. Hence, a possible way forward is to continuously track and update the test cases and test methods to keep pace with the technology trends and advancements in the security area.

The outcome of penetration testing is a penetration test report that explains the tests performed and a detailed analysis of different vulnerabilities and successful attacks. The product development team should analyze this report and improve the software and system to minimize the resulting vulnerabilities and associated risks. Cybersecurity tests are executed on each release cycle of the software, hardware, and system. The cybersecurity tests, especially penetration tests, are initiated once the system reaches a particular maturity level, such as the production version of the system that meets approximately >80%, similar to the final product as for the Start of production (SOP) version. All other test types, like functional testing and fuzz testing, should be integrated into the usual software and system verification and validation test cycles [9.2, 9.31, 9.32].

The scope of cybersecurity tests is not over even when a product is released to the market. The impact on cybersecurity requirements and associated vulnerabilities in the system must be analyzed for any further software updates or adaptations in the system [9.2, 9.3]. This covers the entire life cycle of the product and even at the vehicle service centers as part of service and maintenance. There could also be information regarding the system updates from vehicle manufacturers to their system suppliers. In these situations, system suppliers are expected to provide an impact analysis report to the vehicle manufacturer regarding possible threats and vulnerabilities because of the planned change in the system or its environment. These changes can be as simple as a change in a system interface from CAN to Ethernet or a major one like the migration of the overall vehicle network addressing type from IPv4 to IPv6.

For keeping a system secure over its lifetime, continuous monitoring and updates are also required, which is part of cybersecurity management. With the introduction of the standard ISO/SAE 21434 "Road vehicles—Cybersecurity engineering," there will be a possibility standardization of the cybersecurity processes across different organizations. With the introduction of cybersecurity as an important regulatory requirement as per UN Regulation No. 155 from the UNECE, there will be more focus given to cybersecurity for the development of automotive systems, especially when moving toward autonomous driving.

9.7. Summary

Developing and deploying complex systems which are used for performing safety-critical functions is a complex process. The development should consider the end user, their safety, and security. As ADAS and Automated Driving systems are complex systems, safety and security are of exceptionally high importance. Some standards provide guidelines and recommendations of certain best practices for the development and testing of safety-critical products. Practically, these guidelines cannot be applied as is recommended. Rather adaptations and tailoring are required based on the product, organization, and even the production environment.

Building a safe and secure product requires good and detailed design and analysis. This also requires the need for following good processes in the organization and product development. Processes are acting here as guards for avoiding any ambiguities and ensuring that all required information is generated and available for the development and testing of these products.

This chapter enlightens the reader with an overview of how essential the tools are in safety and security testing and the need for their qualification before even using them for testing a safety-critical system. The different methods now used in the industry for safety qualification and how different test techniques with the concept of test design are applied for testing have been discussed. An overview of testing safety-specific requirements, as recommended by the standard ISO 26262 in hardware, software, and system-level using different test environments with certain examples, helps the reader start with a fundamental knowledge of functional safety testing. The standard methods used, such as diagnostics-based safety testing, explain how safety tests can be executed in a laboratory environment.

Artificial neural networks and their utilization for different functions are growing day by day. Many might not know how these software components are to be tested. This chapter integrates the experience from the automotive industry to summarize the test and performance evaluation process followed in the industry from different organizations. Test methods used for testing neural networks for both their structure and composition and their functionality are explained in this chapter.

Testing cyber-physical systems for guaranteeing security is getting traction in ADAS and Automated Driving. This chapter explains the complete process followed in the industry for testing cyber-physical systems and their execution. In all areas, the verification and validation of AD systems focuses on the types of tests and different test techniques applied to achieve the required test depth. These are discussed in detail in the following chapters.

References

9.1. ISO 26262:2018, "Road Vehicles—Functional Safety," 2018.

9.2. ISO/SAE 21434:2021, "Road Vehicles—Cybersecurity Engineering," 2021.

9.3. SAE Vehicle Electrical System Security Committee, "Cybersecurity Guidebook for Cyber-Physical Automotive Systems," SAE Standard J3061, 2016.

9.4. ISO/PAS 21448, "Road Vehicles—Safety of the Intended Functionality," 2019.

9.5. Bath, G. and McKay, J., *The Software Test Engineer's Handbook*, 2nd ed. (Santa Barbara, CA: Rocky Nook Computing, 2014), ISBN:978-1937538446.

9.6. Clark, J.A. and Pradhan, D.K., "Fault Injection: A Method for Validating Computing-System Dependability," *Computer* 28 (1995): 47-56.

9.7. Casdorph, V., Taylor, B. et al., "Software Verification and Validation Plan for the Airborne Research Test System II, Intelligent Flight Control Program," IFC-SVVP-FOOl-UNCLASS-120100, Institute for Scientific Research, Inc., December 1, 2000.

9.8. ISO 31000:2018, "Risk Management—Principles and Guidelines."

9.9. *AIAG & VDA FMEA Handbook* (Automotive Industry Action Group, 2019), ISBN:978-1605343679.

9.10. International Electrotechnical Commission, "Functional Safety of Electrical/Electronic/ Programmable Electronic Safety-Related Systems," IEC 61508, Edition 2.0, April 2010.

9.11. ISO 14253-1:2017, "Geometrical Product Specifications (GPS)—Inspection by Measurement of Workpieces and Measuring Equipment—Part 1: Decision Rules for Verifying Conformity or Nonconformity with Specifications," 2017.

9.12. RTCA, *DO-330: Software Tool Qualification Considerations*, 1st ed. (Washington, DC: RTCA, 2011)

9.13. Wildmoser, M., Philipps, J. and Slotosch, O., "Determining Potential Errors in Tool Chains: Strategies to Reach Tool Confidence according to ISO 26262," in *SAFECOMP 2012*, Magdeburg, Germany, 2012.

9.14. Gitlow, H.S., Levine, D., and Popovich, E.A., *Design for Six Sigma for Green Belts and Champions* (New Jersey: Financial Times Prentice Hall, 2006), ISBN:978-0131855243.

9.15. Arthur, J., *Lean Six Sigma Demystified*, 2nd ed. (New York: McGraw Hill, 2007), ISBN:978-0071749091.

9.16. IEEE Std 1012-1998, "IEEE Standard for Software Verification and Validation," New York, 1998.

9.17. Hsueh, M.-C., Tsai, T.K., and Iyer, R.K., "Fault Injection Techniques and Tools," *Computer* 30, no. 4 (1997): 75-82.

9.18. Rana, R. et al., "Early Verification and Validation According to ISO 26262 by Combining Fault Injection and Mutation Testing," in Cordeiro, J. and van Sinderen, M. (Eds), *Software Technologies. ICSOFT 2013*, Communications in Computer and Information Science, vol. 457 (Berlin, Heidelberg: Springer, 2013), https://doi.org/10.1007/978-3-662-44920-2_11.

9.19. ISO 14229 : 2020, "Road Vehicles—Unified Diagnostic Services (UDS)."

9.20. Minsky, M.L. and Papert, S.A., *Perceptrons* (Cambridge, MA: MIT Press, 1969)

9.21. Abraham, A. and Nath, B., "Hybrid Heuristics for Optimal Design of Artificial Neural Networks," in *Third International Conference on Recent Advances in Soft Computing (RASC2000)*, England, June 2000.

9.22. Haykin, S., *Neural Networks: A Comprehensive Foundation*, 2nd ed. (New York: MacMillan Publishing, 1999)

9.23. Taylor, B.J. et al., *Methods and Procedures for the Verification and Validation of Artificial Neural Networks* (Boston, MA: Springer US, 2006), doi:https://doi.org/10.1007/0-387-29485-6.

9.24. Bishop, C.M., *Neural Networks for Pattern Recognition* (Oxford, UK: Oxford University Press, 1996)

9.25. Bolt, G.R., "Fault Tolerance in Artificial Neural Networks—Are Neural Networks Inherently Fault Tolerant?," DPhil thesis, University of York, November 1992.

9.26. Lee, E.A., "Cyber Physical Systems: Design Challenges," in *Proceedings of the IEEE Symposium on Object/Component/Service-Oriented Real-Time Distributed Computing*, Orlando, FL, May 5-7, 2008.

9.27. Offensive Security, "Kali Linux Tools Listing," 2019, accessed September 1, 2021, https://tools.kali.org/tools-listing.

9.28. ClusterFuzz, "Google Blog on Blackbox Fuzzing," accessed September 1, 2021, https://google.github.io/clusterfuzz/setting-up-fuzzing/blackbox-fuzzing/.

9.29. Sommer, F. and Dürrwang, J., "IEEM-HsKA/AAD: Automotive Attack Database (AAD)," 2019, accessed September 3, 2021, https://github.com/IEEM-HsKA/AAD.

9.30. Booth, H., Rike, D., and Witte, G., "The National Vulnerability Database (NVD): Overview," 2013, accessed September 3, 2021, https://nvd.nist.gov/.

9.31. Sommer, F., Dürrwang, J., and Kriesten, R., "Survey and Classification of Automotive Security Attacks," *Information* 10, no. 4 (2019): 148.

9.32. Dürrwang, J., Braun, J., Rumez, M., Kriesten, R. et al., "Enhancement of Automotive Penetration Testing with Threat Analyses Results," *SAE Int. J. Transp. Cyber. & Privacy* 1, no. 2 (2018): 91-112, doi:https://doi.org/10.4271/11-01-02-0005.

/Autonomous
/Sensing
/Communication
/Battery
/Navigation
Autopilot mode

10

Verification and Validation Strategy

Verification and validation are two essential processes to qualify a product. In the verification process, it is checked whether the product is implemented correctly according to the requirements. Validation is the process of checking whether the product is developed correctly and whether it serves the purpose for which it is designed. This is about checking the product from an end-user perspective to see if use cases are met for which the product was developed. Having a good strategy for verification and validation of the product is essential as it determines how well a product is designed, considering its performance, quality, reliability, and correctness.

The strategy for verification and validation for product development is usually a reference for all the processes followed and methods applied for testing in a project. Strategy definition for verification and validation is part of the project planning phase. Here an overview is developed on how all the available resources are deployed, considering different processes, methods, and test infrastructures to qualify a product. This summarizes various options considered for planning and utilizing resources, methods, techniques, and processes, considering the boundaries of a particular project. The test plan is the common term used in the industry, usually for the verification and validation strategy. However, in reality, a test plan includes much more than the verification and validation strategy. A test plan is similar to a project plan which incorporates the planning for the complete test life cycle of a project. It includes plans associated with risks and their handling, stakeholders, budget, configuration management, etc. While executing a project, it is less likely that all test iterations move as per the defined strategy. The testing team must apply different tactics to handle irregularities faced while executing the project in these situations.

Different organizations have different approaches for establishing test strategies, and it varies based on the organization and the product under development. Verification and validation strategy is also a reference, which is checked as part of all project assessments.

10.1. Test-Driven Development and Feature-Driven Development

As the automotive industry has transformed into a software-dependent sector, different agile methodologies have found their way in the automotive industry. When the automotive projects become complex and complicated to follow development using classical development methods amid ambiguities, different agile methods help the industry for rapid growth and continuous progress. These days, almost all ADAS and Automated Driving systems are developed and tested using one or the other agile methodologies as the development of these products is complex and highly dependent on software. Two of the typical agile methods used in the field of ADAS and Automated Driving area are discussed here.

As with many new product developments in the automotive industry, the features required for the products are usually defined taking into account certain use cases in the vehicle. The development follows, based on the requirements derived from those use cases. There is a different approach for developing the system by only considering the use cases as the requirement. Test-Driven Development (TDD) and Feature-Driven Development (FDD) are two agile methodologies that help in these situations for product development and are widely used in many organizations nowadays for prototyping and developing automated driving functionalities.

TDD is an agile methodology in which the test cases are defined based on the use cases assumed for the product [10.1, 10.2]. This is a method usually used for developing software components and software stacks. The test cases are designed and developed based on the use cases and functions of the software components that should be developed as the first step. The developed software is further tested against those designed tests, and the misbehaviors and defects are identified. Based on the identified defects, the code is refactored further. This is a continuous process until there is no defect or failure in the software for a specified set of test cases that relates to a particular function or feature and serves the use cases. This method is applied these days in automated driving projects so that the different use cases of the system are defined based on specific scenarios utilizing various simulation tools. The functional model or the software that is undergoing development is tested against those scenarios to check if it serves the use cases. This is commonly followed in developing data-driven software components, which include artificial neural networks used in realizing different ADAS/Automated Driving functions. For testing, scenarios play an essential role in these software components. For the complete system, the calibration and the associated tests for calibration of the features are also performed by similar methods using simulation environments. Repeatability and reproducibility with scenarios are possible in a simulation environment, which is hard to achieve in real-world vehicle testing. TDD approach also helps decompose the software to its components and units and define associated unit software and component requirements and their interfaces.

TDD approach should not be misunderstood for the required testing to evaluate the quality of the product [10.1, 10.2]. This is one of those approaches used in the industry to detail the requirements of the software and system to its lower levels, such as units or components. Another advantage of TDD is that it helps to bring an early check on the software architecture to its smallest components. Tests used as part of the TDD approach are the development tests defined and developed by the software development team. Sometimes a separate test database called an acceptance test suite is used as a qualifier for the iterative release of the software for different functional components [10.3]. A trend seen these days in many organizations is that TDD test suites or databases used for prototyping are automated and used for software unit testing and integration testing later. This is one of the best methods to reuse the work done and improve the testing at lower levels of the software.

FDD is another agile methodology where features of a system are identified and decomposed. An iterative development focuses on different features for its implementation. Feature by phase, or F×P, is a common term used in product development for introducing features in phases. The feature developments are classified into different phases over a complete vehicle program and its release gates. FDD can be considered as a miniature version of that. This is not a new method in the automobile industry. The system is decomposed into different features, and subfeatures and development focus on the individual feature or subfeatures on each development iteration [10.1, 10.4]. In the field of automated driving, higher levels of automation features in the vehicle have their foundation or building blocks, which are the ADAS functions. One can decompose the higher-level automation features to their subfeatures and functions using a feature dependency diagram or any other suitable tools or methods and focus on developing those building blocks and further integrating and achieving features of higher-level automation. As an example, one should not start implementing the Highway Pilot feature without understanding its building blocks such as lane keep assist and adaptive cruise control. Although this is a well-known method, a common mistake made in many projects is not identifying the dependencies and their building blocks in the first phase for developing a feature with higher automation levels, which adds additional risks and effort to the overall project.

Both methods mentioned here are commonly used methods nowadays, although they utilize different test methods to achieve a better system design and architecture. They are not a replacement for the overall product testing, which has to be performed as part of the product development and release. They act as only additional supportive methods that can be used for product development while facing ambiguities with the requirements and functionalities during product development. One should decide and apply these methods depending on the feasibility of the application and the benefit at different levels of the project.

10.2. Purpose of Test Design and Test Depth

The traditional way and one of the most common approaches followed in testing is creating test cases based on requirements and executing them. This has changed with complex projects where there are ambiguities, and the requirements are challenging to be finalized early in the development phase. Hence, a different approach was required to consider for better product development and to ensure its quality. Test design can be simply seen as a way to generate test cases for efficient testing. There are different test techniques, which can be used for defining test cases and designing the overall tests. The application of different test techniques depends on how different input vectors are generated as part of evaluating a system. Different test techniques are the core of any test design for testing a product to guarantee its quality, safety, and security [10.5, 10.6]. Knowing where and how different test techniques are to be applied determines a quality product. Test design is also about knowing how effectively one could test the product, employing different test techniques, and using test environments or infrastructure with different toolchains.

ISO/IEC/IEEE 29119-3 [10.6] is usually considered as a reference in many organizations that define the verification and validation strategy as part of testing a product in most of the projects in the automotive industry. Defining a strategy for testing an Automated Driving systems is challenging. Most of the requirements or even the features might not have been identified during the strategy definition, which adds risk to the project if traditional test methods and approaches are employed and will not help in efficiently testing these products.

Test cases used for verification and validation of a system can be broadly classified into two. First are the tests based on requirements. There are clearly defined requirements as part of the requirement specification, and the scope of the structured requirement-based tests is to evaluate those requirements; Besides these test cases, other tests are performed to evaluate the robustness, reliability, and performance of the system, covering the areas related to faults, error conditions, misuse conditions, security, etc. Sometimes this comes as a usage-specific requirement or hidden set of requirements for any system which needs to meet. Usually, these tests are executed through special test types such as exploratory test types. Exploratory tests are those types of tests that are defined and performed based on the experience of the person performing the tests on similar systems and knowing the possible behavior of such a system in the vehicle.

Specification-based tests that are solely dependent on the requirements are usually classified into three groups: those that cover the functional aspects of a system, tests that focus on the use cases of the system as per requirements, and those that consider the nonfunctional aspects of the system. Different classifications of the test cases are possible in any project, and this chapter discusses one of those classifications with different test techniques considered for testing an Automated Driving system. Different test techniques and methods that can be utilized are well explained in [10.5]. However, this chapter will focus on a few of those which are commonly used in the industry.

Knowing different test techniques will gradually raise questions: How different test techniques are applied? Where should these test techniques be applied? Different test techniques can be applied based on the required depth of testing in each area under test. Table 10.1 shows a classification of different test techniques in structured testing based on a specification, where different test classifications are considered, which were used for the verification and validation of an Automated Driving system. One could consider this as a reference for applying different test techniques to different areas of testing.

Where to apply different test techniques and the depth at which these techniques should be applied is an outcome of analysis performed as part of the test design phase. For safety-critical systems such as ADAS and Automated Driving systems, execution of tests based on the requirement specification is not enough. These systems can experience conditions beyond those which requirement specifications can define during their lifetime. Hence, there need to be special test types that are considered beyond those defined as part of specifications. Most of those tests are covered under the exploratory tests where different test types and conditions beyond the requirement specification are considered. A boundary is defined for testing a system with this approach, thereby improving the overall system performance and quality.

Exploratory tests are highly dependent on the individuals who perform these tests. Even though the tester is the key factor here, other dependencies include the system under test, the scope of testing, the expertise and knowledge of the tester, knowledge about similar products, and end-user-level thinking. As Automated Driving systems are not common

TABLE 10.1 Test techniques and depth classification for specification-based testing.

Specification-based tests (Structured testing using requirements)								
Tests for functional aspects					Use cases	Tests for nonfunctional aspects		
Risk-based classification for prioritizing	Equivalence class technique	Combinatoric logic technique	Logic/ Domain-based tests	State transition tests	Use-case tests (SOTIF, functional safety analysis, and customer usage, misuse)	Configuration-based tests (ECU configuration, feature enablement, vehicle variants, etc.)	Documentation and process based	Stability, robustness, and efficiency-based tests
Low risk	Valid classes tested	Valid conditions (all vectors at the same time)	Testing only best logic conditions (priority based)	Zero-switch method	Testing standard scenario (known scenarios)	ECU configuration for one vehicle variant with single option tested	Peer review for process and documentation	Performing regular tests without covering robustness
Medium risk	Invalid classes tested	Valid condition and Invalid condition (pairwise vectors)	Testing logic conditions with a maximum number of don't care terms	One-switch method and guard testing	Testing known scenarios with exceptions	ECU configuration for the main vehicle with all options enabled and tested	Group review, detailed walkthrough of the code and specifications	Testing for performance, robustness, long-term tests, data flow, stress tests
High risk	Boundary classes tested	Valid & boundary vectors (three vectors at a time)	Testing those including with one don't-care term	Two-switch method tests and MCDC tests	Testing for unknown scenarios (SOTIF analysis)	All vehicle variants with single or multiple options tested	A technical review covering design, architecture, and the test design with specifications and results	Alternate use cases, misuse conditions with the stress test, path load test, data flow stress tests
Safety critical (functional safety, SOTIF requirement)		Valid and invalid vectors tested with all possible combinations (e.g., scenario-based tests)	Testing with no don't-care term (specific application of mutation tests)	Multiswitch tests, including invalid transactions	Testing for safety-critical scenarios	All vehicle and ECU configuration and all associated options covered for testing	Detailed inspection and assessment with more than three samples checked (e.g., functional safety audit)	

yet in the market, having a person with knowledge of a similar product to perform the exploratory test is challenging. The technologies used in automated driving were already in use in other industrial areas. Looking for expertise and skillset from a different industry might help build up your team and expand its skillset. Exploratory tests mainly focus on the weak areas of the software and the system and are usually executed to explore the errors and failures in the system which would not have been found on structured requirement-based test approaches alone. Different test techniques can be used for performing exploratory tests [10.5]. It is also important to know how these test techniques can be used to achieve the required depth for testing the product. Table 10.2 summarizes a classification used for exploratory tests on different testing areas and with different test techniques to achieve a certain depth for testing.

Exploratory tests are commonly performed inside the vehicle or at the system level as part of the validation when the system and its features have reached enough maturity based on different use cases and requirements. Hence different test techniques for exploratory testing should be adapted to the test environment and the available features in the system. It is very common to have exploratory testing performed inside the vehicle. An advantage of performing these tests inside the vehicle is that it will help uncover defects beyond the system, such as bad integration or system misuse conditions. An exploratory test execution does not need to follow structured test steps. An exploratory test, which does not follow the standard test steps or by accessing the system by nonstandard means, will show up how the system will respond to those from a user perspective. Most of the exploratory testing will end up creating defects that would help further to improve the requirements and the associated software and system. This will play a significant role in defining and prioritizing the logical decision-making and state transitions of the system from a user perspective. It is recommended to perform exploratory testing in the vehicles, especially when multiple new systems are introduced or for a new vehicle program.

TABLE 10.2 Test techniques and depth classification for experience-based testing.

Experience-based tests (Not depending on a test specification)				
Risk-based classification for prioritizing	**Fault attack tests**	**Error guessing tests**	**Explorative tests**	**Checklist-based tests**
Low risk	Executing fault attack tests that are customer visible (e.g., specifically on HMI and controls)	Main functionality focused error guessing (multifunction error guessing)—Functions that are regularly used by the customer or frequently visible to the customer	Usage of a single test charter for performing tests	The checklist used only to meet the assessment requirements
Medium risk	Covering major bugs of the system (defect priority-based)	Minor functionality-based error guessing (focusing on the lower prioritized functions, which are not used frequently by the customer)	Usage of 2 test charters for testing the system	Checklist focusing on the functionality (Major and Minor)
High risk	Covering minor bugs of the system and software		Multiple test charters used for testing (>2)	Checklist focused on the failure, fail-operational, and fail-safe features

The analysis performed for testing a domain controller for ADAS and Automated Driving was the basis for organizing and classifying different test techniques, as shown in Tables 10.1 and 10.2, as a mechanism to establish the required test depth. All these test techniques can be applied to different test types across the product verification and validation processes. Readers can consider this as a reference and adapt further for their specific use case and application.

10.3. Developing a Test Suite

Verification and validation of the product are dependent on the test strategy defined as part of that product development. Especially for a new product, building up the test cases and the test suite from scratch is challenging. Usually, for most projects, the test strategy is developed by reusing parts from previous projects and taking into account the inputs and recommendations as proposed by standards such as ISO 29119 [10.6]. One of the challenges faced by many start-up organizations in ADAS and Automated Driving is establishing a solid test strategy and a strong foundation on the required verification and validation with all required artifacts. How can one define and build up the framework of a test suite for an Automated Driving system?

Test suites for an Automated Driving system can be grouped in different ways. Here the focus is those tests that are performed after the software integration has been completed. This can be classified into two, one for the test area, which covers all the activities after an integrated software is available in a laboratory environment, and the other is for the vehicle testing, which is performed in a proving ground or on public roads.

How would one estimate if the test cases are generated and covered for all the applicable areas of the system and beyond? The standard approach followed is to map these test cases to available requirements or requirement groups. It is usually a challenge to achieve maturity for Automated Driving systems until very late in the product development phase because of the iterative product development approach. On the other hand, the infrastructure and the test design required for testing the whole system should consider all the test areas at the start of the test itself. Otherwise, the rework would be expensive, especially for the tools and infrastructure that are required for testing. ISO/IEC 25010 standard is a reference that would provide insight with details on testing the product for the system and software quality [10.7]. The standard classifies the system and software product quality based on eight characteristics, as in Figure 10.1.

This section will discuss an example of creating a test suite for a new product in the field of AD by taking into account different input sources such as system-level requirements, applicable standards, regulations, and feature lists. A test suite should cover test cases in each of those areas for the system and the software. The analysis performed for the requirement specifications can be reused along with the inputs from the standard ISO/IEC 25010 to make sure all those eight characteristics are covered with the test cases. Experience-based tests should also be considered as part of the test suite as this will help in further developing test cases and improving those.

System-level test cases can be analyzed and prioritized to three different levels of testing based on the risk associated with each level (Figure 10.2):

- Level 1: Test group with tests for regulatory requirements, the requirement from standards and safety requirements
- Level 2: Test group associated with feature quality and characteristics
- Level 3: Test group with tests related to system-level use cases, user experience, stability, robustness, and reliability of the system and software

FIGURE 10.1 Characteristics of the system and software quality in a product.

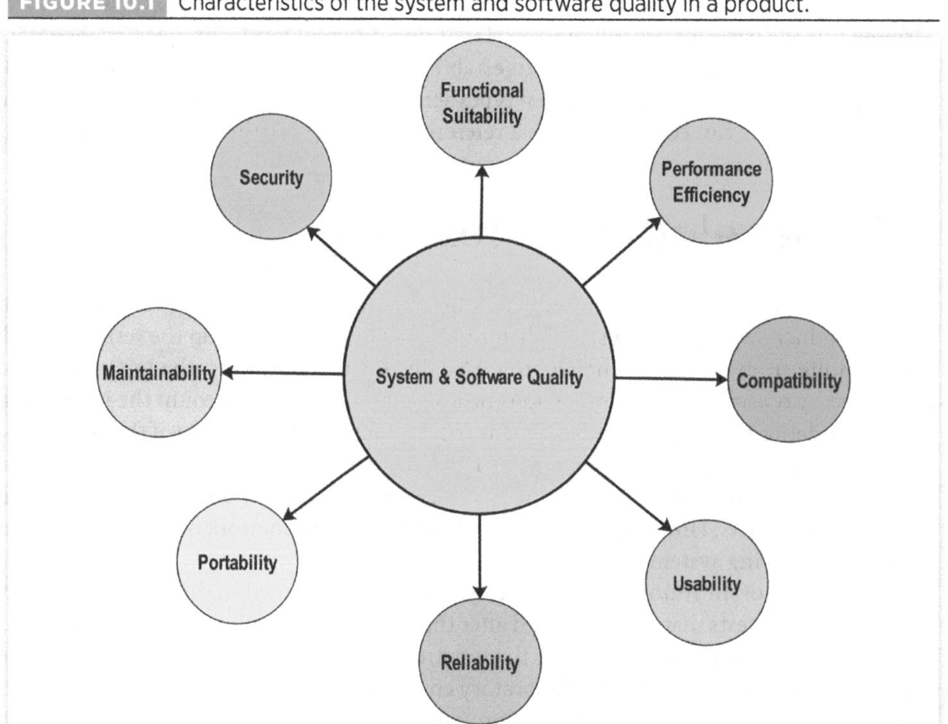

FIGURE 10.2 Test case classification and prioritization.

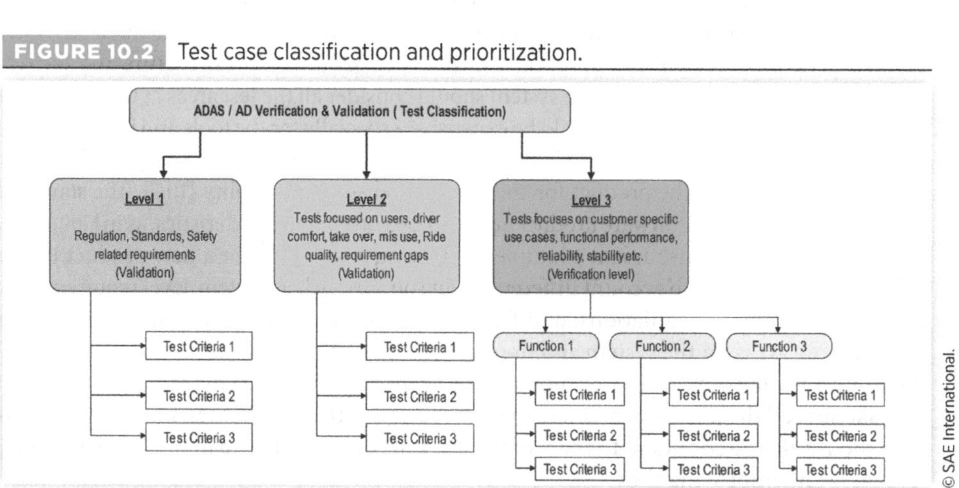

Each of these classifications has specific test criteria defined for which the system and software will be evaluated. Those test criteria are defined either based on the requirements defined for the system and features or even based on certain agreements with the vehicle manufacturer as their acceptance criteria, which covers the criteria related to the system performance and quality, feature stability, robustness, and reliability.

Figures 10.3 and 10.4 show an example of the framework used for building up a test suite for system-level laboratory-based and vehicle-level tests. These classifications take into account the prioritization and risk assessment, as shown in Figure 10.2. Even if different test infrastructures are to be used at different stages of testing, having such a framework will help to identify the areas the team has to work on to generate test cases. These would guide the team when a new product is considered for testing or when dealing with ambiguities early in the product testing phase.

FIGURE 10.3 A sample test suite structure for an Automated Driving system and its functions.

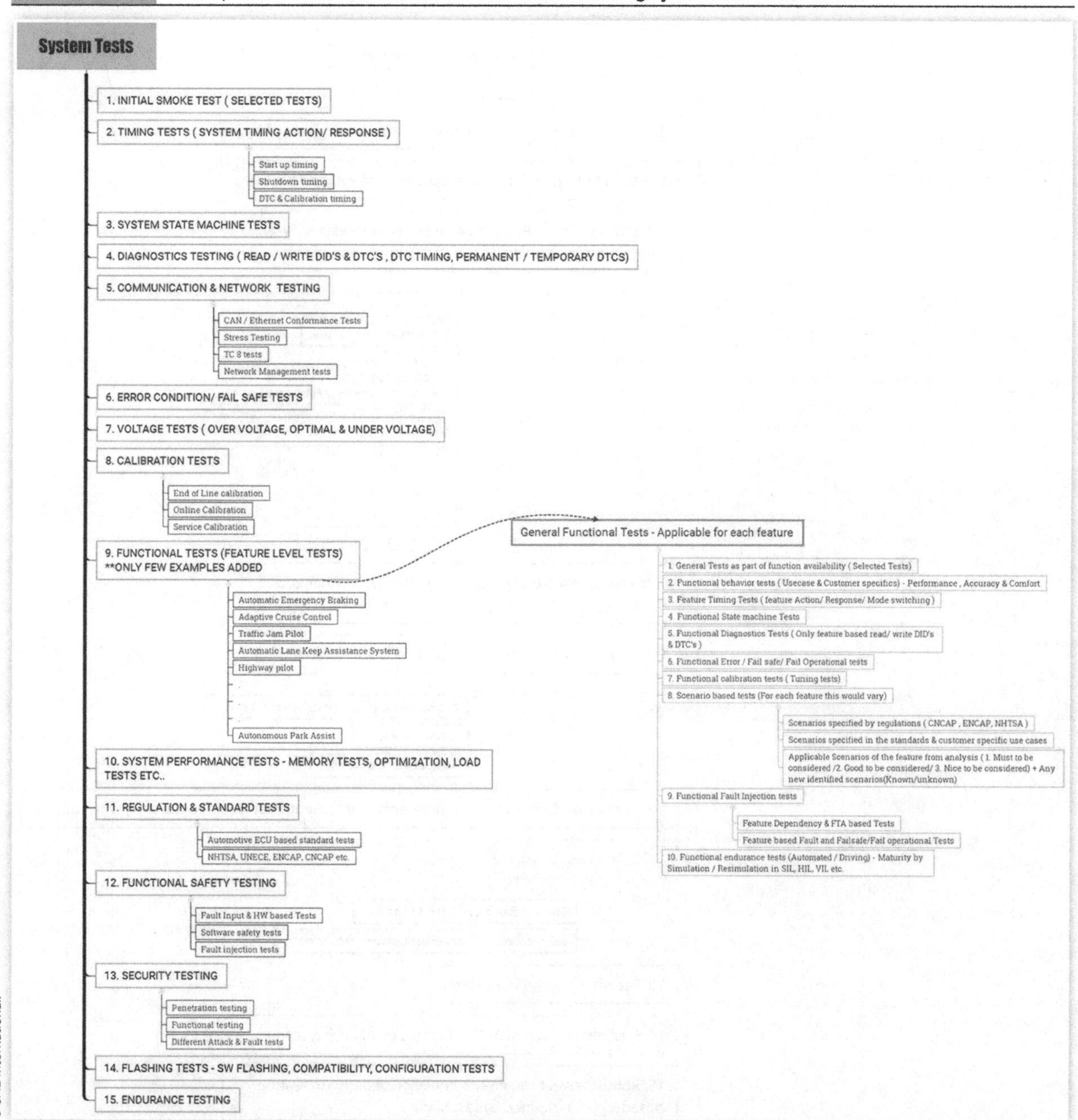

FIGURE 10.4 A sample test suite structure for vehicle testing.

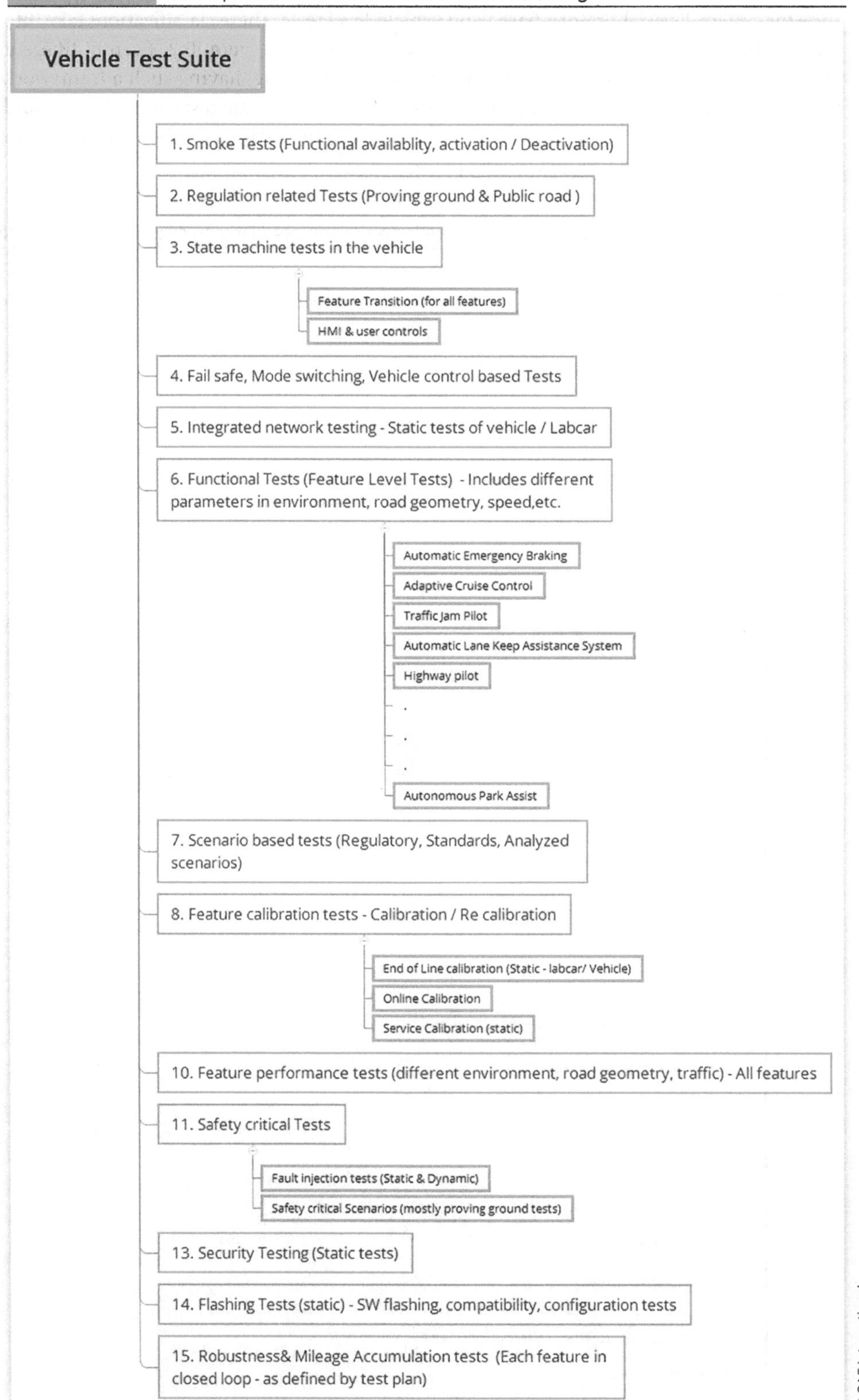

TABLE 10.3 An example of system test maturity evaluation metric.

Test case classification in (system level) test suite	Weightage (risk-based weightage allocation)	Test maturity
General system test cases	1	
Flashing and data flow test cases for the system	1	
Test cases related to regulations (geographic location specific)	1	
State machine test cases (system state machine)	1	
System operations and operational timing test cases	1	
System interface and network with network timing test cases	2	
Test cases related to applicable standards	2	
Diagnostics test cases	2	
Calibration test cases (system and sensor)	3	
Error condition/fail-safe/fail operational test cases of the system	2	
Safety-specific test cases	2	
Cybersecurity test cases	3	

Weightage 1: Critical requirements which are the foundation for the system and feature; **Weightage 2**: Those with medium impact on the system and feature; **Weightage 3**: Those which are not blockers but required as part of final delivery of the system and feature.

TABLE 10.4 An example of feature test maturity evaluation metric.

Test case classification in (feature level) test suite	Weightage (risk-based weightage allocation)	Test maturity
General feature related test cases (Activation/ Deactivation)	1	
Feature-state machine-related test cases	1	
ODD, boundaries, and prerequisite for feature enable test cases	1	
Feature timing test cases (reaction, response timing)	1	
Mode-switching and driver override test cases	2	
Feature-related standard test cases	2	
Feature diagnostics test cases	2	
Feature calibration test cases	3	
Feature interdependency and error/fail-safe test cases	2	
Test cases related to regulations (UNECE, ENCAP, CNCAP, NHTSA)	1	
Scenario-based test cases (scenarios classified to must-have, good to have, nice to have)	2	
Test cases associated with safety goals and feature safety requirements	2	

Weightage 1: Critical requirements which are the foundation for the system and feature; **Weightage 2**: Those with medium impact on the system and feature; **Weightage 3**: Those which are not blockers but required as part of final delivery of the system and feature.

Different test techniques and design methods can be applied in each area considered while designing test cases. Another challenge one would face while performing tests in any new product is about the coverage. Coverage of requirements with the test cases for a particular system and its features are one of the metrics usually used by almost all organizations for acceptance. It is easy to track the coverage employing different tools which are available in the market. A simple mechanism to classify different test groups for generating metrics to track test maturity during project execution is shown in Tables 10.3 and 10.4. This is a handy reference to prioritize and track different test groups and associated maturity achieved with the test cases. The weightage considered for each subgroup of the complete test suite is an outcome of system analysis and risk assessment.

10.4. Test Process

Test execution follows specific processes over the entire verification and validation. Having a robust test process is a sign of better control of the quality of the product during its development. In this section, an overview of the processes is explained and then followed by an example. Having a systematic approach for the end-to-end test activities is mandatory, especially to control the quality and performance of complex products.

It is common in many organizations to have a test framework that covers all test types as part of the test infrastructure. This test framework would integrate different test environments and tools as part of the verification and validation infrastructure. As shown in Figure 10.5, the high-level test process shows the workflow and the interactions across various test environments at different levels of testing. Test selection, execution, and integration of those with different test environments are managed either by a standard interface tool or using multiple tools as part of a toolchain. The use of scenario-based simulation and tests associated with data-driven software development increased the

FIGURE 10.5 An overview of requirement-based and experience-based testing processes.

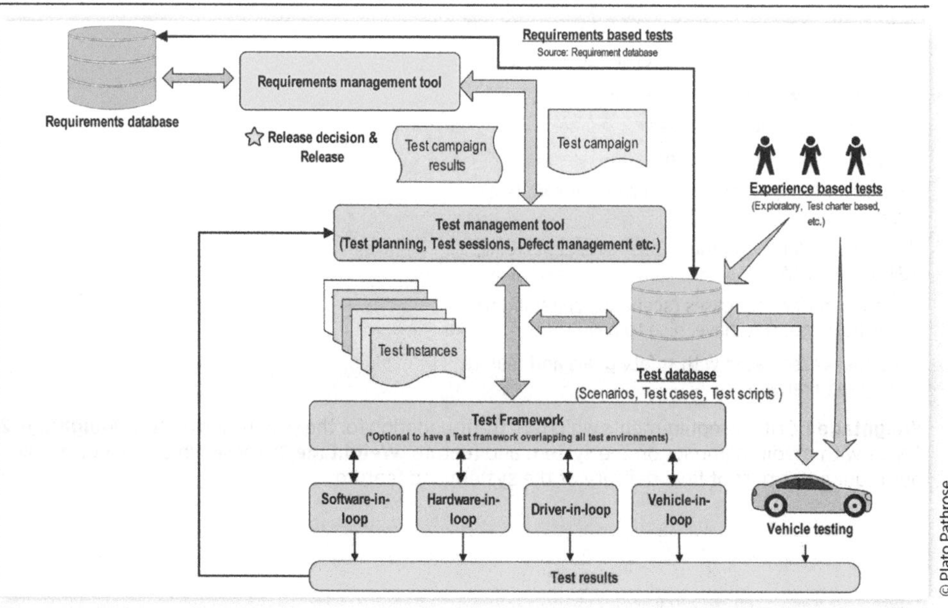

demand for the interface and the infrastructure. This forced many organizations to depend on certain external platforms and tools, mainly the cloud platform providers and associated software packages, to integrate with the test framework.

This is a generic process flow used as an example for system-level verification and validation. This varies with each organization and application, along with the tools and infrastructures used. The test framework acts as the main interface between different lifecycle management tools which are used for requirements and test database management. The test management tools are used to organize and plan test sessions and objectives for each test campaign.

The results from each test campaign are accessible from the test management tool as log files or reports generated locally at different test environments. The health monitoring system of the test framework will help identify the bottlenecks and the glitches in test executions from tools and other infrastructures, including errors by either tracking it independently or collecting information from the health monitoring functions of independent tools and test environments.

Using a health monitoring function in the test framework is advantageous, especially when used across different test environments and infrastructures. As the scope and the scale of testing have increased multiple folds from the traditional test approaches, new technologies and methods were required to meet those demands. That is how the utilization of cloud platforms and cloud technologies gained popularity in the automotive industry for software and system testing [10.8, 10.9]. Cloud-based testing was used earlier in different industries. With a more software-driven approach in the automotive industry, these technologies found their place and became popular in the automotive industry. These technologies, along with the platforms, provide a flexible ecosystem for development and testing. This is beneficial and helps reduce time, money, and effort to move data from one place to another.

These days, the test environments and infrastructure has changed a lot, even for software testing. Large-scale utilization of cloud platforms and cloud technologies is found in many organizations in ADAS and Automated Driving. The primary usage of these infrastructures is in software development, testing, and system verification and validation, as shown in Figure 10.6. Scenario-based testing in different test environments and the development and verification of data-driven software components mainly utilize the cloud platforms. With the support of cloud-based platforms, different types of tests can get executed, and that too on large scales. It is possible to execute thousands of scenarios using multiple instances of execution in a cloud environment. This is one of the major advantages of cloud-based development, integration, verification, and validation activities. It saves the time, cost, and effort required to move the data from one place to another.

Moreover, automation in the test execution also helps to utilize the cloud platforms and applications for testing [10.9]. One of the major challenges in deploying complete test activities in a cloud environment is in having a reliable test framework. The test framework should manage different tools and infrastructure as part of the verification. The validation toolchain should be capable of handling multiple instances of tools and test iterations operating in the cloud. It should also be able to timely detect and report errors and failures. For long-term test runs, it is a common failure that sometimes the tools become nonresponsive, fails to report exceptions or errors, etc. Hence, having a health monitoring function in the test framework is very important. This also adds complexity to the test framework design and development phase for scaling the test executions in a cloud platform.

It can be noticed that in the recent past, many organizations have migrated to utilize data centers of the cloud platform suppliers to establish remote HiL test systems and

FIGURE 10.6 An overview of cloud-based test execution.

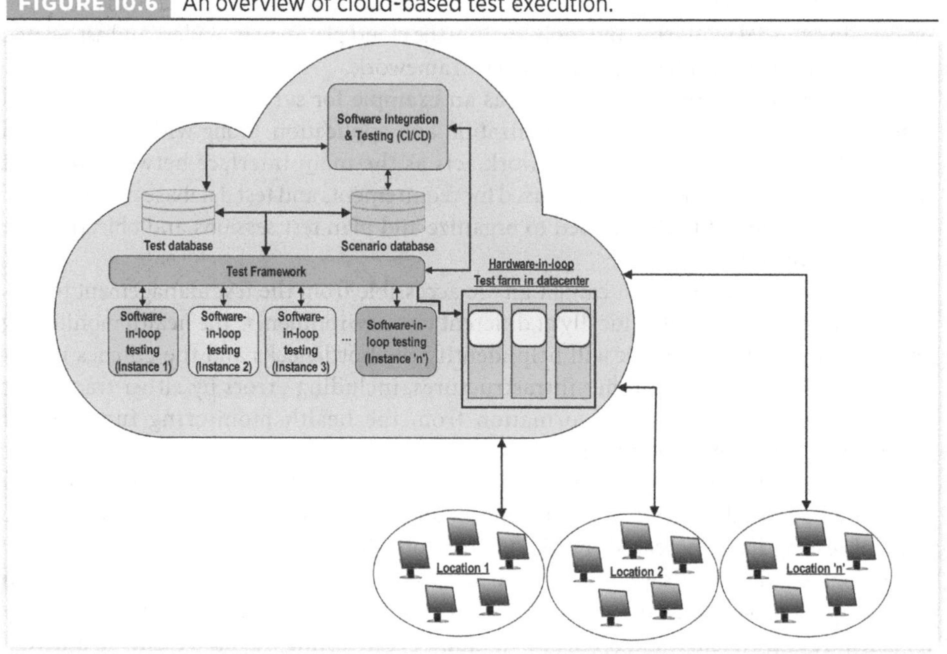

test farms. This is mainly utilized as a platform for performing integration tests as well as system-level tests. This is usually integrated as part of continuous integration and development (CI/CD) in the product development workflow to use time zone advantages by multinational organizations as there is no delay because of data transfer. The traditional large-scale HiL Farms is gradually converted to the data center-based HiL infrastructure by cloud platform providers. This is now in use by different suppliers and vehicle manufacturers.

10.5. Testing in the Vehicle

Testing in the vehicle is unavoidable for any automotive product development. The scale of tests executed and the effort to execute those tests in the vehicle are inversely proportional. The tests executed in the vehicle are mostly part of the validation, where the use cases and user experiences are checked especially rather than pure requirement-based testing. This does not mean that none of the requirement-based tests will be executed in the vehicle. If not appropriately managed, testing in the vehicle is the most complex and expensive testing for any vehicle program. This has grown recently because of the introduction of safety-critical ADAS and Automated Driving features in the vehicle. This is because of both the complexity of the tests involved as well as the time taken to achieve the reliability and robustness required for those features.

A general test execution process followed for vehicle-level testing is shown in Figure 10.7. Readers are free to adapt and use the process depending on the time and strategy defined for vehicle-level tests in a project. Testing in the vehicle for ADAS and Automated Driving features usually starts once when the system and software achieve a certain level of maturity based on the requirements defined for the system. Different entry criteria can be considered for entering the test phase in the vehicle. The effort and time required for executing tests in the vehicle are huge compared to any other laboratory-based test activities such as

FIGURE 10.7 Vehicle test process and workflow.

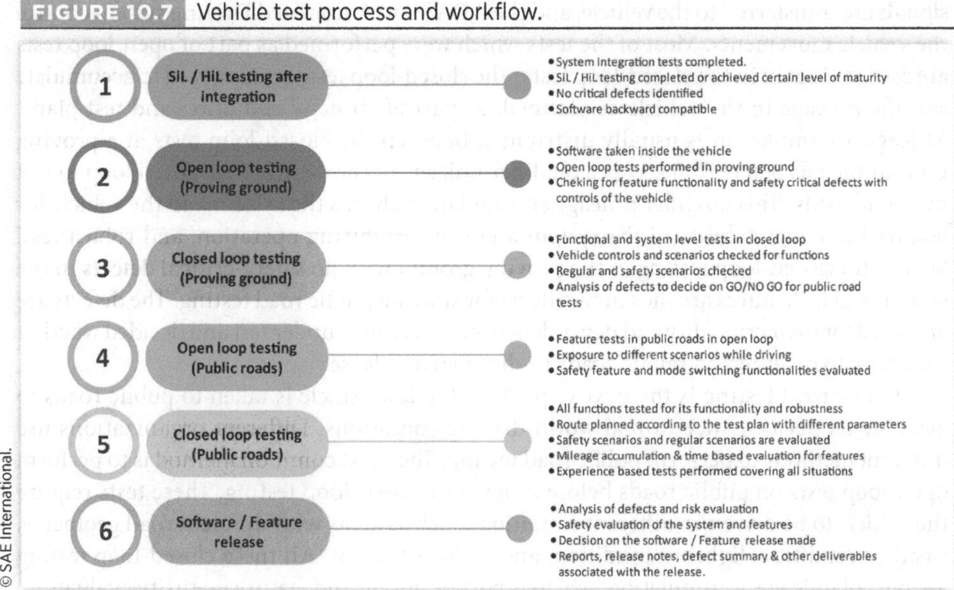

Software-in-the-loop testing or HiL testing. Laboratory-based test executions are usually performed earlier to check the maturity of the software and system before testing it in the vehicle.

Vehicle testing is usually performed in an open-loop configuration where the control signals are only analyzed in the first phase and later in the closed-loop configuration where the control signals are transmitted to vehicle controls and actuators, thereby closing the loop. The environment where vehicle testing is performed can vary depending on whether it is in a controlled environment such as a proving ground or test track, or even in public roads and highways.

Vehicle testing is usually performed using a fleet of vehicles by the vehicle manufacturer. On the other hand, the suppliers who provide ADAS and Automated Driving systems also perform vehicle tests with a limited number of test vehicles. The vehicle test process needs to be adapted based on the project scope and the availability of the vehicles for testing. Usually, the entry criteria for a system to undergo vehicle testing are its successful execution of a software-in-the-loop or HiL tests where the system and its functions are checked in a laboratory environment. The results of these tests are taken as input to check if the system and its features qualify for testing in the vehicle. Open defects, implemented requirements, the status of the features under tests, etc. are checked before proceeding with vehicle tests. The vehicle test plan should define the type of tests and the test depth required based on the software and system release for each test cycle.

Vehicle testing usually starts with open-loop testing in a controlled environment such as a test track where the vehicle is evaluated through data collection and analysis and checked for its behavior while testing different functionalities. These tests are mainly used to check and make sure that there are no safety-critical defects present in the system and the software for different functions and to evaluate certain fundamental functional tests. These tests usually include specific scenarios which are common on public roads while driving. A set of scenarios, along with exploratory tests, are performed inside the vehicle. The system is checked for functional misbehaviors and other defects which are safety critical. Once the open-loop tests are completed, the closed-loop tests are executed where the control signals are transferred to the vehicle, and the ADAS and Automated Driving system control the vehicle movements. Most of the tests which were performed as part of open-loop tests are executed again in the closed-loop tests. The closed-loop tests are also used to accumulate specific mileage in the vehicle as planned, as part of strategy definition and test plans. Mileage accumulation is usually distributed between the closed-loop tests at a proving ground such as 500 or 700 km as a qualifier mileage of the vehicle in a closed loop to test in public roads. This qualifier mileage accumulation checks the system and the vehicle for feature behavior, stability of the system and features during operation, and robustness. Successful closed-loop testing in the proving ground with no safety-critical defects in the system and its features are the entry criteria for starting public road testing. The defects are analyzed for their criticality and dependency on the features under test and decided whether they have to be tested on public roads in the current release.

Public road testing is the next step where the test vehicle is taken to public roads to evaluate its performance in real-world driving conditions. Different organizations use different methods to perform public road testing. The most common method is to perform open-loop tests on public roads before moving to closed-loop testing. These tests require the vehicle to be driven in different conditions such as areas with different road geometry, weather condition, lighting condition, and traffic situation. All these closed-loop testing are part of mileage accumulation, which is part of system and feature maturity evaluation. These tests will continue until a defined time or mileage is reached in a closed-loop test with minimal defects which are noncritical. During these closed-loop tests, the data logs mainly capture the sensor data, middleware data, and vehicle network information, which are stored for further analysis and evaluation of features and the system in the laboratory environment.

Software release and a change in software after each test iteration in closed loop usually occur after four to six weeks. Whether the software needs to be delivered to the end customer is decided based on the test results. For a vehicle manufacturer, these iterations are executed with a fleet of vehicles, and the maturity of the software is measured based on the defects and the feature behavior across all the vehicles in the test iteration. Vehicle manufacturers consider these iterations as continuous improvement of the system and software on each of the quality gates of the vehicle program. A system supplier will consider it as part of the maturity and release gates of the Automated Driving system delivered to the vehicle manufacturer.

10.6. Summary

Verification and validation and different methods of applying those processes to qualify ADAS and Automated Driving systems were discussed with examples in this chapter. The transformation of the verification and validation process to agile through different methodologies such as TDD and FDD was discussed in detail. As the methods and approaches used for the development and testing of the products changed with the adoption of new technologies, test design and application of different test techniques found their way in different products. Especially for complex products like ADAS and Automated Driving systems, the test design becomes a mandatory component for testing and qualification.

Different ways in which the test techniques are applied to design test cases that can address the coverage and depth of the system were discussed with a few examples and references from the industry. Test depth definitions and the adoption of different test techniques help test the system and the software to attain better quality. Test cases must be designed as per the requirements and based on experience, considering possible chances of error. This is mandatory while testing a safety-critical product.

How to define a test suite? This is challenging for someone who does it for the first time. A sample framework of the test suite was discussed with examples in Figures 10.2 and 10.3. Usually, there is a tendency to reuse the structure defined in the requirement specification for the test suite for better organization and differentiation of the overall system. Testing an Automated Driving system involves a lot of processes across different levels of testing. Having such a robust process will help have better control and management in the system development and testing phases. Following processes does not guarantee quality if the technical qualifications and maturity of the products are not considered.

In-vehicle testing is an important phase for any automotive product. ADAS and Automated Driving systems undergo a robust test phase in the vehicle both from the supplier side and the vehicle manufacturer. A high-level vehicle test overview and process give the reader insight into how the vehicle testing is organized and the associated phases in testing automated driving features in the vehicle. Testing without any measurements is not useful. For an organization with a legacy of ADAS products such as adaptive cruise control and AEB, it is easy to apply those learnings in developing and testing Automated Driving systems. For those organizations, the goal is to improve test execution speed and the depth of testing by using various test techniques, methods, and automation. It also includes adopting better tools and technologies. For safety-critical product testing, the tools and the techniques applied along with the process and methods determine the quality of the product.

References

10.1. Crispin, L. and Gregory, J., *Agile Testing: A Practical Guide for Testers and Agile Teams*, 1st ed. (Boston, MA: Addison-Wesley Professional, 2008), ISBN:978-0321534460.

10.2. Beck, K., *Test Driven Development: By Example*, 1st ed. (Upper Saddle River, NJ: Addison-Wesley Professional, 2002), ISBN:978-0321146533.

10.3. Pathfinder Solutions, "Effective TDD for Complex Embedded Systems Whitepaper," April 30, 2012, accessed September 10, 2021, https://web.archive.org/web/20160316153308/http://www.pathfindersolns.com/wp-content/uploads/2012/05/Effective-TDD-Executive-Summary.pdf.

10.4. Palmer, S.R., *A Practical Guide to Feature-Driven Development* (Upper Saddle River, NJ: Prentice Hall International, 2002), ISBN:978-0130676153.

10.5. Batch, G. and McKay, J., *A Software Engineers Handbook*, 2nd ed. (Santa Barbara, CA: Rocky Nook Computing, 2014), ISBN:978-1-937538446.

10.6. ISO/IEC/IEEE 29119, "Software and Systems Engineering—Software Testing," September 1, 2013.

10.7. ISO/IEC 25010:2011, "Systems and Software Engineering—Systems and Software Quality Requirements and Evaluation (SQuaRE)—System and Software Quality Models," 2011.

10.8. Banzai, T., Koizumi, H., Kanbayashi, R., Imada, T. et al., "D-Cloud: Design of a Software Testing Environment for Reliable Distributed Systems Using Cloud Computing Technology," in *2010 10th IEEE/ACM International Conference on Cluster, Cloud and Grid Computing*, Melbourne, Australia, 2010, 631-636, doi:https://doi.org/10.1109/CCGRID.2010.72.

10.9. Tilley, S. and Parveen, T., "Migrating Software Testing to the Cloud," in *2010 IEEE International Conference on Software Maintenance*, Timi oara, Romania, 2010, doi:https://doi.org/10.1109/ICSM.2010.5610422.

/Autonomous
/Sensing
/Communication
/Battery
/Navigation
Autopilot mode

11

Acceptance Criteria and Maturity Evaluation

As with any product developed, there will be challenges and limitations during the development and deployment of Automated Driving systems. Even though all wish to have an ideal Automated Driving system that handles every situation while driving, it is impossible. There are no fully autonomous vehicles for public road usage in large-scale production available, and will not be available at least for some time. With all the limitations of these Automated Driving systems, there come risks in different forms such as safety, security, and functional misbehaviors to its users and society. It is a complex process to deploy safety-critical functions inside a vehicle. When the system is not in ideal conditions, how is it possible to improve the system performance, safety, and reliability?

As with any other system made, there is a goal defined for the performance, which must be met to be taken inside the vehicle. Complex systems like ADAS and Automated Driving systems also undergo measurements across certain defined matrices and goals, collectively called acceptance criteria. This chapter takes you through the acceptance criteria used for certain ADAS and Automated Driving systems that were already deployed or are undergoing development to launch in the market shortly. Readers will understand how various vehicle manufacturers and their suppliers decide on the quality and functionality of complex systems inside their vehicles. When deploying such a complex system, there are also risks involved if they are not mature and not tested enough. A novice in ADAS or Automated Driving can consider this as an informative reference to understand how the vehicle manufacturers and their suppliers deploy systems inside their vehicles and how they measure quality and prove it is safe enough. Moreover, the residual risk associated with the acceptance criteria is an important factor determining the system quality and user safety while using the vehicle.

11.1. Need for Acceptance Criteria

Acceptance criteria are a set of metrics or goals which a product needs to meet for its acceptance by the customer. A set of measurable metrics will be used to estimate how good the system is for acceptance. During the concept and planning phase, vehicle manufacturers define certain functionalities and their use cases in the vehicle. These end up with their suppliers as requirements for the system, which has to be developed. A vehicle manufacturer will define specific, measurable outcomes from the system as functions or with a minimum quality and performance, which is acceptable. These matrices will also be integrated into the requirements for the system development. While the tests are executed, there should be ample evidence to prove that these acceptance criteria are met for the acceptance of the product [11.1]. Acceptance criteria are part of the scope definition for any project to define what is an acceptable quality of the deliverable from that project. It can vary from products or even can be defined for services.

With the transformation of projects and product development from traditional ways to digital and using agile methodologies, there is a misunderstanding across many people with the terms "Acceptance criteria" and "Definition of done." While following agile methodologies, iterative product increments are achieved over the entire duration of the project. Agile testing will help in making sure these iterative increments are released with quality and performance as required. Definition of done is a term described in the scrum guide for those requirements that a product increment should meet to be releasable [11.2]. Definition of done is applicable for any user stories or tasks against which it must be measured. A set of requirements and goals are defined for each task to meet the goal for that task. It brings measurement for completeness of the task or the iterative increment of the product. It also helps in having a clear understanding among the team. This also gives an overview of the quality and performance requirements of these increments.

Acceptance criteria can be considered as a superset where a set of definitions of done are part of it on each incremental release iteration of the product. This is mainly defined as the product quality and performance requirements as a finished product with its features and not for the smaller chunks delivered to a customer. Acceptance criteria is an input to derive tasks and user stories and define their definition of done on iterative product development. When the user stories and tasks are derived based on a product backlog, the user stories are expected to encapsulate the acceptance criteria and the definition of done. Both are required to better understand the product developed by the team as regards what should be achieved for the incremental release and what is expected from the customer from the finished product. This chapter explains the acceptance criteria of ADAS and Automated Driving systems that determine the product-level features and functionalities that are considered while developing and deploying vehicles for production.

11.2. Defining Maturity of the System and Features

Acceptance criteria define the maturity and quality of the system and its functions, which directly influences the performance and quality of the features in the vehicle. Each vehicle program has certain defined quality gates to measure the progress and the quality at different program phases. These gates also incorporate the requirements that must be met to achieve a certain level of performance, quality, and maturity of different features in the vehicle. In a vehicle program, these are usually achieved by adding new features, improving existing features from the previous gates, revising the hardware and software, or even replacing the system. This chapter will discuss how the maturity is defined for ADAS and Automated Driving systems in a vehicle program considering different phases of its development.

The system quality and performance requirements and features at different project phases are being defined based on the eight characteristics of product quality as defined by the standard ISO/IEC 25010 [11.3]. Vehicle manufacturers usually define the acceptance criteria based on the features when working with a system supplier. This is usually defined as part of the requirement specification. The quality and performance requirements will be translated to test requirements and are integrated into the verification and validation strategy as part of the system development. The acceptance criteria are defined for the system-level verification and validation strategy, considering different types and the deliverables from those tests. The metrics defined for each test type should be achieved by testing, and their evidence must be provided as part of the project deliverable.

The acceptance criteria defined include metrics for the system as well as for the features such as adaptive cruise control and highway assist. System verification and validation are the main processes where these matrices are measured and evaluated. The acceptance criteria, in most cases, speak only about the evidence that the system and its features are stable, robust, and reliable over a period meeting all the requirements from the customer as use cases. This is proven by the evidence provided across three different test environments after executing different types of tests as part of system verification and validation, as shown in Figure 11.1. They are primarily software-in-the-loop, HiL for the system verification, and

FIGURE 11.1 An overview of the acceptance criteria definition for system verification and validation strategy.

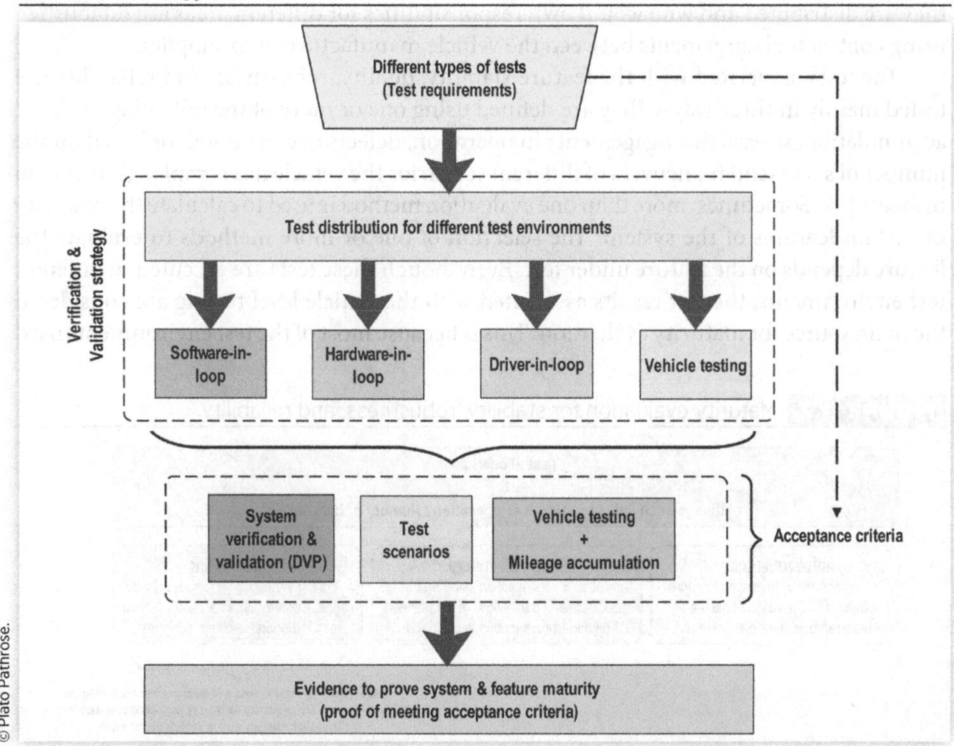

vehicle testing as part of system validation. Even though various organizations use DiL and vehicle-in-the-loop test environments, they are not considered the basis for acceptance criteria commonly because those test environments were not used iteratively as part of regular verification and validation process in many of the projects.

During various tests executed as part of the system verification and validation in these three environments, the stability, robustness, and reliability of the system and features are tested. There will also be tests that evaluate the mode switching conditions where the control transfer between humans and the autonomous system is tested in various conditions while in operation.

Typically, organizations follow different methods to define maturity and acceptance criteria for an ADAS and Automated Driving system for different gates or phases in a vehicle program. Most of them are defined based on the following factors measured against various test environments. They are (i) Coverage and traceability of the requirements with test cases, (ii) Open defects and their criticality evaluation, and (iii) Results of regulatory tests, use cases, and the tests associated with stability, robustness, and reliability along with planned mileage accumulation. The evidence of all these factors is considered for deciding whether the acceptance criteria are met.

The number of tests one could execute in each test environment varies while moving from the software-in-the-loop simulation to the vehicle testing at different levels. The test strategy should define what should be executed as part of tests in each test environment, and the test plan should cover those for each test iteration [11.4]. The tests related to stability, robustness, and reliability are executed across various test environments. The distribution on how much one should test in each test environment for acceptance is usually defined jointly by the vehicle manufacturer and supplier. Sometimes, part of those tests, which should be executed in the vehicle, is taken care of by the vehicle manufacturer and the supplier, or even independently. Although the acceptance criteria cover all these areas, how they are distributed and who would own responsibilities for different areas are established using contractual agreements between the vehicle manufacturer and supplier.

The tests associated with the feature stability, quality, robustness, and reliability are tested mainly in three ways. They are defined using one or more of the following: mileage accumulation, several disengagements in operation, defects over a period, or based on the number of successful maneuvers of different scenarios the vehicle has completed, as shown in Figure 11.2. Sometimes, more than one evaluation method is used to calculate the maturity of certain features of the system. The selection of one or more methods to evaluate the feature depends on the feature under test. Even though these tests are executed in different test environments, the test results associated with the vehicle level testing are considered the main source for maturity evaluation. This is because most of the test environments used

FIGURE 11.2 Maturity evaluation for stability, robustness, and reliability.

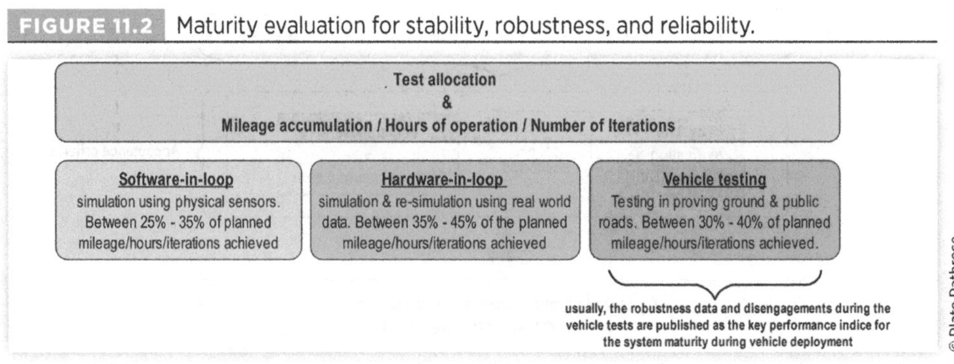

for the vehicle programs by the suppliers vary. The lack of correlation of the laboratory-based test environments to the real world is a concern for the vehicle manufacturer. This could also be because of the unavailability of identical test environments at the supplier and vehicle manufacturers.

11.3. Maturity Evaluation for the System

While developing an ADAS or Automated Driving system, the maturity is evaluated in two parts. One is the maturity of the system, an ECU, or domain controller with all interfaces to the vehicle and the other is the maturity of the features associated with those systems in the vehicle. Even though the features are provided by the system in the vehicle, feature maturity is evaluated separately. For the system-level (ADAS/Automated Driving system) maturity, the evaluation primarily looks at the requirements coverage and traceability and the status of the open defects and their criticality.

The test cases considered as part of maturity evaluation of the system are part of either or both system or software verification. Requirements-based testing and experience-based testing are performed in various test environments to evaluate the performance, stability, robustness, and reliability of these systems. Environmental tests, production-specific tests, endurance tests, load tests, and stress tests are usually carried out as part of the system testing and are also checked for acceptance of the system. The maturity of the system varies at different phases and different versions of the product during the product development.

Evaluation of the system maturity is checked normally in a vehicle program when the system meets at least 70% of the listed system requirements. The tests that are executed for evaluating the stability and reliability of the system for any prior versions do not bring much value to the measurements. Usually, in the product development of an ADAS or Automated Driving system, there can be three or four different versions of the ECU or the ADAS/Automated Driving system over the whole product development. These numbers would change depending on the supplier and the project. Various versions of the system can also relate to prototypes that are only for laboratory use and those that can be integrated into the vehicle. The versions that are integrated into the vehicle should be considered for the maturity evaluation and acceptance at different phases of the vehicle program.

An overview of the acceptance criteria of the system in a vehicle program is shown in Figure 11.3. Here the maturity is evaluated from the field prototype of the system to the production version that is integrated into the sellable vehicles. This is only for reference and can vary with the maturity evaluation by different organizations and projects. Each vehicle manufacturer and supplier has its own maturity evaluation methods.

11.4. Maturity Evaluation for the Features

Like system maturity evaluation and acceptance, the features from the vehicle level also undergo evaluation for their maturity based on their performance, stability, robustness, safety, and quality. This is evaluated based on how good and error-free the features are while in operation. The feature maturity evaluation is a subset of the system maturity, and both are interrelated. Hence, they are evaluated in parallel based on the deployed features for each software release.

Feature maturity evaluation is always performed for the closed-loop test execution in different test environments. There are also certain open-loop re-simulation tests performed

FIGURE 11.3 An example of acceptance criteria and maturity for ADAS/AD system at different phases.

		Test coverage	Defects
System prototype Ver.1	• Field prototype Ver.1 • 60% similar to the production version	• 70% Functional requirement test coverage. • 50% Non functional requirement test coverage. • 100% Functional safety test coverage . • 50% Diagnostic test covergae	• No critical defects present.(such as system release blockers) • No safety critical defects present. • Not more than 40% of functional defects present • Not more than 60% defects on following (Diagnostics,Hardware,Electrical,Homologation)
System prototype Ver.2	• Field prototype Ver.2 • 90% similar to the production version	• 100% of coverage of the functional requirement. • 100% of coverage of non functional requirement test coverage. • 100% coverage of functional safety requirements. • 100% of diagnostics test coverage • 100% of PPAP sample tests covered	• No critical defects should be present which are production blockers • No Functional safety defects should be present. • No Functional defects should be present. • No non functional defects should be present.
System prototype Ver.3	• Field prototype Ver.3 • 100% equivalent to the production version • Sellable to customers as in prototype vehicles	• 100% test coverage of functional requirements. • 100% test coverage of non functional requirements. • 100% test coverage of functional safety (technical safety concept). • 100% test coverage of diagnostic requirements • 100% coverage of PPAP sample tests, End of line tests, about 3000 or more software flashes in system	• No critical defects should be present in PPAP sample Tests and EOL. • No Functional safety defects should be present. • No Functional Defects should be present • Should have only 5 or less failures in 3000 plus flash tests
SOP (Start-of-production) Version	• SOP (Start-of-production) version. • System ready for the fully functional dealer vehicles.	• 100% test coverage of functional requirements. • 100% test coverage of non functional requirements. • 100% test coverage of functional safety (technical safety concept). • 100% test coverage of diagnostic requirements • 100% coverage of PPAP sample tests, End of line tests, about 3000 or more software flashes in system	• No critical defects present for PPAP tests and EOL tests. • No functional safety defects present. • No functional Defects present • 5 or less failures in flashing long term tests. • If any defects exists,all defects should be cleared with 100% coverage of defects with deviation Report agreed.

in the industry for evaluating features by some organizations. Here, the focus is more on the software, especially the perception part of the software. How different tests must be performed and which test environment to be used for testing are defined as part of the verification and validation strategy. Coverage and traceability for the feature-specific requirements, including requirements associated with safety, use cases, and the criticality evaluation of the identified defects, are usually considered as the matrices to evaluate the acceptance criteria and maturity of features at different phases of the vehicle program. Scenario-based testing plays an important role in ADAS and Automated Driving feature evaluation for its functionality, stability, robustness, and reliability. Tests are executed to evaluate the features, and according to their results, the quality of the feature and the usage restrictions for those features are established. Usage restriction simply means defining the boundary conditions where the software can be used. The usage restriction for a particular software version with associated features is the outcome of verification and validation results and the analysis of the identified defects for their criticality and impact on the user. With these classifications based on risk, it is helpful to communicate the risk associated with the software and associated features in the vehicle during its intermediate delivery stages to a vehicle manufacturer or their supplier. Figure 11.4 provides an overview of a sample feature maturity evaluation metric showing the evaluation criteria at different phases of a vehicle program. Testing ADAS and Automated Driving features and the measurements of certain specific predefined parameters help evaluate the feature quality and performance. They are defined as part of the strategy and are considered while performing feature-level testing, requirements-based tests, and experience-based tests. There is another metric that is widely

FIGURE 11.4 An example of acceptance criteria and maturity evaluation for features.

	SiL, HiL, DiL Test coverage	Vehicle test coverage	Defect evaluation
Feature prototype Ver.1 • Initial version of feature available with usage restrction* upto level 1 • 80% feature requirement implemented.	• 100% of feature requirements tested • 2- 5% of the mileage accumulation planned is achieved. • 10% of the standstill Iteration tests completed. • Started of feature regression tests(atleast 2-3 iteration	• No mileage accumulation** in vehicle performed • 90% of the stand still iteration tests are completed for features. • Start of vehicle regression tests (stand still tests).	• No feature blockers should be present. • No fail safe feature spoilers on the feature present • Communication tests completed at HIL,No alive Counter, missing Message related defects are present.
Feature prototype Ver.2 • feature with usage restriction* upto level 3 • 100% of feature requirement achieved (only trained drivers can drive)	• 100% of feature requirement tested. • About 20% of the mileage accumulation planned is achieved. • Regression tests executed for atleast more than 15-20 iterations for multiple software Releases) • Stress and Robustness Tests for features executed more than 2-3 times.	• 60% of the mileage accumulation** planned is achieved • 100 % of the Stand still Iteration tests performed. • vehicle Regression tests > 10 Iteration tests performed.	• No feature blockers should be present. • No fail safe feature spoilers on the feature present • Defect Iteration tests so that feature defects <10 % of the total defect • All electrical,communication and FUSA fail-safe feature operation are without any defects.
Feature - production intended • Function intended for production. • No usage restriction*(Can be driven in roads with normal drivers)	• 100% of feature requirement tested. • About 70% of the mileage accumulation planned is achieved. • Regression Tests in full swing for (cumulative iteration > atleast 100 softwares release cycles) • More than 70 Iteration of stress and robustness feature tests completed.	• 80% of the mileage accumulation** planned • 100 % of the stand still Iteration tests performed • vehicle Regression tests > 100 software release cycles performed	• No Feature Blockers. • No feature level EOL,Diagnostics or stress tests defects present • Robustness defects for feature < 3 • ODD related defects should be < 10% of overall defects.
Feature in production(SOP) • SOP(Series production) level feature maturity • fully functional feature for custtomer vehicles	• 100% of feature requirement tested. • 70% of the mileage accumulation planned is achieved. • Regression tests in full swing for (cumulative iteration > atleast 100 softwares release cycles) • More than 70 Iteration of stress and robustness feature tests completed.	• >90% of the mileage accumulation** Done in vehicle • 100 % of the stand still iteration tests performed • vehicle Regression > 150 software release cycles performed	• Zero defect and Technical sign off achieved for all features • overall driver comfort issues < 5 • No fail safe, fail operational defects and meeting ODD > 95% for all features.

* usage restriction refers to the constraints in using a released software in different test environments including vehicle due to the risks associated and the maturity of the software.
** As an example for the feature maturity, the feature maturity and acceptance criteria based on mileage accumulation only is considered here.

used these days for the features associated with ADAS and Automated Driving. "Safety performance index" is the metric that is coming from the aviation safety manual, which gets used these days for evaluating the safety of the system and features based on the failures or misbehaviors associated with it while repeating or executing certain tests.

A fleet of vehicles is usually used to test and evaluate different features in the vehicle for their functionality, stability, robustness, safety, and reliability over a period. These tests usually start with the first vehicle-deployable prototype version of the ADAS or Automated Driving system (ECU or domain controller) with the features. These vehicle tests are usually split between tests executed on a proving ground and those executed on public roads. Most of the safety-critical features like AEB and emergency steering assist are usually tested, and its performance and maturity are evaluated in proving grounds as there are certain regulatory tests supported here for the acceptance criteria definition and also to avoid risks of executing such tests in public roads. Even though tests are executed in different test environments, the feature acceptance for the production deployment is purely based on the feature behavior in the vehicle. The evidence is collected during each test cycle to prove that the feature is good enough and error free. The accumulated test results and measurements in different test environments are to be considered for the acceptance criteria as defined in the test strategy.

For example, a safety-critical feature that has acceptance criteria of 1 failure in 100,000 km as defined by a vehicle manufacturer is not always tested in the vehicle. It is a collective evaluation of the tests in HiL and vehicle testing, especially for the failures. Some organizations extend the maturity evaluation of the features to software-in-the-loop testing as well, but this mostly focuses on the software robustness and solely depends on the environment and tools used for testing them.

Depending on the feature that needs to be evaluated for its maturity, the test environment is also selected. For example, adaptive cruise control is a comfort feature and also safety critical. The comfort provided to the driver can be evaluated inside the laboratory environment using a DiL or vehicle-in-the-loop test environment. On the other hand, using a software-in-the-loop simulation without having a human or without the influence of the human factors to evaluate the driver's comfort does not make sense.

Vehicle testing has the highest priority over other test environments when evaluating various features. The validation of the features is usually carried out extensively with the tests performed in the vehicle. The strategy for vehicle testing will cover how different tests are executed in proving grounds and on public roads. Unfortunately, many features still have limitations and may not function as intended on public roads when specific scenarios occur for which it has successfully responded in proving grounds. This is a known problem for many vehicles as there are very few scenarios based on which an ADAS feature is qualified for market deployment as part of homologation and approvals. When the vehicle is deployed on public roads, the scenarios that a vehicle experiences are enormous, and many features are neither designed nor robust enough to handle all those scenarios.

Depending on the agreement between the vehicle manufacturer and its suppliers, feature-level acceptance for quality and robustness and the test environments usage to qualify them vary. The qualification of features for acceptance by the vehicle manufacturer can range from using software-in-the-loop, HiL, DiL, vehicle-in-the-loop, and testing in the vehicle. Some vehicle manufacturers define the acceptance only based on vehicle testing, and some others use a mix of HiL and vehicle testing. Along with test environments, the type of tests and the duration of tests to be executed for measurements are defined. This can be defined for kilometers of operation in the vehicle or are re-simulation in an HiL environment. Alternatively, this can be based on hours of closed-loop operation of the feature in the vehicle and an HiL test environment or maybe for special features like automated parking; these definitions can be the number of iterations of different scenarios and maneuvers the vehicle undergoes successfully during testing for acceptance.

As shown in Figure 11.2, maturity evaluation can be distributed across different test environments. This can be specified from any vehicle manufacturer and integrated into the test strategy for a project. Examples of a few acceptance criteria and how the maturity of different features are evaluated in the industry will be covered later in Section 11.6 of this chapter.

11.5. Vehicle Testing and Feature Maturity Evaluation

Maturity evaluation of the system and its features are evaluated in the vehicle during multiple test iterations. Even if vehicle testing is one of those types of testing in product development, it is of great importance. This is mainly because laboratory-based test environments are not mature enough to exactly replicate the environment and the behavior of the vehicle as in the real world. Various processes followed for vehicle testing were already discussed in Section 10.5 of this book.

Testing in the vehicle as part of validation covers the functionality tests and the long-term closed-loop tests where the feature performance, stability, robustness, and reliability are evaluated. That is usually executed by planning driving routes for certain kilometers with different scenes and scenarios that help test various features. Sometimes the tests are also organized to have repetitive test execution for different scenarios in a proving ground. When different automated driving features are planned to be tested in the vehicle, it is mandatory that the building blocks of those features, the ADAS functions, must be tested before testing automated driving features. Many organizations make this mistake in vehicle testing by immediately starting the AD features without making the building blocks stable. Without having a mature ADAS function, which is the building block, the higher levels of automation will not be stable. It will be a waste of effort and time to fine-tune the higher levels of automation than to fix its foundation. To avoid this mistake, analysis and understanding of the functional architecture and the system architecture will help during test planning.

Highway Chauffeur and Traffic Jam Chauffeur are two of those features that are in development and production at many vehicle manufacturers today. These are SAE Level 3 automated driving features as per the standard SAE J3016 [11.5, 11.6]. Most of the vehicle manufacturers perform the tests for these features in the vehicle for its maturity. The evaluation of these features is made using two approaches: The first one is acceptance defined based on the mileage accumulated in a closed loop for a certain number of kilometers with different scenes and scenarios included according to the ODD of the feature under consideration [11.7]. The second one uses a certain number of hours as a metric where the vehicle is operated in a closed loop with the feature enabled and evaluates the vehicle behavior in closed-loop operations. The vehicle behavior is evaluated for different scenes and scenarios for which the feature is designed and its operating limits.

Different vehicle manufacturers set different numbers for these metrics for the kilometers driven and hours of operation while qualifying these features. For example, the feature Highway Chauffeur is evaluated for its maturity and robustness by the number of kilometers with different scenes and scenarios. These can range from 1 million km to 8 million km, covering different geographical locations with different vehicle manufacturers. While testing, the kilometers covered involve road routes of different countries, scenes, situations, scenarios, and weather conditions. All these are part of the vehicle test plan and according to the operational limits of the feature. Driving on straight roads and accumulating kilometers planned for maturity evaluation is not enough to evaluate the feature for its robustness and maturity. While creating a vehicle test plan, the route plan for vehicle testing should consider covering different scenes, scenarios, and different situations. This is challenging and should be planned with care.

Some vehicle manufacturers consider both the kilometers driven and number of hours for which the feature is in closed-loop operation for the maturity evaluation. This is a different approach from what has been used for Highway Chauffeur feature evaluation. Certain vehicle manufacturers and suppliers use this approach to evaluate features such as Highway Assist, Highway Chauffeur, Traffic Jam Assist, and Traffic Jam Chauffeur features. Rather than evaluating each feature separately using independent metrices, a metric that covers all the features is defined and are evaluated against it. This evaluation criterion is considered because of the feature architecture used to implement these features. These features have common components which act as the building blocks, and the vehicle manufacturer might be reusing those to implement higher levels of automation features. Because of legacy reasons, the vehicle manufacturer decides not to test those building blocks again on their full scale. Each of these evaluation processes is iterative, and the feature is evaluated and improved through software updates over each test iteration until it is production-ready.

Traffic Jam Chauffeur feature, a Level 3 feature as per SAE classification, undergoes its maturity evaluation, including stability and robustness, by the number of hours the feature is enabled and in operational condition in the closed loop. The operational condition of the feature is evaluated in various traffic jam conditions on different highways or highway-like roads. Different vehicle manufacturers consider this evaluation with different numbers for the required hours for the feature in operation. Some vehicle manufacturers would split the overall required operational duration to each software release cycle. This acceptance metric of hours of operation in closed loop for qualifying Traffic Jam Chauffeur feature varies from 500 to 2500 h across different vehicle manufacturers. Some vehicle manufacturers and suppliers even distribute these planned hours for maturity evaluation as a metric for each software release cycle which ranges from 50 to 75 h of operational testing where the quality and performance of the feature is evaluated. The traffic situation on public roads varies day by day, and so does the test route planning as part of vehicle testing. The traffic situation is not always predictable, especially when planning a fleet test across multiple countries. One of the most common approaches used in the present for vehicle test planning is to make use of live map information, such as information from Google Maps, TomTom maps, and HERE maps, for analyzing the probability of traffic jam conditions on specific routes for scheduling vehicle tests.

Certain features such as automated parking and valet parking are neither measured using kilometers driven by the vehicle nor the hours in which the feature is operational in a closed loop. These features are evaluated by scenario-based tests, which are repeated as multiple iterations for various scenarios. For example, a parking feature is designed to be operational for certain defined parking maneuvers like parallel parking and perpendicular parking. The evaluation of the feature for its maturity is based on how successfully the vehicle can execute those maneuvers in different scenarios. This includes repetition of parking maneuvers with different situations or foreign influences like pedestrians and other vehicles. The parking features are mostly evaluated for their maturity in closed and controlled environments, like a proving ground. This approach usually limits the functionality of real-world situations if the proving ground tests are not robust enough. The maturity evaluation metric used for parking features is the measure of successful parking maneuvers in various scenarios. This ranges from 500 to 2000 iterations of successful parking maneuvers in various parking scenarios.

An interesting approach taken by one of the European vehicle manufacturers and its supplier to evaluate parking features during vehicle testing was to get into an agreement with a supermarket chain. This facilitated the vehicle fleet when deployed for testing across different locations, utilizing some of the parking lots of this supermarket store to execute certain parking tests. This was more effective in testing the parking feature than performing the proving ground tests alone. It gives more real-world conditions where these features are used in the real world, such as faded lane markings and influence from other vehicles and pedestrians. The building blocks of automated driving features are the ADAS functions, which would also undergo robust testing in different test environments such as software-in-the-loop, HiL, and the vehicle. The acceptance and maturity evaluation of those features are measured using metrics discussed above with different scenes, scenarios, and misuse conditions while testing them.

Different vehicle manufacturers publish the maturity and reliability of the features in their vehicles as a measurement of mileage they have accumulated as part of test and validation. This can range from one million kilometers to multiple millions of kilometers. When should the mileage accumulation start? What should be the measurement associated with the defects identified over this time? Does this mileage accumulation cover the tests performed in a proving ground and on public roads? These are some of the details which are not published or not available to the public. Almost all vehicle manufacturers consider their mileage accumulation strategies differently. Some measures it only for public road testing, and some include both proving ground and public road testing in a closed loop as part of mileage accumulation. These measurements are usually calculated from the first vehicle deployable prototype of the Automated Driving system with multiple rounds of software updates until production-ready.

How good or bad the automated driving features behave while in operation despite the planned mileage accumulation is mostly measured based on the disengagements of the features during their operation. For example, assuming that the Highway Chauffeur or Traffic Jam Chauffeur feature is engaged in the vehicle and expected to behave all the time as designed during its operation, while testing, there might be situations where these features behave differently. This might be because of a defect or a disengagement due to some other reason. It should be analyzed to identify the cause of the malfunction, or the disengagement is either because of a system fault exiting from the defined ODD or because of some other reason. These disengagements usually occur across the whole mileage accumulation campaigns. Identification of the root cause of these disengagements is possible by utilizing detailed data analysis. Acceptance criteria are usually defined for the error rate and the number of disengagements for different features during the mileage accumulation phase. This will provide the KPIs on the feature readiness for production. The decision to allow certain disengagement causes to move over to the production phase is also made by evaluating the risks and the probability of occurrence of those while the feature is in use in driving conditions. Table 11.1 gives an example of how those disengagements and the errors are measured for different features under test during driving test campaigns.

The acceptable error rates for each feature are defined as part of the requirement of feature quality and performance, and it varies across vehicle manufacturers and suppliers. The decision on what is an acceptable error rate for a feature is developed by some of the vehicle manufacturers utilizing the concepts from Six Sigma. Even though Six Sigma methodologies are originally applicable and designed for production quality environment, these are used to evaluate the feature quality and its improvement by measuring defects per million kilometers. Different vehicle manufacturers and suppliers consider this method as practical and know that none of these features will behave ideally over their entire lifetime.

Table 11.2 shows the standard Sigma levels and associated Defects Per Million Opportunities (DPMO). These measurements were converted to defects per million kilometers for the defect calculation over million kilometers for the driving test campaigns and mileage accumulation. Depending on the planned mileage accumulation for each feature-level testing, a certain Sigma level is allocated to achieve the defects and is used as the KPI for feature performance and quality. Sigma levels between 3 and 4 are the most commonly used Sigma levels for acceptance of various ADAS features such as adaptive cruise controls, lane keep assist, and highway assist. For automated driving features such as Highway Chauffeur and Traffic Jam Chauffeur, the Sigma level is kept as Sigma Level 5. Achieving Sigma Level 5 for automated driving features is challenging and is not

TABLE 11.1 An example of disengagement measurement for features.

Traffic Jam Chauffeur—Error rate (example)					
Hours driven	**Total handover requests/ misbehaviors**	**Handover scenarios**	**Error handovers/ misbehaviors**	**Acceptable error rate (10%)**	* The evaluation of Traffic Jam Chauffeur here is based on the number of hours the feature is in operation (closed loop)
8	10	8	2	1	
Highway Chauffeur—Error rate (example)					
Kilometers driven	**Total handover requests/ misbehaviors**	**Handover times**	**Error handovers/ misbehaviors**	**Acceptable error rate (10%)**	* The evaluation of Highway Chauffeur here is based on the number of kilometers driven with the feature in operation (closed loop)
10	10	10	1	1	

* Different vehicle manufacturers use different acceptance error rates. Only for example here is it considered as 10%. Usually, 5% or less is considered by multiple vehicle manufacturers for different features.

TABLE 11.2 Sigma level and associated DPMO (DPMO chart).

Sigma level	DPMO
1	690,000
2	308,537
3	66,807
4	6210
5	233
6	3.4

practical even though, theoretically, it can be described. Hence, the Sigma levels are allocated to the features based on certain scenes and scenarios instead of the complete feature operation. With this approach, it is practical to meet a specified Sigma level in various operational conditions, scenes, and scenarios for the feature. Theoretically, there are proposals to consider the automated and autonomous driving features to achieve Sigma Level 6 and above considering the safety. Unfortunately, it is not practical with the current technology and the infrastructure situations to consider various features with this maturity. Other methods and approaches are also used in the industry to evaluate the feature quality during vehicle test campaigns which solely depend on the vehicle manufacturer and their suppliers.

Not all features are evaluated for production and road deployment in public roads. Certain safety-critical features are evaluated for certain scenarios as per the regulation, which is evaluated mainly on the proving grounds. Emergency steering assist, AEB, etc. are some of those features where the tests are primarily executed in proving grounds for specific scenarios as required by the regulations. Testing these features and scenarios on public roads is risky and can be hazardous; they are mostly tested in the open loop on public roads. The testing and qualification of various ADAS and Automated Driving features based on regulatory scenarios will speed up and make the feature deployment easier for any vehicle manufacturer. Unfortunately, this brings an additional risk that the vehicle is only tested for the feature behavior on a minimal set of scenarios, and there is no guarantee that the vehicle behaves the same way on public roads. This is a common issue faced in the industry with a lot of vehicles when the manufacturers focus only on vehicle deployment and not much on the safety, performance, and robustness of the features.

11.6. Case Study on How Various ADAS Features Are Deployed

Different ADAS features have been integrated into vehicles by various manufacturers from many years back. They were developed and deployed using various methods and processes that are manufacturer or supplier specific. These technologies are proven through successful market deployment and with years of fault-free operations in different models of vehicles. This section provides information about how various manufacturers deployed some of the ADAS features in the market and how that varies while moving to the development and deployment of automated driving features. Various ADAS features up to SAE Level 2 [11.5] were deployed by different vehicle manufacturers with validation strategies of qualifying them both in a laboratory environment and the vehicle as part of road testing. Primary laboratory testing involves HiL simulation and open-loop and closed-loop re-simulation of vehicle sensor data and vehicle networks to the ECU or by feeding the ECU with synthetic sensor data with the help of simulation tools.

Re-simulation methods in HiL are widely used to deploy some of the safety and comfort features in the vehicle. Table 11.3 gives some examples of the acceptance criteria and maturity used by certain vehicle manufacturers and their suppliers for various ADAS feature deployments in the vehicle.

DPMO of the Six Sigma method was used to evaluate the quality and performance of the features by a few manufacturers and suppliers. The defect rate of those features varies from Sigma Level 2 to Sigma Level 4 compared to the reference mileage of 1 million km of

TABLE 11.3 Examples of a few ADAS feature deployments from different manufacturers.

Features	SAE level	Vehicle manufacturer and supplier acceptance strategy
Adaptive cruise control	Level 2	OEM and supplier (year 2010)—600,000 km of joint mileage accumulation and evaluation. OEM and supplier (year 2017)—1,000,000 km of mileage accumulation and evaluation
Lane departure warning and Lane keep assist	Level 2	Supplier (year 2017)—1,500,000 km of mileage accumulation by the supplier and 800,000 km by the OEM
Traffic jam assist	Level 2	Supplier (year 2017) 500 h or 300,000 km in operation. OEM 100 h and 100,000 km
Automatic parking assist	Level 1/2	Supplier (2017, 2019) 1000 iterations of each parking maneuver. OEM 200 iterations of selected common parking maneuvers
Highway chauffeur	Level 3	Supplier and OEM (planned for 2022 and 2023) 4,000,000 km of closed-loop operation
Traffic jam chauffeur	Level 3	1000 h and 700,000 km of operation

closed-loop vehicle testing. Most of the features were tested only in the vehicle for approximately 1 million km or even less. The failure rates are compared with the Sigma levels for evaluation by transforming the values achieved during the planned vehicle testing or simulation to defects per million kilometers.

The utilization of environment simulation tools was not that popular in earlier days like today. Hence, either the data capture and re-simulation or testing directly in the vehicle was considered as part of validation. The data capture and re-simulation approach requires prototype or production vehicles with all sensors on them. On the other hand, each vehicle program mostly requires a new driving campaign for data collection as sometimes they vary in the vehicle attributes or the sensor positions, or even the sensors. There are also dependencies with the route selection for the driving campaign and data collection for a particular geographical location. The limitations would be that the features will have poor performance and quality for specific geographical locations because of their environmental characteristics as they were not taken into account for the testing and training of the algorithms that are the key components of various features.

Compared to the time that was previously needed for a vehicle program, it has now been drastically reduced by several years. This demands using automation and more efficient methods and processes for testing and qualification, including the maturity evaluation of the system and its features. Thus, tools, methods, and processes used these days are much better and more efficient than those used 5 to 10 years back for deploying various ADAS features in the vehicle. With the help of laboratory-based simulation and methods used these days, it is possible to achieve efficient test and qualification of the features and system with automation and simulation without fully depending on vehicle testing.

The major challenge these days are the nonstandard products and features deployed in the market that create a different expectation about automation in the vehicles. Certain vehicle manufacturers define features that are nonstandard and publish them as autonomous features, and that motivates the vehicle users to try beyond the limitations of the features and the vehicle mainly because of wrong interpretations. This must be addressed through strict regulatory requirements and standardization. The vehicle manufacturers are running ahead with the technology adoption in their vehicles than the regulatory bodies are prepared for. On the other hand, many working groups and regulatory bodies are on their way to establishing requirements and standardizing features inside the vehicles. It gives hope that it will provide a better ecosystem for the vehicle manufacturers and for the users to experience automated and autonomous driving features that are safe and secure with better quality and performance.

11.7. Summary

One of the most important parts of product deployment is defining the system acceptance criteria and maturity the system has to achieve. The key takeaway from this chapter is about how different vehicle manufacturers and their suppliers define performance and quality in the ADAS and Automated Driving area. How different is the definition of done and acceptance criteria were discussed? Which were the most misunderstood terms when the automotive industry adopted agile methodologies for product development? The importance of defining acceptance criteria as part of the scope definition will help both the vehicle manufacturer and their suppliers achieve goals and have defined performance and quality with the features in the vehicle. The maturity evaluation is integrated as part of the verification and validation strategy, and how it is planned during the development and deployment was discussed in detail. There are many other sources from where requirements for quality and performance are derived and tested, which are further traced back, and the defects are analyzed.

Various organizations use a stepwise approach for the maturity and quality evaluation of the ADAS and Automated Driving system and its features. They are usually linked with the development gates of the vehicle program. A general overview of system maturity and feature maturity was covered in this chapter. It helps the readers have their first insight about the improvement planned for each gate of the product development. A combination of different test environments is used to evaluate the software and the system over the complete product test life cycle, out of which testing in a vehicle plays a major role in the final deployment of the system and feature in the vehicle. Testing different features in the vehicle is still the major part of testing in any vehicle program. For measurements, many features are still qualified for their performance, safety, stability, robustness, and reliability based on their long-term behavior. The long-term behavior is a measurement of the total mileage accumulated by the vehicle, the number of hours in which the feature has been in operation, or even the number of iterations that the feature completed a particular maneuver. This varies according to the features that are considered. Even for the software release cycles by system suppliers, certain minimum hours of long-term operation of the features in simulation and the vehicle are required.

Many ADAS features were deployed in the market for more than 15 years. Here, vehicle manufacturers used different methods to evaluate those features for their maturity in performance and quality. Some examples of the maturity evaluation used by certain manufacturers and their suppliers for deploying ADAS and Automated Driving features are also discussed, along with the measurement methods used.

The next chapter takes you through one of the most complex and trending topics in the automotive industry, which is data collection and management. Vehicle testing and data collection have become an integral part of the overall system development and testing in automated and autonomous driving. Different methods used in the industry and the challenges faced while applying those methods and how they are resolved are covered by specific examples from the production programs.

References

11.1. Crispin, L., *Agile Testing: A Practical Guide for Testers and Agile Teams*, 1st ed. (Boston, MA: Addison-Wesley Professional, 2008), ISBN:978-0321534460.

11.2. Schwaber, K. and Sutherland, J., "The Scrum Guide—The Definitive Guide to Scrum: The Rules of the Game," November 2020, accessed October 23, 2021, www.scrum.org, https://scrumguides.org/docs/scrumguide/v2020/2020-Scrum-Guide-US.pdf.

11.3. ISO/IEC 25010:2011, "Systems and Software Engineering—Systems and Software Quality Requirements and Evaluation (SQuaRE)—System and Software Quality Models," 2011.

11.4. ISO/IEC/IEEE 29119, "Software and Systems Engineering—Software Testing," September 1, 2013.

11.5. SAE International, "Taxonomy and Definitions for Terms Related to Driving Automation Systems for On-Road Motor Vehicles," SAE Standard J3016_202104, April 30, 2021.

11.6. ERTRAC, "European Road Transport Research Advisory Council Roadmap for Connected Automated Driving," 2019, accessed October 1, 2021, https://www.ertrac.org/index.php?page=ertrac-publications.

11.7. PAS 1883:2020, "Operational Design Domain (ODD) Taxonomy for an Automated Driving System (ADS)—Specification," 2020.

/Autonomous
/Sensing
/Communication
/Battery
/Navigation
Autopilot mode

12

Data Flow and Management in Automated Driving

Automation in vehicles has changed a lot in the way in which the vehicles are developed and tested. Classical methods and processes are changing, and more software-oriented approaches and systems are driving the automotive industry these days. The importance of data has also become more relevant these days as many software components use data in their development and testing phase for better performance and quality. As discussed in the previous chapters, the different methods used for qualifying and maturity evaluation of the software during the development and features in the vehicle are all data dependent. How the data are collected, analyzed, and interpreted will be discussed in this chapter. As Clive Humby said, "Data is the new oil" [12.1]; it is getting true while moving toward autonomous driving and building complex systems inside the vehicle that is controlled mainly by software units that have a dependency on AI. Automated Driving systems are such systems where the utilization of AI components has a significant stake.

The scale at which the data is collected, analyzed, and used for various purposes in vehicle development is very complex. There are a lot of dependencies and challenges that a vehicle manufacturer and its suppliers have to face while collecting, analyzing, and deriving decisions from those enormous amounts of data. In this chapter, readers will learn about the data utilization and the processes followed in the industry in various phases of development and testing and the challenges associated with those. It also discusses how various challenges are addressed these days and how the industry is getting future-ready in handling big data. Big data is a standard term used these days in the automotive industry. It is a term that describes a large amount of data of different types which may or may not be structured. Analyzing these enormous amounts of data will help derive actions and decisions for improving the product and its features. When moving toward Autonomous driving, a vehicle utilizes sensors for capturing the environment around it, and they generate a massive amount of data [12.2]. This needs to be processed, and the actions have to be derived that controls the vehicle. Along with having a high-power system with a huge processing capacity, there should also be a mechanism to analyze those data which are generated and processed. This processed information decides the vehicle controls and motion and is critical for various features in the vehicle, that are highly automated.

The automotive industry lacks the relevant expertise in extensive data collection and management. The lack of skillset and experience added additional challenges when dealing with huge amounts of data while considering the development of Autonomous Vehicles. This change forced the industry to extend classical skills from the automotive industry to the new technology skills such as big data management and data analytics. Data science is one of the key areas right now, which involves structuring, analyzing, and interpreting huge amounts of data using statistical methods. Different sections of this chapter will take the reader through end-to-end processes followed in the industry for the collection, storage, and analysis of data, mainly for ADAS and Automated Driving system development.

12.1. Importance of Data in Automated Driving

Highly automated driving vehicles generate a considerable amount of data. The development of Automated Driving systems also utilizes a huge amount of data generated during its regular operation. The main reason for this is the process involved in the development of those systems. The software components used in the Automated Driving systems play an essential role in migrating the automotive industry to an ecosystem where data is the core, and the software quality and performance depend on the data. This dependency is mainly because these software components, such as computer vision algorithms and machine learning algorithms, generate output based on the input data and the data used to train them during their development. This increases the demand for data during the development phase of these complex systems. How these demands will be met and the practical ways to handle these demands will be discussed as the primary goal of this chapter.

Unlike other industries, such as telecommunication and smart manufacturing, where huge amounts of data are collected and analyzed, the main challenges in data collection and analysis in the automotive industry are influenced by three main factors. These are called the triple "V" constraints. Data collection and analysis in automotive is complex mainly because of these three constraints. They are volume, velocity, and variety, as shown in Figure 12.1. The amount of data generated and the need to analyze them in an autonomous vehicle with multiple sensors are enormous. For example, assume a generic sensor setup in the vehicle with 14 cameras, 6 radars, 3 lidars, etc. With an automated driving feature such as Highway Chauffeur, the vehicle should drive above 100 km/h. In these conditions, the sensors need to process the environmental data, which act as input, so quickly that they can detect objects at higher speeds so that the response action can be planned by the vehicle. A camera should work at least at 30 or even 60 frames per second to process the input, and if they transfer raw images to a central processing ECU, the size of the data will be bigger than the processed data from cameras. This is applicable for all the sensors in the vehicle, even though cameras act as a source of a significant chunk of the overall data in the vehicle from the external environment. The variety of data that needs to be processed includes data from all the sensors, vehicle networks, software data, processor data, etc. All these are of different data types and may be structured or nonstructured. The primary source of

FIGURE 12.1 Triple V constraints of big data collection and management.

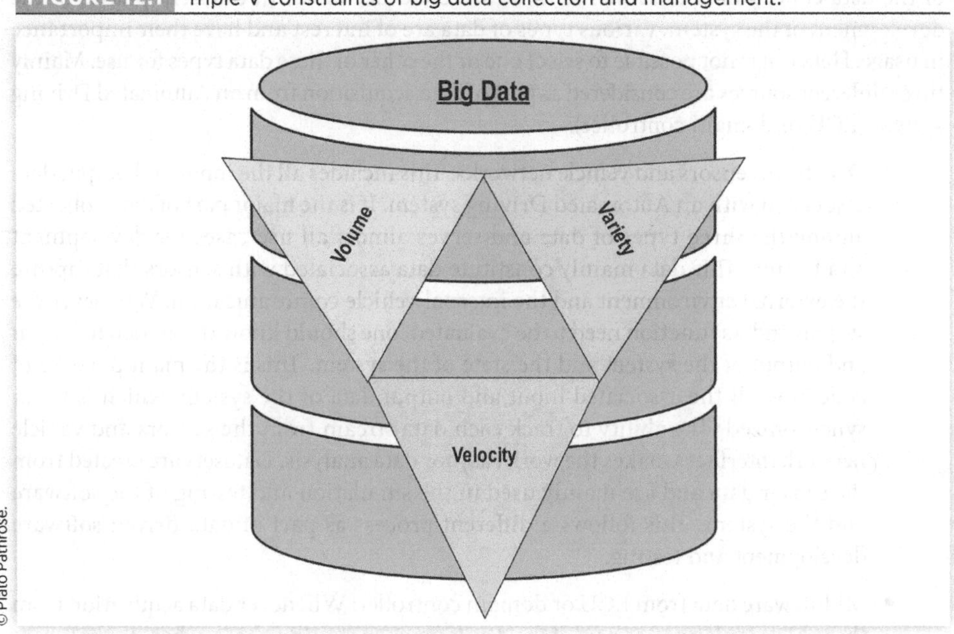

unstructured data is during the development phase of the data-driven software components. For training and validation purposes, large-scale data collection campaigns are organized. There are a lot of third-party data purchases and open-source databases that might not be standardized. Here one could expect to deal with huge amounts of nonstructured data.

Most of the software components of the automated and autonomous driving system depend on various sensor data as inputs, which are required for their development, operation, and testing. It is a complex task to collect data from various sources [12.2, 12.3, 12.4]. Post data collection comes to the cleaning process where only useful data is extracted from the vast input from sensors and then structured for different use cases. Based on the use cases, one could use these data and analyze them for better interpretations and decision-making with those analysis results.

Different types of data are associated with a vehicle during its development and deployment phase. The data in huge amounts, or big data, has found its usage in three main areas of data-driven software development and testing. They are (i) Data for training, verification, and validation of artificial neural networks, (ii) Data for re-simulation as part of verification and validation in a software-in-the-loop or HiL test environments, and (iii) Data collected as a database which is used as a reference for the vehicle level testing of features as part of the validation of ADAS and Automated Driving features. Even though all these are interrelated, these have their specific use cases, and the associated data are maintained separately. The next sections will discuss how the data is collected, stored, and analyzed for these use cases in the industry.

12.2. Types of Data Collected

Data collection for ADAS and Automated Driving use cases should be well planned as this is one of the main areas where many organizations burn a lot of money. The primary purpose of the data collection is to serve the three aforesaid use cases. In different phases of the development of the system, various types of data are of interest and have their importance in usage. Hence, it is not possible to select one or the other of these data types for use. Mainly three different sources are considered as part of data acquisition from an Automated Driving system (ECU or domain controller):

- Data from sensors and vehicle networks: This includes all the input and output data associated with an Automated Driving system. It is the major part of data collected among the three types of data and serves almost all use cases for development and testing. This data mainly constitute data associated with sensors that capture the external environment and the internal vehicle communication. Whenever the system and its function need to be evaluated, one should know the associated input and output of the system and the state of the system. This is the main purpose of collecting all the associated input and output data of the system, which is time-synchronized. The ability to track each data stream from the sensors and vehicle network interfaces makes the work easy for data analysis. Datasets are created from the sensor data and are mainly used in the simulation and testing of the software and the system. This follows a different process as part of data-driven software development and testing.
- Middleware data from ECU or domain controller: Whenever data acquisition from the vehicle is planned as part of the development or testing phase, where the software is not mature enough and still requires improvement, the middleware data from the system play an important role. Middleware data act as an interface between the application layer and the lower layers of the software up to the physical layer in an Automated Driving system. For software development and testing, it is very important to know the input and output of each software component and how they influence the overall behavior of the system and its features. Middleware data are captured and mainly used as part of the software and integration testing during the development. Specific sensors used in the vehicle, which has built-in processors, only provide processed data to the Automated Driving system and are used by the software in the system to process it further. Logging and analyzing the middleware data will help to identify software defects and other issues associated with the system and its subsystems. Usually, middleware does not support external logging unless configured to do so in the production systems. For capturing the middleware data, a separate logging interface has to be integrated with the system software as a wrapper that connects to an external physical interface such as Ethernet. This software interface component is usually integrated only during the development phase of the system until the system and software achieve a certain maturity level for its quality and performance. This interface software component used for logging the middleware data is removed from the production systems.
- SoC or processor data from the ADAS/Automated Driving system: Capturing processor data from the ECU or a domain controller where multiple processors are integrated to provide required processing power. These data support massively in the development phase to analyze and evaluate the processing performance and quality of the software running in each of those processors. These are usually captured using specific debug interfaces available in the processor. These debug interfaces are exposed to the external environment with additional hardware and software components integrated into the system hardware during the development phase.

These prototype versions of the hardware are used for data collection, analysis, and debugging purposes. This is usually performed in a laboratory environment or even inside prototype vehicles. Processor data capture from the Automated Driving system and its analysis are commonly performed for those systems where new microprocessors or microcontrollers are introduced. This might be done as part of a new system design or as a strategy to replace an existing processor with a new one for better performance. The captured data over these debug interfaces are used to analyze the performance and status of the microprocessors during their operation. This is one of the methods used to evaluate the performance of artificial neural networks in specific processors and optimize them. The analysis of this data helps in various other ways, such as optimization for the execution of various low-level software threads, optimizing memory handling functions, and data access and exchange functions by the processor with other processors in the system. Debugging interfaces of an ECU are mainly used for the hardware and software integration tests. These tests are executed with simulation environments such as HiL environments or even directly in the prototype vehicles. When multiple ECUs or domain controllers are used for redundancy as part of fail-operational or fail-safe architecture in the vehicle, fault injection tests are executed to evaluate their functionality. The data captured from the processors will help assess the behavior of those processors in the presence of various types of faults. In the product development phase, the processor data logging and analysis are usually performed by the system suppliers rather than the vehicle manufacturer.

Capturing all this data requires complex systems with multiple interfaces that fit the system and subsystem interfaces. The data captured during operation should be time stamped and synchronized so that data logging can continue without any failure over the entire period of its operation. An average data collected with all the sensors and vehicle networks from a vehicle with semiautomated driving or automated driving features would be in the range of terabytes in a few hours. This data rate varies depending on the number of sensors and the format in which the data is captured. Many vehicle manufacturers use cameras that generate raw data. This adds additional data load and bandwidth requirements. This also depends on the resolution, and the format of the image captured using different cameras. There are various ways in which all these data are collected and analyzed. Vehicle manufacturers and their suppliers use data loggers with different interfaces to collect data that are integrated into a vehicle. Figures 12.2 and 12.3 show examples of two types of data-logger architectures mainly used to capture vehicle data. This includes the sensor data, vehicle network data, middleware data, and processor data from an ADAS/AD system.

Various types of data loggers available in the market from different suppliers provide flexibility to integrate into the vehicle and plug into different network interfaces to log data in different formats. A major trouble faced during data logging from various sensors in the vehicle is with the camera interfaces. Even though the sensors transfer the environmental data to the system, it is a bidirectional communication channel through which the ECU and the sensors communicate that is not standardized and is developed according to the system suppliers. The automotive cameras are of various types and usually have specific configurations that vary with different camera suppliers. Integrating cameras into data loggers is usually troublesome unless precise specifications are available for the design and the data flow between the ADAS/Automated Driving system and these camera modules. These include the information related to sensor configuration, calibration by writing certain data into the registers of the camera module on every startup phase, timing, and encoded

FIGURE 12.2 A data-logger architecture with centralized logging.

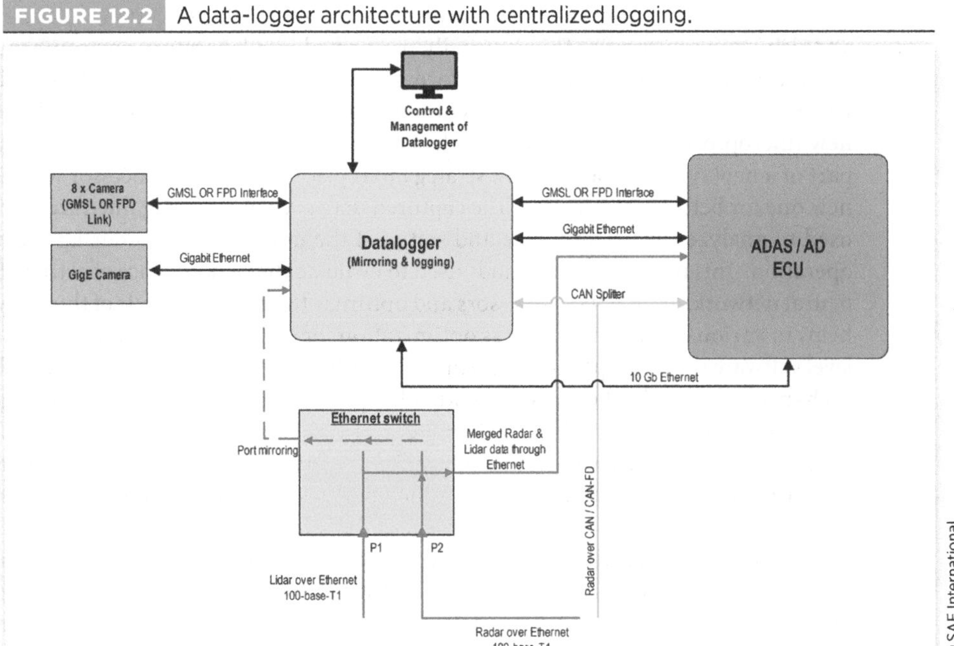

FIGURE 12.3 A scalable data-logger architecture with various capture modules.

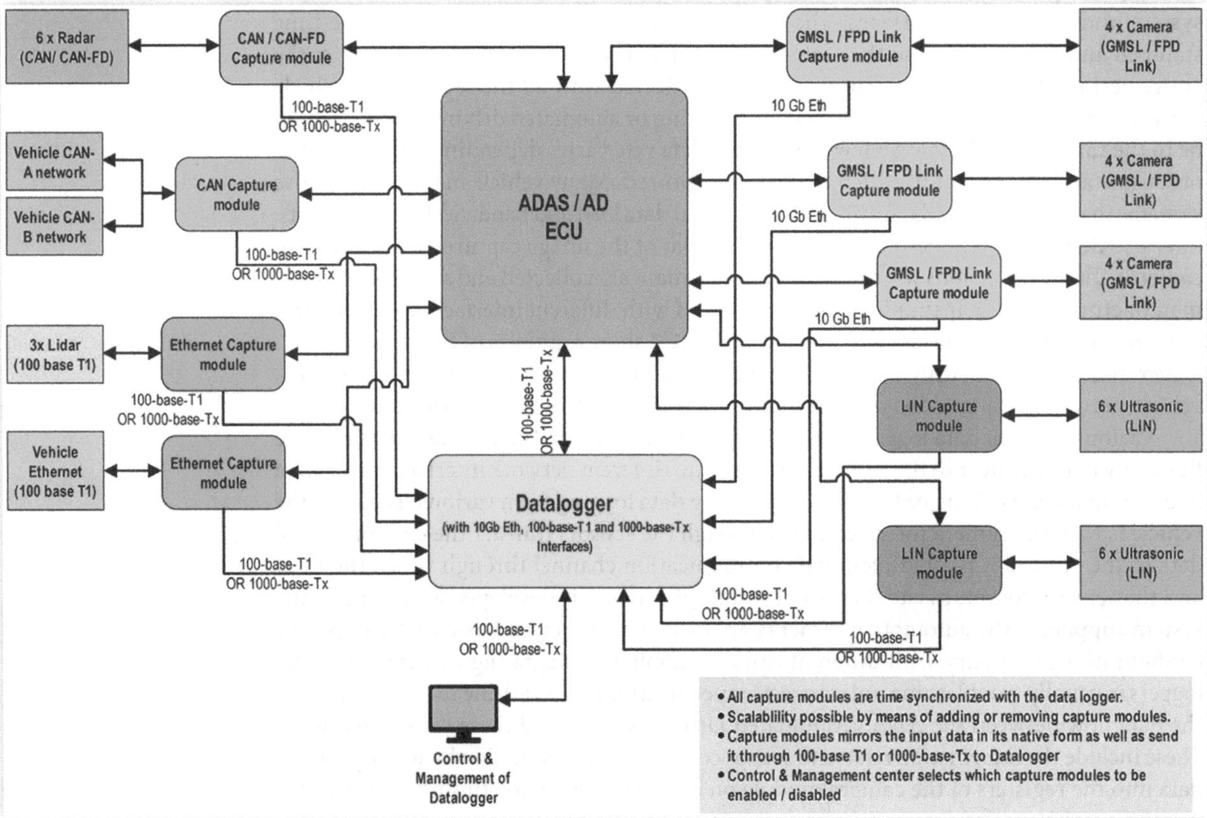

data format in which the connection is established. Depending on the number of channels and the data-logging duration, the data loggers vary in size and cost and may get complex. Sometimes a series of data loggers are interconnected and used to scale the number of data channels in the logger for data logging. An alternate approach to scale the interfaces is adding additional interface modules according to the demand for more data channels, as shown in Figure 12.3. In this case, all these interface modules synchronize while logging into each other and with the data logger.

12.3. Data Acquisition Strategy and Data Loggers

Collecting a vast amount of data that is generated at higher rates will continuously challenge any process, infrastructure, and mechanisms used from the traditional automotive industry. These days, many organizations have even established different departments and upgraded their infrastructure and processes to handle these challenges. Establishing a data strategy is vital to cover the whole life cycle of data collection, storage, analysis, and interpretation. This is a common challenge faced by many organizations as they start directly with the data collection activity without establishing a strategy and process, being unaware of the scale of the challenge they have to face. This will gradually hit them back in no time.

The data strategy involves defining the workflows, deploying required resources and infrastructure, and establishing processes that would help collect, store, organize, and analyze data in a minimum amount of time. This strategy should cover the whole data life cycle from generation until archiving it after its specific use. The data strategy should define and classify different sources from where the data is generated for each use case in the ADAS and Automated Driving system development.

There are many different ways in which large-scale data acquisition occurs in the industry. The data source can also vary according to various use cases and the suppliers providing them. This can be as simple as data collected with a single dashboard camera from the vehicle or a complex one such as a data logger to capture the complete data associated with the vehicle and the Automated Driving system. There are many situations where the data is generated from environment simulation tools. These are mainly used for the development and testing of data-driven software components. Another source of data is buying it from suppliers. These are usually processed data in datasets for specific geographical regions or specific classes from suppliers. All these different sources should be considered while defining the data strategy as each of them is different in terms of data format, data type, and structure. Converting these different varieties of data into a usable data format and using it for various use cases, deriving inferences based on analysis, and archiving it is the complete process that should be defined as part of the strategy. Out of these, usually the data collected directly from the vehicle sensors constitute the major part of the overall data collection.

Data collected from the vehicle as part of the Automated Driving system development using a data logger need to be planned according to the usage and how the data can be stored. They are classified into different types of data and are further used for generating datasets employing manual or automatic labeling methods and tools. When the data collection campaigns are planned with a fleet of vehicles, the capacity of the loggers, the format in which it is stored, and how the data is classified and managed for different use cases should be predefined. When all vehicle sensor data to the Automated Driving system are captured, according to the type and number of sensors in the vehicle, the data size can quickly scale to gigabytes (GB) within a few seconds of logging. Storage and management of data in this scale will be complex if it is not separated into smaller chunks and labeled while logging. Many data loggers use a mechanism for automatically trimming data to different files after a few seconds or after a defined file size while logging. Most of them create a new file for storage once a file reaches a particular size. These individual file size in the data logger varies from 1 GB to 5 GB or even more. In certain data loggers, the file splitting is configured based on time. After some logging period, such as 30 sec or 1 min, a new file will be created. As the file data size depends on the logged sensor type and the data flow, using this method can result in unevenly distributed file sizes. Both these mechanisms are usually configurable via data-logger software, and hence it is selected during the integration of data loggers for data capture. Data loggers need to be designed in such a way that the data channels are selectable for data logging. This functionality will help select particular data channels or sensors for data logging based on need and is not necessary to log all the interfaces. On the other hand, the possibility of selecting data channels for logging gives more flexibility to scale the interfaces and usage of data loggers.

Selecting the format of the data that needs to be logged in the data-logger storage is important considering the different use cases and the toolchain used for its analysis. Various data formats are used in the industry for data logging. A few of those commonly used data formats are rosbag, .mdf (Media Descriptor File), HDF (Hierarchical Data Format) [12.5], ADTF DAT (Automotive Data and Time-Triggered Framework) file format, etc. Depending on the toolchain and the planned data storage and analysis workflow, one of those data formats can be selected. Some data loggers support the storage of data in multiple data formats, which can be chosen prior to logging. The data formats and file sizes are mostly controlled by the data-logger software, which is configurable and usually optimized and developed according to specific requirements. The most challenging part of data loggers is synchronizing different data channels and the time stamping of the data on each channel. There are other protocols used for time synchronization, such as PTP (Precision Time Protocol) [12.6], NTP (Network Time Protocol) [12.7], and PLP (Probe Logger Protocol) [12.8]. Data loggers make use of one or more of those protocols for the time synchronization of the data or even based on a custom-generated clock. The tolerance limits in the accuracy of time stamping and the data-logger architecture also play a significant role in selecting one of those protocols for synchronization.

One of the most valuable features available in a data logger is the capability to tag specific events that occur while running data collection campaigns or vehicle testing. Different methods are used to tag the data specific to events in data loggers, and those are partially driven by the capability of the data-logger control and management software. Tagging of the data while logging can be implemented through a simple button press to a complex neural network-based voice recognition algorithm that supports voice tags and converts that to specific data-oriented tagging in the recorded data. Data tagging for events is usually possible while driving after the event has occurred. This is handled in some of the data loggers using additional temporary memory (Random-Access Memory [RAM]), which can hold data up to 15 to 30 sec of data recording before moving it to the permanent memory. The size of temporary memory to be used can vary and can be decided according to the need for event capture. Adding these features in the data loggers increases the cost and the complexity.

While the data is logged, there are many challenges that the user has to face, such as the data logger not logging from various sources, situations like missing frames from the camera channel, frozen input channels, disconnection, or even short-circuits. Overheating of the data loggers is also a common issue because of prolonged usage while in operation. All these conditions have to be considered and compensated to achieve the optimal performance of data loggers. Similar to an ECU or a domain controller, these faults and failures must be identified and addressed so that the data logger and the data collected are not lost. A health monitoring or a diagnostic mechanism integrated with the data logger can be used to identify and capture these faults and failures. It will also act as an interface to analyze and perform software updates to different modules of the data logger. It is not very common to have a live health monitoring system in data loggers available in the market. As there is no standardization available and these devices are custom developed, the monitoring is only performed in the software. Any error or failure in the system that affects the stored data can only be identified when the collected data is analyzed.

The data-logger design should consider a fail-safe or a fallback mechanism from its regular operation in the event of a detected failure. Detection, Warning, and Safe shutdown are the processes that are taken care of in the event of a failure in data loggers. The captured data is not lost and is recoverable, at least by exchanging the data storage disks. There should be considerations given while designing the data logger to detect missing camera streams or frames, time synchronization loss of data and checking against a reference to re-synchronize, detecting the disconnects, overheating of the system, etc. All these require separate sensors in the system or sensing mechanisms in the software of the data logger that provides the status of the data logger in operation as part of the data collection campaign to the user who is driving the vehicle. In the event of a failure, the data logger should detect it and inform the user, and in the case of no response from the user within a certain period, the data logger needs to be shut down to save the data and effort. A visual and audible notification should be provided to the user when no logging is performed because of a system fault or when the memory is unavailable. These are only a few common failures and detections considered in data loggers that are commonly available in the market. The data loggers can also become complex when further requirements are considered, such as designing the diagnostics in the data-logger system with specific reference IDs for each failure, transmitting those to the fleet management center automatically, and even organizing replacement loggers in case of critical faults.

Data logging of the vehicle sensors is usually performed with built-in sensors of the vehicle. This might be specific to a vehicle program so that each of those sensors and their positions are considered while data is collected. On the other hand, data logging is also performed to use the data to evaluate and improve neural networks and algorithms for specific sensor positions such as front-facing camera and lidar as part of data-driven software development, which is not focused on a specific vehicle program. Because of the particular use cases, there is no dependency on all the sensor positions in the complete vehicle. Instead, the focus is only on certain specific sensors for development and testing. For making use of data from other vehicles as a reuse strategy, certain tools are available in the market that are used to consider the shift of sensor position and regenerate the captured data to a new sensor position, which helps in compensating and reusing the captured data from the complete vehicle sensors up to a limit. A significant drawback with this approach is that the geometric shift of the already captured data by software has limitations in the transformation to new sensor positions beyond certain angles. It is a time-consuming activity that involves comparison with the data captured from the changed sensor positions from the vehicle. We hope that someday such a mechanism would help reduce the additional effort of recapturing the whole drive data.

12.4. Data Reuse Strategy

Usually, vehicle manufacturers and their suppliers spill a lot of money during the development of ADAS and Automated Driving systems for data collection campaigns and vehicle testing. Even though these are an essential part of product development and testing, there is a lot of repetition seen in the industry with different manufacturers and suppliers where the same data is being collected at different instances. Having a data strategy in which the collected data can be shared and reused will avoid repetitive work related to data collection campaigns and ground truth generation for data-driven software development.

When a data reuse strategy is developed, even with the existing data capture and storage methods, there is the possibility to reuse the collected data for the development and test purposes in multiple vehicle programs. The only vehicle data that cannot be reused and must be repeated are those collected as part of the vehicle test campaigns for qualifying different features in the vehicle. Many organizations have considered the data reuse strategy in the data collection campaign and vehicle testing to avoid multiple drive campaigns [12.3]. Vehicle manufacturers and their suppliers rely on different ways of data collection. Data collection methods can vary with different suppliers as well as with different vehicle models. This ranges from using a custom vehicle rooftop sensor box to capturing data directly from the integrated sensors in the vehicle.

Large-scale data collection and its usage are seen with those organizations which develop software components that are part of data-driven software development used in ADAS and Automated Driving features. These software components can be anything from a simple detection algorithm to complex perception or path-planning integrated software with multiple algorithms. This is also widely performed by the system suppliers where the data is used for integration and test activities such as re-simulation in HiL [12.2, 12.3]. If the software development is not undertaken at the vehicle manufacturer's side on a large scale, the data collection for algorithm development and re-simulation will be limited at the vehicle manufacturer's side. The development part can utilize the collected data from each vehicle program to process it and generate ground truths or datasets. This will help in the training and validation of neural networks.

The data collection for algorithm development can also be performed by certain off-the-shelf products from different suppliers called Ground Truth Systems (GTS). These devices are integrated into the vehicle rooftops and can be used to generate ground truth data in runtime or by utilizing offline methods. Many suppliers are in the market that provides different types of devices that can be integrated into the vehicle. These devices come with different types of sensors that can be fixed on the rooftop of cars or commercial vehicles to collect, store, and generate ground truth data. These collected data can be processed and used for software development and improve the performance and quality of the software. This acts as an asset for the development, verification, and validation of artificial neural networks. Usually, this type of data can be reused to develop and evaluate the software, even for different vehicle programs. The advantage of this method is that it is not dependent on any vehicle or its sensor setup. Instead, it captures the external environment by a reference set of sensors from the rooftop position of the vehicle. The only constraint in using these GTS is that they should be qualified and generate qualified datasets that are trustable and with higher accuracy and precision.

The sensor data from the vehicle can be captured directly and used for generating ground truth while performing vehicle testing. This is a standard method widely used for data collection and reuse for development and testing purposes. This requires the data loggers used in the vehicle to adapt each time for a different vehicle program based on various sensor interfaces. This is challenging, especially from a system supplier's perspective when they work with different vehicle manufacturers simultaneously. The sensor positions and the sensors in the vehicle will influence the quality of the data collected. For a vehicle manufacturer, data collection and reuse might vary according to different types of vehicles and their sensors. On the other hand, for a system supplier, reusability is possible for specific sensor positions. Out of the whole vehicle sensors, the data from front-facing sensors (camera, radar, and lidar) are used for the software development and testing irrespective of projects.

As described in this section, the data collection and reuse are solely dependent on the strategy defined by an organization based on their products and use cases. The effective use of data will control the cost of data collection and ground truth generation.

12.5. Data Analysis and Data Flow

Data collection, storage, and analysis in any organization involve a lot of methods, processes, and workflows. These are challenging when it is big data and in the scale of terabytes and petabytes. Having a mature approach and workflow would help handle these enormous sizes of data and optimally utilize them for different use cases. The analysis of these massive amounts of data involves more complexity as the requirements for analysis vary depending on the product, organization, and its use case. Hence a generic analysis approach cannot be derived for the whole data management.

A generic data flow for an Automated Driving system development and its verification and validation is shown in Figure 12.4. This can be adapted based on the organizational needs and the process followed. The different phases of the dataflow are (i) Data collection phase: where the data is collected by various means and sources; (ii) Data encapsulation phase: where the data is structured and stored; (iii) Data usage phase: where the data is accessed for analysis, and the evaluation of the stored data takes place [12.2, 12.3].

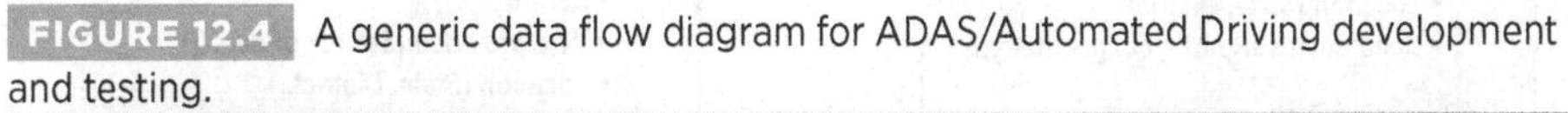

FIGURE 12.4 A generic data flow diagram for ADAS/Automated Driving development and testing.

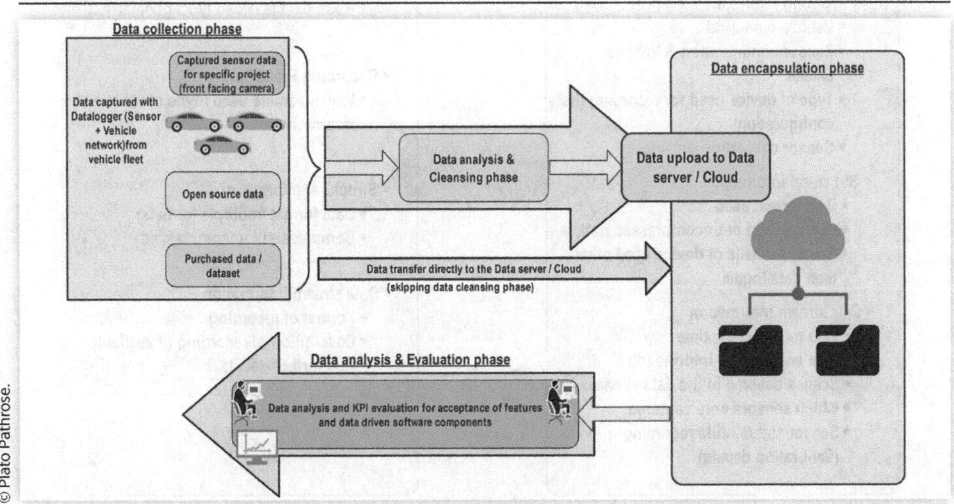

Depending on the processes followed by an organization, certain phases might overlap as per this data flow diagram. The main goal of any strategy should be to make the data available for the end user with minimum delay. The data collection phase involves data gathering from different sources. These data might not always be available as structured and in the same data formats. The various sources from which the data is collected are (i) Vehicle data, including data from the integrated sensors, middleware data of the Automated Driving system, the processor from the Automated Driving system (ECU or domain controller); (ii) Data that comes from external organizations that cannot be influenced; (iii) Data procured and collected from the market, which could include purchased datasets, collected data using any nonstandard devices or methods, any relevant data that is available as open source and free to use. In the data collection phase, overall data can be classified as data that can be controlled for its quality and the way it is generated called controllable data and noncontrollable data where data cannot be influenced and are available to use in different formats and structures. This is usually coming from other organizations or collected using nonstandard methods. An overview of the characteristics and attributes which should be considered of controllable and noncontrollable data is shown in Figure 12.5. This information will help define the labels and tags for different types of input data stored as part of the data encapsulation phase. These tags must be defined as part of the configuration management and should cover the complete data life cycle. This is helpful not only in a single project but also in multiple projects with the possibility of data reuse.

The tags and the labels used in Figure 12.5 are a few examples that can be used as part of configuration management while storing data from different sources. They would help in identifying the data as well as optimally using it for other use cases. The method described

FIGURE 12.5 An overview of the contents and tags used in controllable and noncontrollable data files.

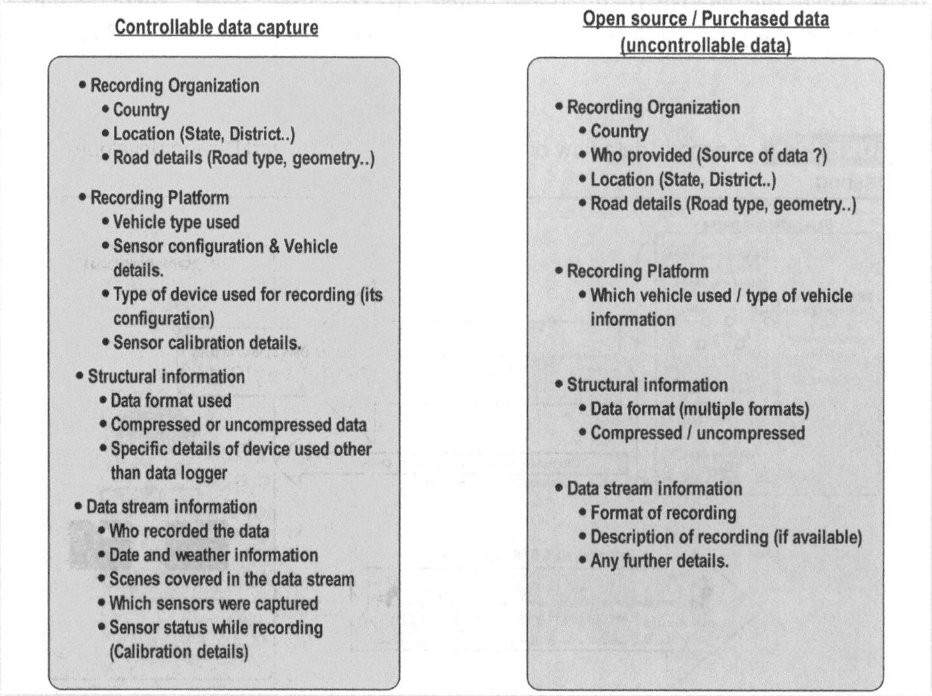

here is about organizing the data. Even if data comes in different formats and structures, it is recognizable and usable by the end user. The controllable data can be influenced by the data collection methods and guidelines defined as part of the data strategy of an organization. Usually, they are collected directly by an organization or contracted to be collected by another organization. Here the data content, structure, and type of data are defined by guidelines. Primarily, these data will be structured with more information available such as its source, platform used for data collection, and the content.

The noncontrollable data includes open-source third-party data, which is procured from other organizations, where the data is prepared and available in specific formats by the supplier. The supplier might have prepared the data for some other use case in a different structure and form, and now they are selling that data. The tagging and the labels associated with these data files should be adapted according to the end user and their purpose. The goal of any data management system should always be to provide flexibility and ease to the end user in searching and using the correct data from the storage.

The analysis of the stored data varies with different teams involved in the project. Even though the collected data can be used for analyzing different KPIs at the vehicle and feature level, the principal use case of the collected data is to analyze and evaluate the performance and quality of data-driven software components. Other use cases of stored data include:

- Generation of ground truth datasets for training and verification of neural networks
- Extraction of various scenarios and scenes from the collected sensor data which can be used for re-simulation and simulation-based tests
- Analysis of vehicle dynamics by extracting data associated with vehicle motions and different maneuvers such as the longitudinal or lateral motion of the vehicle in various conditions

There are specific tools available in the market that are integrated with various data storage platforms that support data analysis and can extract scenes, scenarios, and vehicle motions from the collected sensor data. These tools can help in extracting what is needed from the collected data and can use these data files for re-simulation or to check the performance of system and software components.

There is a misunderstanding with many organizations that the analysis of the collected vehicle data can be performed with manual effort. Hence the utilization of any data analytics tools is usually not planned earlier in the data strategy. These organizations will face many challenges if they do not employ specific data analytics tools as part of their data management strategy. The end user, who might be an engineer or a tester, might not identify and evaluate the collected data that might help in finding a defect in the system or a feature. Without using data analysis tools, an engineer might be required to manually go through the whole data stream to identify the defect of interest.

Various analytics tools help the end user and the data supplier in providing required information about the stored data files and the possibility to extract, preview, and analyze the area of interest from each stored data file. The data analysis in detail includes evaluation, comparison with reference algorithms and reference data, measuring the accuracy and precision for various events, detection and localization algorithms, etc. Usually, the analysis tools used in the data platform are integrated with a test framework and other tools that help further analysis and processing of the stored data. These analytics tool frameworks include tools for training algorithms, tools for labeling data streams from different sensors for ground truth generation, tools for performance comparisons using reference algorithms, etc. An overview of the data analysis and management tool that provides specific interfaces at the user and the data backend to the data storage server or the platform is shown in Figure 12.5. The exchange of information at the user end and the data backend can be configured according to the use case and the necessity of the analysis from the users. When different teams from the project access the data stored in a data platform, data access differs based on different requirements. Hence a configurable user interface will help to address those demands. Many organizations utilize a Web interface where a data file stored in the data platform, such as a data server or cloud storage, can be previewed with all associated information.

It is also beneficial to consider a command-line interface and access to the data files and the Web interface. When different types of data are accessed, these end-user Web interfaces can be configured to include additional information and profiles for each user type based on the teams and their demand for data access. This helps the users to have data access with minimal complexity. The command-line access to the data storage will allow users to access the data even with minimal display and system resources. In most cases, reference algorithms that replicate different ADAS and Automated Driving functions are deployed in the backend. They are used to evaluate the performance of the vehicle level functions based on the captured and processed data from the vehicle. This type of data analytics has huge demand these days from the vehicle manufacturer and the system supplier side as such tool frameworks are not readily available to purchase from the market and are usually custom developed by various organizations that need them.

The purpose of any tool is to reduce the effort of its user and make work easy. This should be the goal while designing an analytics tool or deploying it from any tool supplier. Suppose the tool is not supporting and is adding additional burden to the process defined as part of the data collection, management, and analysis; in that case, the tool should not be considered for deployment. These evaluations must be done thoroughly before deploying any tool for data access, management, and analysis of big data for multiple users.

12.6. Data Storage and Management—A Case Study

Managing data generated with high velocity, volume, and variety from the vehicle is challenging. Data handling requires a well-defined strategy, infrastructure, well-defined processes, and workflows. If not, it is hard to achieve efficiency in handling the data. This will affect the data usage, causing wastage in effort, resources, time, and money. A case study about how the data collection, management, and analysis were performed for one of the projects is discussed in this section. This will help the reader gain knowledge about one of the methods to handle data in automated driving projects of this scale and to understand various challenges associated with it. When large-scale data is to be handled for different use cases, it is good to know about certain critical areas and understand how the data from a fleet of vehicles are collected, stored, and processed.

This case study covers an example where the data was collected from three different locations with a smaller fleet of vehicles. The data was collected to improve the performance and quality of software components and in HiL re-simulation and testing various automated driving features as part of vehicle testing. Approximately 12 terabytes of useful data were generated from each location per day after the logged data from data loggers were cleansed from respective sites. The collected data included data from all the vehicle sensors and the middleware data from the Automated Driving system or the domain controller. The data was captured in the rosbag format in data loggers integrated into the vehicles. As there was a possibility of laboratory access after each day of data collection and testing, the logged data from each vehicle were taken to a cleansing process where unwanted data were trimmed before uploading these to a cloud data platform. This might not have been possible if the fleet size was big or the data collection was performed by drivers who could not complete the cleansing of the data or if there was no daily access to the laboratory or a data ingestion center for uploading the data. In such cases, the cleansing part in the data flow path could have been skipped, and alternate methods for data ingestion, such as the manual transmission of the data disks or using a logistic service for data disk transfer, could have been considered. This was a viable option when the data collection was performed in remote areas where access to a laboratory environment or the availability of high-speed data transfer was challenging.

The data ingested from all three locations were used across different sites by engineering teams for the development, verification, and validation and for performance evaluation of the system and the software of the different automated driving features. The data storage and management system includes a cloud platform with a certain toolchain, applications, and services that support the data storage, availability, and security across various locations of the same organization.

A high-level architecture of the data system is shown in Figure 12.6.

This architecture focuses on three main areas of data collection and management:

1. Data storage in huge volumes and velocity
2. Data availability at different locations for multiple users at the same time
3. Data security for all the locations involved

The data ingestion part includes collecting data from vehicles with the help of data loggers, processed data from the laboratory environment, and procuring data from multiple suppliers. They are extracted, labeled, and uploaded to a cloud data platform as part of the

FIGURE 12.6 An overview of minimal functions and interfaces required for analytics tools for data handling.

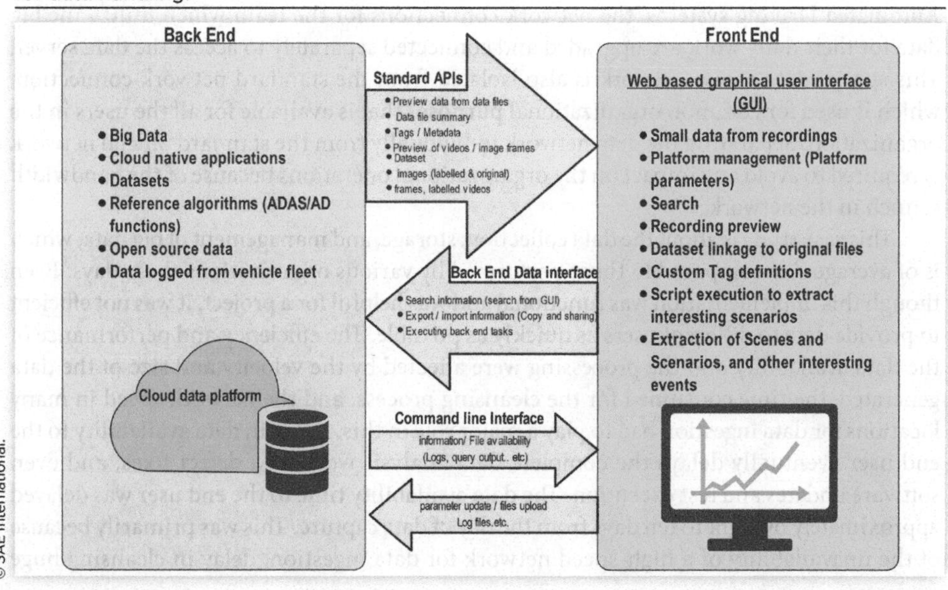

data ingestion. Many services are available with which intelligent data caching has been established at each location to speed up data availability. This also helps in establishing data caches based on the accessibility and needs of the end user.

A security layer is established with a firewall across the organization to maintain data security for all data access locations. The data available to the end user is a challenge when the data ingestion continues for some time. Sometimes, the cleansing and ingestion take multiple days, and the availability of the data to the end user might be after five to seven days or more from the time of data collection. Whenever there is a sudden demand to access specific data files which are not part of the intelligent data caches at each location, it is a challenge on data availability. This challenge is addressed by establishing a three-layer data area. The data area layers are defined based on the speed with which the data files can be accessed from each layer. The three layers of the data area are Hot, Warm, and Cold layers, where the classification is purely defined based on the time taken to access the data files from each layer. The hot layer provides immediate access to the data files, which are also part of the intelligent data caches at each location. The warm layer provides access to the files with a few hours delay, which is the time to bring in the data to the local data caches from the archive, which might be a physically different storage area. The cold layer requires a few days to access the data files. These data might even be archived in a different storage area, even in a different location. All the files, irrespective of their place in the data area layers, can be searched based on a centralized look-up table of data logs across all locations. The application used to control and manage the data in the data platform also facilitates data access to the end user. Thus, it allows the end user to search, update data information, preview data files from the cloud environment, etc. The previews and replays of video streams or creating a copy of those files are possible without any delays depending on the location of the data file in the data area layer, i.e., hot, warm, or cold.

The data management service that monitors, controls, and manages the data caching at each location has a central command center operated by a data administrator. In many organizations, similar monitoring, management, and security functionalities for the data servers are usually handled by the organizational Information Technology (IT) infrastructure team. These data platforms and services must align with the organizational IT policies. Because of the considerable data inflow and outflow for the development and testing of Automated Driving systems, the network connections for the team which utilize the big data for their daily work are upgraded and connected separately to access the data server. This special data access network is also isolated from the standard network connection, which is used for common organizational purposes that is available for all the users in the organization. Separating the data network individually from the standard official network is required to avoid any impact on the organizational operations because of the bandwidth crunch in the network.

This case study is about the data collection, storage, and management of big data, which is of average size compared to those performed by various organizations these days. Even though this implementation was functional and was helpful for a project, it was not efficient to provide data to different users as quickly as possible. The efficiency and performance of the data availability and the processing were affected by the velocity and size of the data generated, the time consumed for the cleansing process, and the network speed in many locations for data ingestion had to play a major role in this. Delay in data availability to the end user eventually delays the complete data analysis workflow, defect fixes, and even software updates and test execution. The data availability time to the end user was delayed approximately by eight to ten days from the day of data capture. This was primarily because of the unavailability of a high-speed network for data ingestion, delay in cleansing huge amounts of data, lack of network availability in remote locations, etc. Delay in data availability was one of the major bottlenecks in the complete workflow when collecting data for analysis and testing. The cleansing part, which is followed directly after the data collection phase, requires a set of engineers analyzing the captured data, trimming it down by identifying valuable data, and uploading it to the cloud data platform with required data labels and information associated with those data files.

The available storage space and the possibility of scaling it depend on the cloud platform provider and the type of contract established. As per the established contract with the cloud provider, additional space requirements in the cloud platform will increase the cost. This was partially addressed by the cleansing process in the data flow path where the data selection happens, thereby preventing unwanted data transmission into the system, which could bring errors that are human dependent. This approach is similar to the process improvement method of controlling the causes which would induce errors in the system if they are allowed to pass through the system. The cleansing process is not only performed for the data collected from the vehicle. This is also performed for any dataset generated or procured from any third-party suppliers or internal to the organization. This is found to help establish reference datasets and a scenario database that acts as the golden sample for releasing a particular software.

There are also cases where the vehicle manufacturers and their suppliers have to work with multiple cloud platforms. Data migration and access to the data files for multiple cloud platforms are required. Even though it is a complex data architecture, several projects are executed in this fashion across various organizations. The supplier cloud data platform where the data ingestion takes place differs from the vehicle manufacturers' cloud data platform where the data is analyzed and used for different purposes. Usually, data migration and transformation are planned for these situations, as indicated in Figure 12.7. These are typically performed by expert organizations outside of the automotive industry specialized in data migration and transformation services (Figure 12.8).

Data security is an essential factor for the collection and handling of data. The German Association of the Automotive Industry (VDA) recommends that the organizations involved

FIGURE 12.7 A data storage and management architecture.

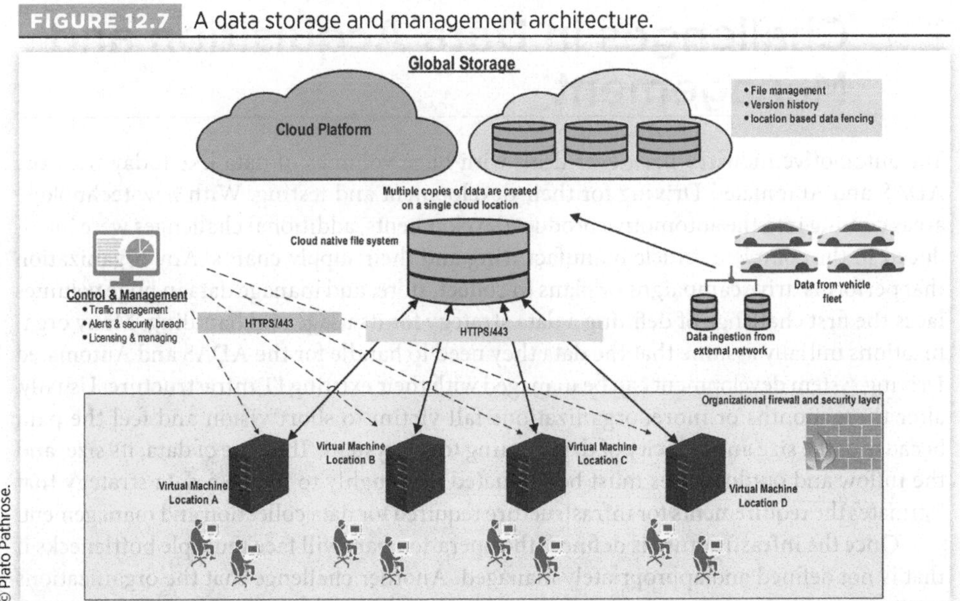

FIGURE 12.8 A data storage and management architecture involving multiple cloud platforms.

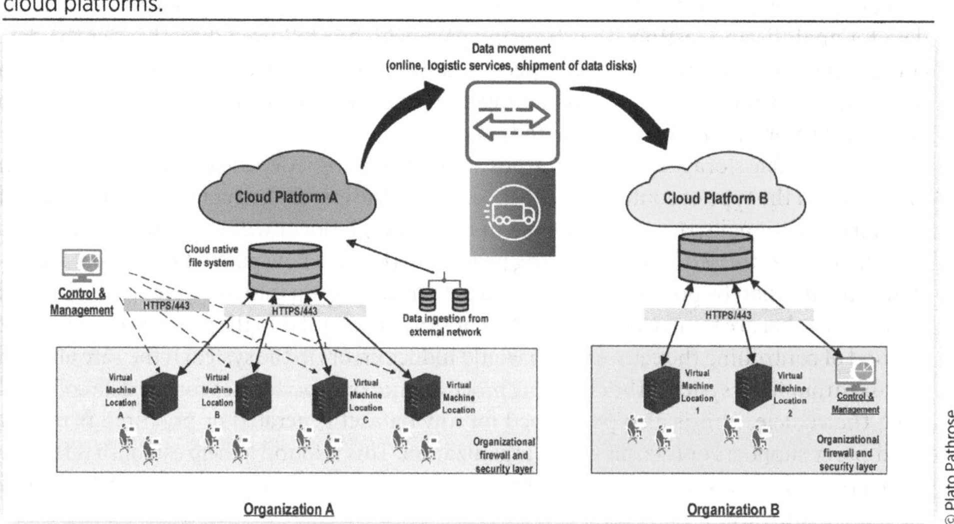

© Plato Pathrose.

in data transmission and collection undergo and meet the requirements of information security assessment TISAX (Trusted Information Security Assessment eXchange) [12.9]. The information security assessment from VDA tries to harmonize the information security over the complete vehicle and its supply chain. This covers both prototype development and large-scale production programs involving multiple suppliers. Now many organizations are far away from establishing these guidelines in their organization. Most of them have their own organizational IT security policies and guidelines because of legacy reasons or because of the IT infrastructure dependencies. It can be expected that this will change in the future to comprehensive information security policies established throughout the automotive industry.

12.7. Challenges in Data Acquisition and Management

The automotive industry has never dealt with huge volumes of data like today with the ADAS and Automated Driving for their development and testing. With new technology areas coming into the automotive product developments, additional challenges were introduced in the complete vehicle manufacturing and their supply chains. Any organization that performs drive campaigns or plans to collect, store, and manage data in huge volumes faces the first challenge of defining a data strategy for its usage and handling. Many organizations initially assume that the data they need to handle for the ADAS and Automated Driving system development can be managed with their existing IT infrastructure. Usually, after three months or more, organizations fall victim to short vision and feel the pain because of the size and velocity of data coming to the system. The type of data, its size, and the inflow and outflow rates must be evaluated thoroughly to define a data strategy that estimates the requirements for infrastructure required for data collection and management.

Once the infrastructure is defined, the operation part will face multiple bottlenecks if that is not defined and appropriately managed. Another challenge that the organizations involved in data collection campaigns is to manage the data availability with minimal delays [12.2, 12.3]. This ranges from the process defined for data ingestion to the infrastructure until the output is generated and archived. Organizations handle this in different ways: Some follow the process of bringing the data and uploading it to a data platform from different locations. Some cloud platform providers run data movement as a service, collect the data drives from various locations, and upload it directly to their data center. Some organizations even ship hard drives over the logistics service providers to a destination where the data is locally loaded and processed. Whatever be the method used for data movement, the data availability time for the end user is a challenge that requires different technology solutions.

Data analysis is handled differently by various organizations. Many organizations rely on manual effort to go through the collected data and trim out unwanted data from it. This is a waste of time as well as effort. Instead, areas of interest from the captured data should be predefined, and those can be identified using automated scanning with specific data analytics tools [12.2]. While performing manual data extraction and cleansing, there is a risk that the prediction of data that is not useful might go wrong because of human error. This requires all the data to be stored even after the cleansing phase if the data is intended to be used for different use cases. This adds up the needed infrastructure and the cost of keeping the data. Moreover, the cleansing process is not adding any value to the data flow in the case of a manual process.

There is a general doubt across the industry regarding how long the data needs to be stored? Or how long should it be made available to the team? Keeping anything that is of no use is a waste. Of course, there has been a lot of investment done to capture and process the data to make it usable. But if it is not used for long and there is no intention of using it, it must be removed to save storage space and money. The data used for the data-driven software development and the HiL re-simulation are usually stored for a defined time as per the project agreement between the supplier and the vehicle manufacturer, taking into account the support and maintenance duration. The datasets are usually stored forever in many organizations as they are assets for developing algorithms, training, and their evaluation. Some move the datasets to separate storage areas when they are not used for one year or less. Some organizations utilize specific mechanisms to identify the most used and not used data files by tracking the access from users. This will help determine the data files that are to be retained for quick access and those that can be archived.

Most of the raw vehicle data captured during drive campaigns are useless after one or two years of translating to datasets or using them as re-simulation database files. Even though the data reuse strategy for different vehicle programs is not successfully executed, it should be considered in the future. The opportunity to reuse the data for the development, training, and evaluation of various features in the vehicle and for the data-driven software components will help every organization involved in collecting and managing the data. Now it is still a challenge for all to reuse the data.

12.8. Summary

The importance of data and how different types of data have found their usage in product development and testing was discussed in this chapter. Big data, which is of enormous volume, velocity, and variety, has become part of the ADAS and Automated Driving system development and testing. Data acquisition, ingestion, analysis, and management become challenging as the automotive industry was not used to handling data of these sizes before.

The skills for handling vast amounts of data are not available in the automotive industry, and those skills need to be harnessed from the software industry where it is available. This chapter discussed common data-logger architectures used in the industry for various types of data logging from vehicles. The scalable version of the data logger is widely used because of its advantage of being adaptable and usable for multiple vehicle types, irrespective of the number of sensors from where data needs to be captured simultaneously. The importance of having a data strategy for the ADAS and automated product development and testing along with various use cases helped define the workflow and processes.

An overview of various processes associated with the data flow from the data acquisition phase to the data analytics and management phase was discussed with examples. From this, the readers will understand various factors that should be considered while setting up a data system for handling huge volumes of data. A detailed overview of the types of data and how they can be managed by means of establishing good processes and configuration management were discussed. Using data analytics tools will help the end user in speeding up the process of data management and analysis. The case study from the industry explains the experience and the challenges faced while establishing a data system that intakes terabytes (TB) of vehicle data per day. Even though it is not up to the range of the volume of data managed these days, which are in the scale of petabytes (PB), this example helps the readers to gain a first impression about handling vast amounts of vehicle data and the challenges faced while establishing such a data management system.

Various challenges associated with data acquisition and management were discussed toward the end of the chapter. That will enlighten the reader with information to prepare for the future of the automotive industry and identify solutions to address some of those challenges. Data analytics is a new technology area in the automotive industry. Even though many organizations perform end-to-end data acquisition and analysis processes, it is still not optimized and has very low efficiency in operations. Most of the challenges associated with data acquisition and management can be addressed by adopting better technologies and infrastructure and defining better processes.

References

12.1. Arthur, C., "Tech Giants May Be Huge, But Nothing Matches Big Data," *The Guardian* (2013).

12.2. Deloitte, "Big Data and Analytics in the Automotive Industry Automotive Analytics through Piece," 2015, accessed November 1, 2021, https://www2.deloitte.com/content/dam/Deloitte/uk/Documents/manufacturing/deloitte-uk-automotive-analytics.pdf.

12.3. McKinsey & Company, "Unlocking the Full Life-Cycle Value from Connected-Car Data," February 11, 2021, accessed November 1, 2021, https://www.mckinsey.com/industries/automotive-and-assembly/our-insights/unlocking-the-full-life-cycle-value-from-connected-car-data.

12.4. Silberhorn, H., "Porsche's Data Strategy: Managing Data in the Digital Era," Porsche AG, February 25, 2021, accessed November 1, 2021, https://medium.com/next-level-german-engineering/data-in-digital-transformation-c377d1e0bb67.

12.5. Release Notes, "HDF5 Version 1.10.7, 2020-09-16," accessed November 1, 2021, https://support.hdfgroup.org/ftp/HDF5/releases/hdf5-1.10/hdf5-1.10.7/src/hdf5-1.10.7-RELEASE.txt.

12.6. IEEE 1588-2008, "IEEE Standard for a Precision Clock Synchronization Protocol for Networked Measurement and Control Systems," 2008.

12.7. Mills, D.L., *Computer Network Time Synchronization: The Network Time Protocol* (Taylor & Francis, 2010), ISBN:978-0-8493-5805-0.

12.8. ASAM, "Capture Module Protocol V1.0.0 (Formerly Probe Logger Protocol)," accessed November 1, 2021, www.asam.net/project-detail/probe-logger-protocol/.

12.9. ENX Association, "TISAX Participant Handbook," January 20, 2021, accessed November 1, 2021, https://portal.enx.com/tphen.pdf.

/Autonomous
/Sensing
/Communication
/Battery
/Navigation
Autopilot mode

13

Challenges and Gaps in Testing Automated Driving Features

ADAS and Automated Driving features have become the highlights of any new vehicle that attract customers, promising better safety and comfort. On the other hand, we are still in a world where vehicles are sold in certain geographical locations where the basic safety systems such as airbags are sold as luxury features in vehicles. The development and deployment of ADAS and Automated Driving features bring massive responsibility to the vehicle manufacturers and their suppliers. Almost all vehicle manufacturers procure various ADAS and Automated Driving systems and functions from different suppliers for their vehicles. Under these circumstances, the vehicle manufacturer usually has access only to minimal information regarding the design and operation of these complex systems obtained from the system suppliers. This adds risk in using the system in a vehicle if those systems are not rigorously tested. Tests are to be executed by the system suppliers during the development and by the vehicle manufacturer for the quality and performance of the functions.

Verification and validation play an essential role in deploying various ADAS and Automated Driving features in the vehicle. Whenever any vehicle manufacturer tries to project new and delighting safety-critical features, there should be evidence that enough testing was performed before deployment in the vehicle, or else it affects the safety of the driver and other road users. Many different challenges and gaps exist in the industry in the development and deployment of Automated Driving systems. Even though the ADAS features have been deployed in vehicles for more than a decade, there are still challenges as regards performance and safety in many of those systems. This gets complicated when moving toward automated driving, which is complex and involves situations where the vehicle takes control under certain conditions. The introduction of new technologies and methods tries to address some of those challenges. However, the technology is still not mature enough to move toward a fully autonomous driving vehicle deployment on a large scale.

13.1. Challenges due to Infrastructure Quality

Unlike ADAS features, the vehicle will take control in certain circumstances in automated driving when the feature is enabled and in operation. The ADAS functions like lane keep assist and traffic sign recognition are the basic functions of higher automation features. Accurate and precise environmental data is important for these features as the vehicle path planning and control solely depend on its environment. The infrastructure should provide the most accurate information about all the static components in the environment and act as a reference for the vehicle to plan and execute certain maneuvers. Having a reliable reference for the vehicle solely depends on the infrastructure and quality [13.1, 13.2, 13.3]. When the infrastructure cannot provide the basic information about the environment around the vehicle, such as lane markings in the roads, traffic signs, etc., the only possibility for the vehicle is to assume that the environment is not supporting and the automated driving features might not get activated. Otherwise, it will wrongly interpret the environment and end up in hazards.

For experiencing various safety and comfort features in modern vehicles, the infrastructure should be available with the required quality. As per the current global scenario, many countries are still struggling without primary road networks beyond the main cities. The quality of the infrastructure required by these safety features, like lane marking, traffic signs, etc., is not available even in some bigger cities. Building quality infrastructure and road networks are the fundamental requirements for integrating advanced technologies in the vehicle [13.3]. Using automated driving features in the areas where the infrastructure does not support it is hazardous. The infrastructure should not only provide the capability to utilize advanced technologies but also need to make sure they are safe and secure. Cybersecurity threats are higher on the infrastructure elements such as traffic infrastructures that automatically send messages to vehicle controls, especially when more connectivity in the vehicles is considered. Hence, safeguarding the data and information exchange over various infrastructure elements is also a challenge these days. Cybersecurity is one of the key areas which should be considered in the modern infrastructure development which supports autonomous vehicles.

13.2. Challenges in the Design of Automated Driving Systems

Automated Driving systems integrate one or more SoCs with higher processing power in the scale of hundreds of teraflops. These types of systems were not commonly used in the automotive industry before. They found their way into the automobile industry because of the demand for higher processing power, especially in infotainment and automated driving. Most of these high-end SoCs have their roots in the consumer industry. Unlike older days, many systems in the market these days incorporate SoCs and other components that are not designed for usage in the automotive industry. These days, the qualification processes followed in many organizations are not robust enough, which adds to the risk of vehicle recalls and the safety of the users. There are many cases in which the features are deployed by various vehicle manufacturers inside the vehicles without being mature enough and have faced backlash. The Automated Driving systems are even developed nowadays using newly introduced SoCs from various manufacturers where the analysis and testing are minimal and have no data to prove the safety. This will drive the technology forward and pose a risk to the safety and quality of the system when used in automotive environments.

Many reasons oblige vehicle manufacturers to take these approaches that were not done earlier in the automotive industry. First, vehicle technology has changed these days, with a lot of electronics and software taking over most of the vehicle functions [13.1, 13.4, 13.5]. The deployment of electric vehicles in the market and moving away from the classical internal combustion engines have reduced the overall moving components inside the vehicle, thereby reducing noise and vibration, which were the most harmful by-products of internal combustion engines. Robust system and component qualifications are required when any new components or systems from other industry areas and use cases are used to develop safety-critical systems. The functional safety standard provides the flexibility to adopt SEooC. Almost all ADAS and Automated Driving systems use SEooC components, but the qualification is not robust enough and increases risk.

The duration for developing and deploying new vehicle models has been reduced drastically from six years to two or three years. A rapid prototyping and production approach is widely seen these days in the industry with many new vehicle manufacturers also introducing new technologies. Unfortunately, it is a hidden truth in many organizations that fundamental system design methods and processes are not followed at their best in many areas of development. Many production projects with various vehicle manufacturers and system suppliers even fail to have matured system requirement specifications toward the SOP of the vehicle.

Adopting new technologies and methodologies that deviate from the classical automotive product development is acceptable, but that does not prevent performing required analysis, qualification, and verification and validation of safety-critical systems such as Automated Driving systems. TDD and FDD methods are widely followed in the development phase of many new automotive products. Unfortunately, they are not executed well in many organizations to build the technology and improve the system and software in many projects. As with any product, the design and the qualification of the system will be influenced by the costs involved. This is also a concern in the development and deployment of complex products, where the cost associated with testing is considered as an area to control, which affects the overall quality of the system and software.

Design principles and good practices from the systems engineering and proven methods must be followed even when the products are developed from scratch using new technologies. The practices and methods are adaptable and should be adapted whenever there is a benefit, either because of new technology or a different method. Rapid prototyping of systems typically does not follow structured methods or processes, but when these prototypes are to be reengineered to build a production-ready system, using the methods and best practices from the industry should be considered. The automotive industry still considers the classical V-model as a reference for different development and test processes even though the iterations or the improvements while developing a product are not reflected or considered in these processes. Different agile development methodologies used in the industry these days, such as FDD and TDD, are not tightly bonded with different automotive processes like Automotive SPICE and functional safety. The automotive industry should consider this change in ecosystem and methodologies and adapt references from the classical V-model-based approaches.

These days many organizations are in the transformation phase to follow agile methodologies and digitalization. There are challenges associated with these transformations and new operations and processes as these usually affect the product development when the development engineers misinterpret these processes. One of the common misunderstandings in this regard is the belief that "following an agile method does not require providing documentation and following the change management process."

13.3. Challenges in Performing Simulation-Based Testing

Simulation is widely used for the development and testing in ADAS and Automated Driving systems. Scenario-based testing is the most common simulation, which is popular nowadays for ADAS and Automated Driving. Other simulation-based test methods are used to verify and validate different domains at different levels of testing. Hardware-based system simulations are widely used in the automotive industry as qualified component models are available. The actual hardware can be modeled, simulated, and evaluated in a PC environment even before finalizing the PCB (Printed Circuit Board). The use of new high-end SoCs affects the possibility of performing in-depth hardware-based simulation and evaluation as part of the design. As the new SoC models might not be available with various toolchains used for simulation, they cannot be evaluated in the design employing simulation. This is a common challenge nowadays, and some organizations even skip the simulation and tests because of tool constraints. Many organizations utilize prototype development boards from SoC suppliers for the design evaluation and tests, which are rarely used to test and evaluate like the simulation tools used for hardware simulation and testing.

Mechanical simulations and tests are still followed as part of production readiness and qualifications. Different types of simulations are performed with models, software, system, and the vehicle in the development processes. Establishing a qualified toolchain and test framework is a major challenge these days. Even though the software and system tests are usually performed with various toolchains in a simulation environment, scenario-based testing is still a challenge because of its complexity to establish a reliable environment with multiple tools integrated into a test framework. When the simulation environment involves multiple simulation tools for co-simulation, qualifying the simulation environment itself is a great challenge. Because of the complexity in qualifying the toolchain, many of the simulation tests performed in various organizations use custom test frameworks and toolchains that are not qualified. In many organizations, the test framework used for verification and validation of Automated Driving systems is not evaluated for the confidence level; instead, they are adopted because of legacy reasons. Adding new tools to this framework requires qualification, which is usually not considered.

Not having a qualified toolchain and test framework raises the question about the reliability and trustworthiness of the test results. This also brings risk in the verification and validation of safety-critical applications of the system. The current situation in the industry is that the environment simulation tools and most of the simulation tools used for scenario-based testing are not qualified by their suppliers. There are simulation tools used even by vehicle manufacturers which are not developed following standard development processes or are not qualified. Even though scenario-based testing is a vital test type for Automated Driving systems, not having tools and infrastructure which are qualified is a risk for reliable measurement and testing of Automated Driving systems.

The only possible way ahead is to spend time and effort to qualify tools and test infrastructures that are used to evaluate the Automated Driving system. The correlation of the tools and infrastructure with the real world will help establish confidence in the toolchain and trust in the measurements.

13.4. Challenges in Laboratory-Based Tests and Vehicle Tests

System verification and validation are the main qualification processes for any automotive product, and the outcome of these processes becomes vital in product deployment if they are safety-critical products. The system verification and validation processes executed in the industry still have a lot of gaps, such as methods followed at each test level and test depth planned and achieved for each test cycle, correlating with the test strategy and the test execution. How much testing is required? This is a common question raised during discussions about testing in projects. The amount of testing that should be performed must match the product and project under consideration. The end user should always be considered while evaluating the quality and performance of a system. The hazards triggered by the failure of these systems will make the verification and validation complex in any product development.

Unfortunately, in many projects, system verification and validation are performed by skipping many intermediate steps. This is not recommended and is not a good practice in the development of safety-critical products. The time available for these projects and the vehicle launch is usually the reason behind these malpractices. An alternate plan is laid in those cases to provide updates and fixes later once the products are deployed. This is a good strategy, but the execution of these strategies is not as good as those planned for as this requires the complete supply chain of the vehicle manufacturer. The maintenance and support from them might not be good enough to implement the aforementioned strategy.

One of the major challenges in system verification and validation is defining the acceptance criteria and achieving those for the product and its functions. Significant drawbacks in many projects are those acceptance criteria that were not defined or clearly specified as part of the requirements. This creates a difference in interpretation of the supplier and the vehicle manufacturer's product performance and quality needs. If the requirements and the expectations are not the same, evaluations will be made at both ends differently. Even the tools and environments might vary with different evaluation criteria for the vehicle manufacturers and suppliers if it is not specified. Even though this looks like a silly mistake, this is one of the common mistakes repeated almost in every project by various vehicle manufacturers and suppliers across the industry.

Defining clear acceptance criteria for the system and the feature in the vehicle with applicable operational design domain and the environmental conditions as part of the requirements and functional definitions helps to have a common understanding among all stakeholders involved. There should not be any excuse for not defining a function or a feature in detail and communicating it to the suppliers by a vehicle manufacturer. Although simulation is considered a supportive evaluation mechanism for the system, features, and their robustness, because of lack of correlation between the simulation toolchain with the real world, certain in-vehicle driving tests are still required to be performed. This can be avoided to an extent by qualifying the test infrastructure used for simulation. Vehicle tests are part of system verification and validation where the system functionality and its reliability are tested in the vehicle. Qualifying ADAS and Automated Driving functionalities in the vehicle requires certain acceptance criteria such as a successful operation for millions of kilometers driven and the hours for which the features are in operation successfully. These are not standardized; instead, each vehicle manufacturer and supplier establish acceptance criteria that are mutually acceptable and evaluate the risks associated with the feature robustness. This is later considered for vehicle deployment. Having very few measurements for acceptance criteria in the vehicle tests adds risk to its users because the features are not being tested enough or not evaluated for their robustness and reliability. This is a common problem with any new automated driving feature deployed in vehicles.

The tests that are part of the homologation of the vehicle type approvals are always tested in the vehicle in proving grounds. With the new features and complex systems inside vehicles, the regulatory requirements and tests are significantly less and do not cover the entire range of features supported by most vehicles. Many vehicle manufacturers utilize these gaps in the regulatory requirements and the tests to get approval for a vehicle with specific features for market deployment. These features might not be required to be tested either because it is not stated in the regulation or the tests covering those features recommended by the regulations are not enough to guarantee the minimum quality and performance along with the safety of the people exposed to it. In the current scenario, certain vehicles are tested and approved with features that are beyond the vehicle type approval requirements laid by the regulations. These gaps in the regulations create certain dangerous scenarios where the quality and performance of those delighter features from the vehicle manufacturers are not safe and are dangerous to the users and society. In many cases, various features with the same names available across multiple vehicle manufacturers are not standard or uniform, and they vary drastically.

With various complex features integrated into the vehicle, the homologation tests for the vehicle type approvals are minimal when executed in the proving grounds. They cannot cover most of the scenarios that a vehicle frequently experiences on public roads. For extending the testing and quality evaluation for approvals, a more extensive set of scenarios are to be tested for various features inside the vehicle. This is possible only by utilizing alternate means of testing, including various test environments, including various simulation environments beyond the vehicle. Recently the UNECE worked toward establishing different levels of testing as part of vehicle evaluation and certification. This includes simulation-based evaluation along with vehicle testing. This would be considered as the future way for the vehicle qualification and approvals from the next version of homologation requirements (Figure 13.1).

FIGURE 13.1 Different types of tests considered for the vehicle qualification and approvals.

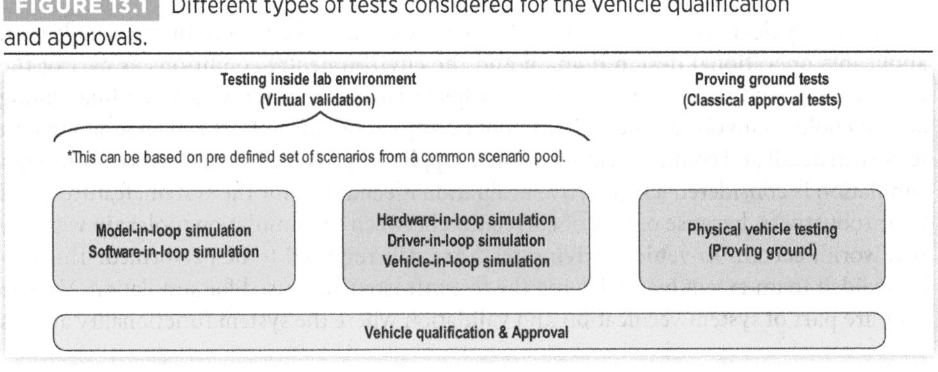

13.5. Challenges in Using AI

With the new technologies, AI has found its way into almost every system inside the vehicle. The performance of various systems inside the vehicle and their functionalities were improved multiple folds by using different applications of AI such as machine learning and computer vision. Although the use of AI has a lot of advantages with the performance and processing speeds, its trustworthiness in all of its operating conditions is still unclear. Now there are many software components of ADAS and Automated Driving systems, such as computer vision and machine learning algorithms, which are mainly used for environmental data processing and to control the vehicle during various events. There are a lot of AI-based algorithms used for the internal sensing features deployed in the vehicles, such as driver monitoring system, occupant monitoring system, and driver drowsiness system.

The scale at which the AI algorithms are used in the vehicle increases tremendously when autonomous vehicles are in focus as the next-generation mobility delighters. Unfortunately, most of the AI-based algorithms that are used in the industry are not tested enough. Even though the evaluation of those algorithms is mainly for the performance, the safety evaluation or trustworthiness evaluations are not performed. This is mainly because of the lack of standard working procedures and standards that can be applied in these areas to qualify artificial neural networks for their safety and reliability. The main focus during the development phase of an ADAS or Automated Driving system is to improve the system performance by optimizing the neural networks running in a system. Unfortunately, the verification and validation part of these software components is not much in focus. It is difficult to understand how deeply these software components are tested in different organizations. In many organizations, a defined method or process and the required toolchain to qualify neural networks do not even exist. The investment done to improve the performance always overrides the effort and investment to qualify these neural networks to be safe in almost all organizations, even where the artificial neural networks play the core component of their product. Luckily, there are many initiatives to establish specific processes, methods, and tools for qualifying AI for its safety and usage in the automotive industry [13.7, 13.8]. This gives hope along with promoting technology when the industry is evolving so that there will be certain metrics used for the evaluation of AI algorithms for their performance, safety, and trustworthiness, which will be used uniformly across organizations. One of the famous initiatives in the safe use of AI is the collection of German research project KIFamilie (www.ki-familie.vdali.de), a group of four research projects that focused on the approaches to build and use AI safely for automated driving. They are KI Absicherung (Safe AI for Automated Driving), KI Wissen (Automotive AI Powered by Knowledge), KI Delta Learning (Scalable AI for Automated Driving), and KI Data Tooling (A Data Kit for Automotive AI).

One of the major drawbacks of data-driven software development is the quantity and quality of the data used for its development, performance improvement, and testing. In most of the projects, an obvious challenge is that significant effort is spent in data collection, preparation, and analysis. This is later used for the software development, performance improvement, and evaluation of artificial neural networks. Once the project is finished or when the vehicle is launched, there is confusion about what to do with these collected and prepared data and spending a lot of effort and money. This is primarily a challenge for the system suppliers in ADAS and Automated Driving when they work with various vehicle manufacturers. Different vehicle manufacturers have different sensor types, sensor positions, etc. There are many variabilities, and hence, the collected data from a particular project with vehicle manufacturers might not be relevant for use in other projects.

The possibility of reusing and synthesizing the data for different vehicle programs from a supplier perspective is a great challenge, and millions of dollars are wasted in data collection and analysis for each vehicle program. There needs to be a strategy and plan on how the data can be reused and utilized in an optimal way to improve the system performance or evaluate the systems. There is always a crunch with the data available for evaluating the neural networks than its training. Hence, if there is a strategy defined and a plan is there to reuse the data, the verification and validation dataset of the neural networks can be improved and can be made comparable with quality and size similar to the datasets used for training these algorithms.

Data reuse might not be as easy as we think. There have been multiple research projects executed to evaluate the data reuse possibilities. This is purely dependent on the vehicle sensors. Any captured data needs to be adapted for different vehicle programs with respect to the sensor position and the sensor type used in those projects. One of the reasonable approaches of data reusability is seen by using a reference system for the data collection and qualifying the measurements of the systems in the vehicle to those reference systems. The reference data remains the same, and the data shift from the changed sensor positions and the vehicle height can be adapted up to a limit. While this seems to be an easy approach, the practicality of this approach for large-scale data transformation and utilization is yet to be proven.

13.6. Challenges in Scenario-Based Testing

Scenario-based testing is one of the critical test types used for the simulation and qualification of ADAS and Automated Driving features. These simulations are repeatedly executed at different levels and in different test environments. This ranges from model-in-the-loop simulation to vehicle-in-the-loop simulation. Scenario-based tests are also executed as part of vehicle testing in proving grounds and on public roads. Every organization executes these tests in different ways and uses different toolchains. The need to have a qualified toolchain to evaluate safety-critical systems was discussed in detail in earlier chapters. It is still a challenge to establish confidence and trust with the current tools and toolchains used in the industry for executing scenario-based tests.

Moreover, different organizations use different scenarios for the development and qualification of different features in the vehicle during the development and testing. Beyond those scenarios, which are part of regulation and vehicle certification, no standardization or a unified set of scenarios are considered by the vehicle manufacturers or their suppliers to ensure that the system is capable of handling those scenarios which the vehicle experiences during its operation on public roads. Different vehicle manufacturers define their features with different ODDs. This creates a difference in features with the same name, which creates a variation in the scenarios used for feature operation and testing. This makes the vehicle manufacturers and suppliers operate independently rather than having features with the same or even different names but providing different functionalities. For example, ADAS features like AEB and forward collision warning behave differently in various scenarios while driving when tested across different vehicle manufacturers. The quality of detection, the ability to apply brakes, types of objects detected, etc. vary even though all vehicle manufacturers must meet the minimum set of scenarios for vehicle approval. The functional performance, quality, etc. vary drastically among different vehicle manufacturers. This is challenging and dangerous, which end up in hazards when we move toward automated driving features where the vehicle takes control and performs certain maneuvers, and there is a great chance that the user expectation of automated driving features might not match with the capability of the features in the vehicle. There is no standardization across various automated driving features from different vehicle manufacturers, and in most cases, the vehicle manufacturers highlight certain delighter features that they launch so that it creates misunderstandings among the users about its capabilities and limitations. This is quite dangerous and can lead to potential hazards.

For better safety, specific initiatives have been devised in the industry to address these challenges. The effort has been taken to unify a set of scenarios as a scenario pool that can be used as a reference to qualify the automated driving features in the vehicles (www.https://www.safetypool.ai). These initiatives will help vehicle manufacturers and their suppliers to use a standard set of scenarios to qualify their features in various test environments. The regulatory bodies can also utilize this scenario pool for homologation and approvals. Using such an approach to qualify the features in the vehicle will make sure that various features integrated into the vehicle meet particular performance and quality in specific scenarios and help the users to have a consistent experience across different vehicles.

13.7. Challenges in Testing for Functional Safety and Cybersecurity

Testing for functional safety and cybersecurity has a vital role in the overall system verification and validation processes. Most functional safety and cybersecurity activities are planned and executed as part of system design and development through analysis and generation of requirements. In the verification and validation part, the main goal is to ensure that the requirements are met, traceable, and no critical defects are present for the functional safety and cybersecurity requirements. Unfortunately, there is no means to guarantee that the functional safety and cybersecurity requirements are complete or mature enough. The evaluation of the system based on a set of test cases that checks only for requirements is not enough to prove that the system is safe and secure for its users. Many organizations use this common approach to stick to a set of functional safety and cybersecurity requirements and only focus on testing those requirements.

Testing for functional safety and cybersecurity of the system beyond the listed requirements is complex and time consuming. This is always a constraint for production programs. This will also limit the depth and amount of testing required for the system to achieve a better level of safety and security for its deployment and operation. Because of time constraints in most of the projects, there is a massive push to focus only on the verification and validation of the available requirements. Even if it is not exposed to the outside environment, this is how the organizational dependencies and project plans are prioritized to guarantee the safety and security of the products. This is good for the project to be finished within the scope of time, budget, and planned resources.

On the other hand, considering the long-term operation and life of these complex safety-critical systems inside the vehicle, the tests must be executed beyond those listed as part of the requirements. This is one advantage of using agile methodologies for testing.

A lot of flexibility was brought into the industry by these methods in product development and testing.

Cybersecurity threats are always focused on Automated Driving systems, especially their design and development. There is also the possibility of cybersecurity threats and attacks on the infrastructure used for the development and testing of automated driving vehicles. Nowadays, because of the large-scale transformation and digitalization of various industries and adopting cloud infrastructures for development and testing, these infrastructures are also vulnerable and are exposed to various cybersecurity threats. It is a general concern around this topic, especially when more and more data are migrated to cloud platforms. How prepared are we for the data leakage and cybersecurity attacks on the cloud platform? The liability associated with the data leakage is very high, especially when vast amounts of critical data from various organizations are deployed in the cloud.

Continuous monitoring and maintenance should be planned for all those safety- and security-related tests as part of verification and validation. Today, the threats are evolving in different areas; this is an area of concern for automated driving vehicles. If continuous monitoring and maintenance are not planned and executed, this will pose a risk to the life of its users. The latest regulation UNECE R155 (E/ECE/TRANS/505/Rev.3/Add.154) enforces the cybersecurity processes and requirements in the vehicle manufacturer's product development and their supply chains. These will ensure more focus in the area of cybersecurity and help to design and develop products that are safe and secure [13.9].

Now almost every system available in the market has some open-source software components integrated into them. The automotive industry also uses a lot of software components that are open source or their derivatives. Even though the open-source software components are significantly different from any other custom-developed software components, one factor makes them more vulnerable. Since most open-source software components are commonly used with different products, the threats associated with these software components are very high. These software components have a higher probability of cybersecurity attack as it is vulnerable because of their availability across the industry than any other custom software component. If the vulnerabilities of any of these open-source software components get exploited, it is not just a single product that gets attacked. That could even facilitate cybersecurity attacks across multiple products in various industries. Hence, the threats and vulnerability chances are higher in commonly used open-source software components. This is because the return of investment for any attacker in attacking these software components is very high compared to utilizing the vulnerability of a custom software component specific to a single product.

Open-source software is also challenging when it is used for safety-critical applications. Usually, the code needs to be reengineered, mainly a change from the open-source licenses, which is a complex situation. Since most vehicle manufacturers are unaware of the open-source components within their vehicles provided by various system suppliers, this is a challenging environment to manage and maintain the safety and security of the overall vehicle and systems. Still most of the systems are sold to vehicle manufacturers as black boxes without much information provided to the vehicle manufacturers by their suppliers. A robust open-source software component management is required on the part of vehicle manufacturers to control, manage, and maintain the safety and security related to the use of open-source software components in products.

13.8. Challenges with Legal Aspects, Liabilities and Its Economic Impacts

The challenges for deploying an automated driving vehicle in the market by a vehicle manufacturer for public road usage are very high from the perspective of legal constraints, associated liabilities if something happens, and the impact those liabilities would bring to the vehicle manufacturer. The verification and validation of the automated driving features are essential to check if those features are capable of behaving as defined and designed. There are many examples in the industry where the vehicle manufacturers highlight certain automated driving features in the vehicle in such a way that it creates a wrong interpretation as regards the vehicle capabilities, which can even create hazards because of wrong understanding and use of the features in the vehicle.

Different countries have different laws which cover product safety and product liability. The product manufacturer is liable to compensate for the losses if the product is defective and causes any damage, injuries, or even deaths. How this is treated varies by the laws defined in each country [13.10]. A vehicle manufacturer is liable if the Automated Driving system and the features provided in the system are unsafe and are defective. A thorough verification and validation strategy and stringent processes to release the products supported by required assessments and audits help the vehicle manufacturer to deploy safe products. Many regulations mention specific clauses such as "using state of the art technologies for implementing advanced features in the vehicle," which means a technology that is proven. The technology supplier and the vehicle manufacturer should provide evidence to prove that any new feature integrated into the vehicle is tested and is proven to be safe for its users and free from defects that cause any hazards to its users or people around the vehicle.

The financial liability for the vehicle manufacturer comes when there are proceedings, sanctions, or the vehicle manufacturer needs to recall their products from the market because of some critical defects. This involves cost and incur losses for the vehicle manufacturer and affects the branding and trust of their customers. Nowadays, the selection of a particular brand and a vehicle is usually a decision based on trust and goodwill. When such recalls or any criminal proceedings occur because of poor product quality or defective product, it will influence the customer's decision to select a brand and vehicle.

Vehicle insurance is another challenge both manufacturers and users face when integrating and deploying automated driving features into their vehicles [13.11]. This indirectly costs more for the customer for their vehicle insurance. Usually, vehicle manufacturers tend to absorb some costs to make it affordable for the end user. Very few insurance companies in the market provide insurance for automated driving vehicles. They employ specific applications that supervise and track the vehicle when the automated driving features are "responsibly" engaged. This collects the driving data and is taken for analysis for insurance coverage.

Legally, new regulations are being enforced to have an event data recorder as a mechanism to track the automated vehicle in operation and record the events whenever they occur. This, in other words, is called the "Automotive black box." These are data loggers or recorders that will record several data and various input parameters of the active onboard safety and accident-avoidance systems in the event of a collision. The proposals for European Union Event Data Recorders ECE/TRANS/WP.29/2020/123/Rev.1 are in place, and the group is working for its approval. The collected data may not necessarily reveal the identity of the driver. There are strict data protection norms that the vehicle manufacturer has to follow in handling the data collected through event data recorders. As per the European Regulation 2019/2144, all new vehicles in the European Union will have to be equipped with a so-called Electronic Data Recorder by July 2022. This follows the already existing US and Chinese regulations to have the Event Data Recorder (EDR) in all new vehicles. This adds to the vehicle manufacturer the additional costs related to the additional system inside the vehicle and the need for infrastructure to analyze the data captured by those systems. On the other hand, this helps the vehicle manufacturers and regulatory bodies have a watch on the automated driving features deployed inside the vehicle and understand its behavior, and take responsibility to provide a safe and secure environment to the users of the vehicle.

13.9. Summary

There are a lot of challenges and gaps that need to be addressed in the current methods and approaches for the development and deployment of ADAS and Automated Driving features in the vehicle. This starts from the design phase of those systems, followed through every phase of development, and later to the launch of those products. Designing and developing a complex and software-driven system and integrating it into a vehicle is not easy. This involves a lot of technologies, methods, and processes to build and deploy safe and secure products for the users. The drawbacks and gaps associated with different phases of development were discussed, and a prospective approach on how these gaps can be fixed was discussed in various chapters. Product design and development are vital in bringing new technologies to the market.

Additionally, proper care and importance must be given to the verification and validation of these systems. Verification and validation is the phase where these systems are checked for their correctness and to see if it serves the purpose for which it is designed and developed. Testing includes many simulations and tests executed in a laboratory environment with real-world scenarios. Even though the technology has developed with better tools and methods for simulation and testing, establishing a reliable and stable simulation environment that correlates with the real world is challenging. This forces the tests to get executed in the real-world using vehicles.

The use of AI and complex software and hardware architectures creates a complex and critical ecosystem for automated driving features. They bring significant challenges in the qualification of these products along with guaranteeing the safety and security of the software and systems. These software components should not be inducing any faults or failures in the system that would result in various hazards. The data-driven software components such as artificial neural networks undergo testing and evaluation of different methods and also use scenarios. Scenario-based testing is used as one of the standard test types for evaluating the vehicle ADAS and Automated Driving features. Tests range from model-in-the-loop simulation to vehicle tests in a proving ground or on public roads. Many intermediate tests use various test environments that cover testing based on scenarios. Having no standard scenario database for the qualification of different features in the vehicle creates custom and nonstandard qualification and performance evaluation processes and methods for those features in the vehicle. This is a challenge that is still open for the user and the regulatory bodies. This challenge can be addressed to a certain degree once the scenario pool is available, which would act as the reference set of scenarios that should be met to deploy these features in the market [13.12].

Safety and security are major concerns for all, which increase with the complexity of the systems and the software. When these complex systems utilize open-source software components for their operation and are unknown to a vehicle manufacturer, the risk associated with that system is manifold. Having better open-source software management for the complete supply chain at the vehicle manufacturer's end would help address these challenges. This is further supported by the introduction of the regulatory guidelines and standards for cybersecurity in the automotive industry. The challenges associated with the legal liabilities and the economic impact on the vehicle manufacturer while deploying automated driving features that are not mature and not tested adequately are enormous. The end user or the customers of these vehicles with automated driving features should not be affected by poor quality, performance, or safety. Hence, more robust regulatory requirements and guidelines are necessary that will supervise the vehicle manufacturers and their suppliers in deploying quality products that are safe and secure. The automotive industry is racing to adopt new technologies and innovations to improve the driver and passenger experience. The regulatory bodies and governments across the globe are supporting the technology evolution in vehicles with infrastructure and providing easy access to the market and ensuring the availability of new technologies to the people. There should be no situation where investments and technological advancements affect the safe and secure living of society. The evolution of technology inside vehicles plays a significant and inevitable role in human life, and safety cannot be spared when moving toward automated driving. This is only possible through proper verification and validation of products and stricter regulations for the deployment of vehicles with higher automation levels.

References

13.1. McDermid, J., "Autonomous Cars: Five Reasons They Still Aren't on Our Roads," July 30, 2020, accessed November 1, 2021, https://theconversation.com/autonomous-cars-five-reasons-they-still-arent-on-our-roads-143316.

13.2. Maurer, M., Gerdes, J.C., Lenz, B., and Winner, H. (Eds), *Autonomous Driving Technical, Legal and Social Aspects* (Berlin, Heidelberg: Springer, 2016), doi:https://doi.org/10.1007/978-3-662-48847-8.

13.3. Duvall, T., Hannon, E., Katseff, J., Safran, B. et al., "A New Look at Autonomous-Vehicle Infrastructure," McKinsey & Company, May 22, 2019, accessed November 1, 2021, https://www.mckinsey.com/industries/travel-logistics-and-infrastructure/our-insights/a-new-look-at-autonomous-vehicle-infrastructure.

13.4. Yurtsever, E., Lambert, J., Carballo, A., and Takeda, K., "A Survey of Autonomous Driving: Common Practices and Emerging Technologies," *IEEE Access* 8 (2020): 58443-58469.

13.5. Bagloee, S.A., Tavana, M., Asadi, M. et al., "Autonomous Vehicles: Challenges, Opportunities, and Future Implications for Transportation Policies," *J. Mod. Transport.* 24 (2016): 284-303, doi:https://doi.org/10.1007/s40534-016-0117-3.

13.6. Campbell, M., Egerstedt, M., How, J.P., and Murray, R.M., "Autonomous Driving in Urban Environments: Approaches, Lessons, and Challenges," *Philos. Trans. R. Soc. A* 368 (2010): 4649-4672, doi:https://doi.org/10.1098/rsta.2010.0110.

13.7. DKE German Commission for Electrical, Electronic & Information Technologies of DIN and VDE, "German Standardization Roadmap on Artificial Intelligence," accessed November 1, 2021, https://www.din.de/resource/blob/772610/e96c34dd6b12900ea75b460538805349/normungsroadmap-en-data.pdf.

13.8. ACEA, European Automobile Manufacturers Association, "Artificial Intelligence in the Automobile Industry," November 2020, accessed November 1, 2021, https://www.acea.auto/files/ACEA_Position_Paper-Artificial_Intelligence_in_the_automotive_industry.pdf.

13.9. UNECE, "UN Regulation No. 155—Cyber Security and Cyber Security Management System," E/ECE/TRANS/505/Rev.3/Add.154, 4.03.2021, 2021, accessed November 1, 2021, https://unece.org/sites/default/files/2021-03/R155e.pdf.

13.10. European Commission, "Council Directive 85/374/EEC of 25 July 1985 on the Approximation of the Laws, Regulations and Administrative Provisions of the Member States, Concerning Liability for Defective Products," Brussels, 1985, accessed November 1, 2021, https://eur-lex.europa.eu/legal-content/EN/TXT/?uri=CELEX%3A52018SC0157.

13.11. Wiggers, K., "Autonomous Car Insurance Drives New Opportunities," August 23, 2021, https://venturebeat.com/2021/08/23/autonomous-car-insurance-drives-new-opportunities/.

13.12. Watzenig, D. and Horn, M., *Automated Driving—Safer and More Efficient Future Driving* (Cham, Switzerland: Springer, 2017), doi:https://doi.org/10.1007/978-3-319-31895-0.

Index

Q

R

S

About the Author

Plato Pathrose studied Electronics and Communication Engineering at the University of Kerala, India. He started his career as a software design and development engineer and has worked on safety-critical automotive and avionics products. Later, he changed his career to system engineering and led multiple systems engineering teams in requirement generation, integration, verification, and validation. Working at different automotive industry levels, from vehicle manufacturers to their suppliers, he gained exposure to advanced technologies and other operating models. He worked on multiple projects where Advanced Driver Assistance Systems (ADAS) and Automated Driving systems were developed and deployed in different passenger and commercial vehicles.

He has been supporting various organizations as a consultant, including government organizations, in defining regulatory requirements and guidelines for the deployment of automated driving vehicles. He has been an advisor to multiple organizations as technical and strategic expertise for developing ADAS and Automated Driving systems. He has also been helping a few organizations in positioning their services and products around the ADAS domain in different markets.

Beyond the technical background, he is a process and methods specialist. He is a lean Six Sigma expert and a TRIZ practitioner who helps organizations with coaching and consulting in their process and methods improvement in system development and testing. He is a certified Automotive SPICE Assessor, Project Management Professional, and Agile Coach who helps organizations transform to an agile ecosystem with digitalization. He is also a technology speaker for different events and had a few technology papers published in his name. He was born and brought up in Trivandrum, a city in the southern state of Kerala in India, and lives in Germany with his wife and son. He is a professional flutist who is passionate about music and a traveler.

email: platopathrose@gmail.com
website: www.platopathrose.com